Macromolecular Symposia | 265

Modern Polymer Spectroscopy

Selected Contributions from the conference:
"17th European Symposium on Polymer Spectroscopy (ESOPS 17)"
Seggauberg, Austria, September 9–12, 2007

Symposium Editor:
Peter Wilhelm
(Graz University of Technology, Austria)

Macromolecular Symposia — Related Titles

M. Laus, D. Ferri, P. Lomellini (Eds.)

From Polymer Structure and Rheology to Process Modeling EUPOC 2007

Vol. 263

Th. Heinze, M. Janura, A. Koschella (Eds.)

Structure and Properties of Cellulose

Vol. 262

S. Zhu, B. Li (Eds.)

Advanced Materials and Polymer Reaction Engineering Hangzhou International Polymer Forum

Vol. 261

B. Liu, M. Terano, V. Busico (Eds.)

Heterogeneous Ziegler-Natta Catalysts

Vol. 260

H.-U. Moritz, W. Pauer (Eds.)

Polymer Reaction Engineering - 9th International Workshop

Vol. 259

M. Hess (Ed.)

Polychar-15 World Forum on Advanced Materials

Vol. 258

Macromolecular Symposia | Vol. 265

Modern Polymer Spectroscopy

Selected Contributions from the conference:
"17th European Symposium on Polymer Spectroscopy (ESOPS 17)"
Seggauberg, Austria, September 9–12, 2007

Symposium Editor:
Peter Wilhelm
(Graz University of Technology, Austria)

ISBN 3-527-32438-0

WILEY-VCH

Full text and further information: www.ms-journal.de

Macromolecular Symposia
is published 14 times a year

Annual subscription rates 2008
Macromolecular Full Package
(All seven *Macromolecular Journals*; 101 issues in total):

Europe	Euro	8,999	9,899
Switzerland	Sfr	14,995	16,495
All other areas	US$	11,895	13,085
		print only **or** electronic only	print **and** electronic

Postage and handling charges included.
All Wiley-VCH prices are exclusive of VAT.
Prices are subject to change.
Individual subscriptions, single issues and back copies are available.
Please ask for details at: service@wiley-vch.de

Orders may be placed through your bookseller or directly at the publishers:
WILEY-VCH Verlag GmbH & Co. KGaA,
P.O. Box 10 11 61, 69451 Weinheim, Germany,
Tel. +49 (0) 62 01/6 06-400,
Fax +49 (0) 62 01/60 61 84,
E-mail: service@wiley-vch.de

Copyright Permission:
Fax: +49 (0) 62 01/6 06-332,
E-mail: rights@wiley-vch.de

For USA and Canada: Macromolecular Symposia (ISSN 1022-1360) is published with 14 volumes per year by WILEY-VCH Verlag GmbH & Co. KGaA, Boschstr. 12, 69451 Weinheim, Germany. Air freight and mailing in the USA by Publications Expediting Inc., 200 Meacham Ave., Elmont, NY 11003, USA. Application to mail at Periodicals Postage rate is pending at Jamaica, NY 11431, USA. POSTMASTER please send address changes to: Macromolecular Symposia, c/o Wiley-VCH, III River Street, Hoboken, NJ 07030, USA.

Printed on acid-free paper.
Typesetting: Thomson Digital (India) Ltd., India
Printing: Strauss Offsetdruck, Mörlenbach
Binding: Litges & Dopf, Heppenheim

Macromolecular Symposia: Vol. 265

Cover: The 17th European Symposium on Polymer Spectroscopy (ESOPS17) was held on September 9–12, 2007 in Seggauberg, Austria. The biennial series of ESOPS meetings brings together scientists specialized in various spectroscopic methods.

The cover picture, is taken from the contribution by Chernev et al. p. 272 and shows infrared images of multilayer polymer film.

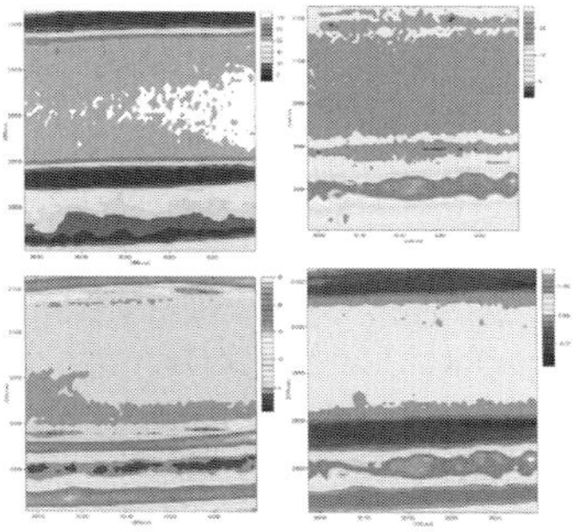

Modern Polymer Spectroscopy

Seggauberg, Austria, September 9–12, 2007

This volume contains selected presentations from the 17th European Symposium on Polymer Spectroscopy (ESOPS17), presented either as invited, short oral or poster contributions. The meeting was organized by the Austrian Centre for Electron Microscopy and Nanoanalysis (formed by the Institute for Electron Microscopy of the Graz University of Technology and the Graz Centre for Electron Microscopy). ESOPS 17 was held from 9th to 12th September 2007 and took place in the ancient castle of Seggauberg, which is situated in the Styrian wine region, near Austria's borders to Slovenia and Hungary.

The biennial series of ESOPS meetings brings together scientists specialized in various spectroscopic methods, from both academic institutions and industry. 103 scientists and students participated in ESOPS17, coming from 27 nations, mainly from Central and Eastern Europe, but also from America (Canada, USA, Venezuela), Africa (Algeria), Asia (India, Japan) and Australia. They discussed the latest developments in the spectroscopic characterization of polymeric materials. Methods, like infrared and Raman spectroscopy and imaging, NMR and ESR spectroscopy, dielectric spectroscopy, also in combination with light and electron microscopy and near-field imaging, were covered by 10 invited lectures, 30 short oral and 45 poster contributions.

Professor Heinz Siesler, member of the International Advisory Board, presented poster awards to Dennis Aulich (ISAS Institute for Analytical Sciences, Berlin), Matt Parkinson (Borealis Polyolefine GmbH, Linz) and Christian Vogel (University of Duisburg-Essen); all three posters dealt with characterisation of polymeric samples by different spectroscopic approaches.

An exhibition of new instrumentation, presented by 10 companies, and new publications on polymer analytics, courtesy of 3 publishers, completed the programme. During the social programme the participants could enjoy local dishes, wine and music, and visit the city of Graz.

The editors would like to thank all the contributors and reviewers, as well as the staff of *Macromolecular Symposia* for their co-operation in preparing this proceedings volume. Additionally, we wish to thank all those who contributed to the success of our symposium, namely the sponsors and the members of the International Advisory Board and the local organising committee.

Peter Wilhelm

Participants from the "17th European Symposium on Polymer Spectroscopy (ESOPS 17)"

DOI: 10.1002/masy.200850501

Micro-Raman and Tip-Enhanced Raman Spectroscopy of Carbon Allotropes

Günter G. Hoffmann,[*1,2] Gijsbertus de With,[1] Joachim Loos*[*1,3]

Summary: Raman spectroscopic data are obtained on various carbon allotropes like diamond, amorphous carbon, graphite, graphene and single wall carbon nanotubes by micro-Raman spectroscopy, tip-enhanced Raman spectroscopy and tip-enhanced Raman spectroscopy imaging, and the potentials of these techniques for advanced analysis of carbon structures are discussed. Depending on the local organisation of carbon the characteristic Raman bands can be found at different wavenumber positions, and e.g. quality or dimensions of structures of the samples quantitatively can be calculated. In particular tip-enhanced Raman spectroscopy allows the investigation of individual single wall carbon nanotubes and graphene sheets and imaging of e.g. local defects with nanometer lateral resolution. Raman spectra of all carbon allotropes are presented and discussed.

Keywords: carbon allotropes; graphene; raman spectroscopy; single wall carbon nanotubes; TERS

Introduction

Carbon is known to exist in a number of allotropes, which range from the hardest of all known material, the pure and single crystalline diamond, to the soft, mainly amorphous, and very impure carbon in the form of soot or glassy carbon. Three of those, diamond, graphite, and of course soot are known since ancient times, while the monomolecular forms of carbon, carbon nanotubes and the Buckminsterfullerenes have been discovered only some years ago.[1,2] These forms can be classified as three-dimensional (diamond and graphite, Figure 1a and b), one-dimensional (single wall carbon nanotubes, Figure 1d), and zero-dimensional (Buckminsterfullerene, C_{60}, Figure 1e). Only recently, in 2004, Gaim and Novoselov[3] produced the two-dimensional form of carbon, graphene by "simply" removing sheet after sheet from graphite to gain a single graphene layer (Figure 1c); and attention is paid to this form of carbon because of its extraordinary functional properties, and potential low production costs.[4]

Since it is possible to produce the allotropes of carbon in large quantities, making it feasible to use them for numerous applications,[5] it has become more and more important to analyze these materials being able to establish, ultimately, structure-property relations at the nanometer length scale. Raman spectroscopy, and in particular micro-Raman,[6] surface enhanced Raman spectroscopy (SERS)[7] or tip-enhanced Raman spectroscopy (TERS) recently introduced[8,9] are characterisation techniques that are able to provide chemical as well as functional information on these materials.

The latter one, TERS, is a characterisation technique combining the power of Raman spectroscopy to reveal chemical composition and molecular structure with the ultra-high spatial resolution of scanning probe microscopy (SPM). Theoretically, TERS allows spectroscopic analysis of any kind of macromolecular material (as

[1] Eindhoven University of Technology, P.O. Box 513, 5600 MB Eindhoven, The Netherlands
E-mail: g.g.hoffmann@tue.nl
[2] University of Duisburg-Essen, Schützenbahn 70, D-45117 Essen, Germany
[3] Dutch Polymer Institute, P.O. Box 902, 5600 AX Eindhoven, The Netherlands
E-mail: j.loos@tue.nl

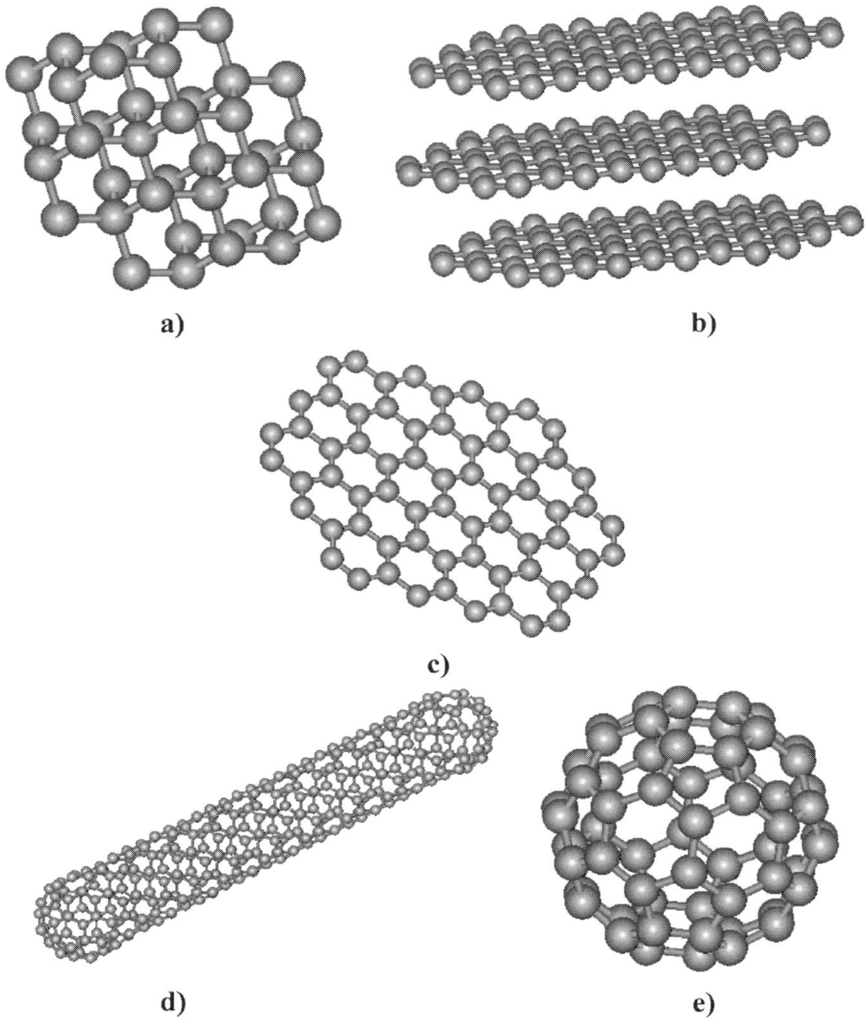

Figure 1.
(a) Diamond lattice and (b) graphite, both three-dimensional structures, (c) graphene (two-dimensional), d) single wall carbon nanotube (one-dimensional), and (e) Buckminsterfullerene (zero-dimensional).

well as inorganic materials like Si) with nanometer resolution, merely depending on probe quality. Only recently TERS imaging was demonstrated with lateral resolution far better than 50 nm and made it possible to identify local defects along one individual single wall carbon nanotube. The group of Novotny and Hartschuh[9–13] realized a resolution of 15 nm while reaching an enhancement of 4, whereas our group resolved about 30 nm while enhancing the G^+-line 256 times.[14]

The potential of TERS is enormous: TERS on biological macromolecules such as proteins and ribonucleic acid (RNA) as well as on various organic dyes has been demonstrated and resulted in spectra that are enhanced compared to conventional confocal Raman spectroscopy by factors 10^6 to 10^{14}, as claimed by the authors; on the other hand, from theoretical considerations even sub-nanometer spatial resolution, resolution below the curvature size of the actual SPM tip, should be possible to be realised by tuning the interaction between tip and sample.

It is the purpose of our study to introduce recent results obtained on various carbon allotropes by micro-Raman spectroscopy and TERS. Main focus is to demonstrate that all carbon allotropes show variations in their Raman characteristics, which allows us, ultimately, to locally identify lateral variations of the composition or defects in carbon-based materials with nanometer resolution by means of TERS.

Experimental Part

A general view of the multi-purpose scanning near-field optical spectrometer NTEGRA SPECTRA^{TM} (NT-MDT) is shown in Figure 2. This back-scattered geometry based configuration for analysing a specimen on transparent substrates allows us to record: atomic force or shear force microscopy (AFM/SFM) images (topography, phase contrast, etc.), confocal optical images, confocal far-field Raman and fluorescence spectra and images, tip-enhanced near-field Raman spectra and images (fluorescence enhancement/quenching). A cw linearly-polarized laser beam from a He-Ne laser operating at 632.8 nm ($E = 1.96$ eV, TEM_{00}) enters the spectrometer through a single-mode optical fibre. The laser output passing through the plasma line filter is expanded and converted to a mode with a given polarization (linear, circular, radial, azimuthal). After that it is reflected by the edge-filter and goes into the inverted optical microscope (Olympus IX70) through an unpolarised beam-splitter cube (10/90) and a pinhole. A 100x oil immersion objective

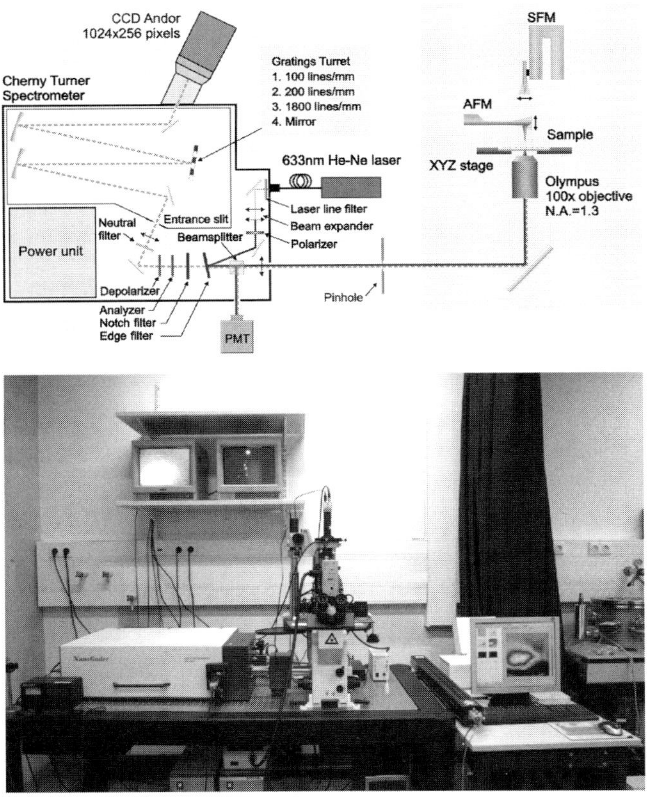

Figure 2.
(top) Optical setup of the TERS instrument used, and (bottom) photography of the TERS setup.

(Olympus, n.a.=1.3, refractive index of oil $n = 1.516$) focuses the laser beam into a spot with a size of less than 300 nm and the power at sample is about 100 μW. The tip is positioned into one of two longitudinal lobes near rims of the diffraction-limited laser spot to locally enhance the electromagnetic field beneath its apex. This system allows one to lock the tip position inside the laser spot to maintain optimum illumination conditions. A near-field Raman image is established by raster scanning the sample with a *xy*-scan stage equipped with a close-loop operation system. Scattered and/or reflected light is collected with the same objective and directed back to the spectrometer through the pinhole. An additional Kaiser notch-filter is installed into the optical path to suppress the Rayleigh scattering. In the laser confocal and spectral modes the light transmitted by the beam-splitter cube is detected with a photomultiplier (Hamamatsu, PMT943-02) and a thermoelectrically cooled charge-coupled detector (ANDOR, DV420), respectively. All Raman spectra were recorded within a spectral range of 150–2500 cm^{-1}. A 200 lines/mm grating provides a spectral resolution of better than 15 cm^{-1}. The pinhole size was equal to 40 μm because of making use of the 100x oil immersion objective.

Sample Preparation

We utilized purified single wall carbon nanotubes (SWCNTs, HiPCo, Carbon Nanotechnology Inc.) as a proper one-dimensional object with strong Raman active spectral lines for testing ultrahigh spatial resolution. A dispersion was prepared by mixing 0.05 g SWCNTs with 20 g dichloromethane in a flask and then sonicating the resulting dispersion for about 1 hour. The sonication was carried out using a horn sonicator (Sonic Vibracell VC750) with a cylindrical tip (10 mm end cap in diameter). The output power was 20 W and, therefore, delivering energy was 1100–1200 J/min. The flask was placed inside a bath with ice water during sonication in order to prevent rising of the temperature. A droplet of this dispersion was spin coated at 300 rpm for 2 min on a microscopy cover glass slide. The glass slide was preliminarily cleaned in a "piranha" solution (H_2SO_4:H_2O_2) at ambient conditions for 30 min, then rinsed in CH_3OH and finally dried under a continuous N_2-gas flow. A similar procedure was applied for preparation of graphene sheets but starting from highly oriented pyrolytic graphite (HOPG, NT-MDT), and for the C_{60} material (Aldrich).

Transmission electron microscopy (TEM) investigation of graphene after dispersing in aqueous solutions were performed using a Tecnai 20 (FEI Co.) operated at 200 kV. The samples were prepared by dipping a copper TEM grid in the graphene dispersion and subsequent drying.

Results and Discussion

Diamond and Amorphous Carbon

The Raman spectrum of diamond has been reported to consist of a single sharp line at 1332 cm^{-1} (e.g. Knight and White[15] and Figure 3). On the other hand, glassy carbon is a material composed of varying amounts of graphite and amorphous carbon. Two broad lines are observed at 1340 and 1590 cm^{-1}. These results are consistent with the turbostratic structure, which has been suggested for glassy carbon (with a particle size L_a of approximately 30 Å) by Nathan *et al.*[16] The corresponding spectrum is shown in Figure 3. Diamond layers deposited by vapour deposition (CVD) find a growing number of applications e.g. in electronics technology due to their hardness, insulating properties and thermal conductivity while being transparent to broad ranges of UV, visible and infrared radiation.[5] In these layers small diamond crystallites are surrounded by graphite and amorphous carbon in the grain boundaries. Raman spectroscopy e.g. is applied to easily discriminate these non-diamondoid impurities by their variations of the Raman spectrum, and thus it is employed to judge the quality of such coatings.

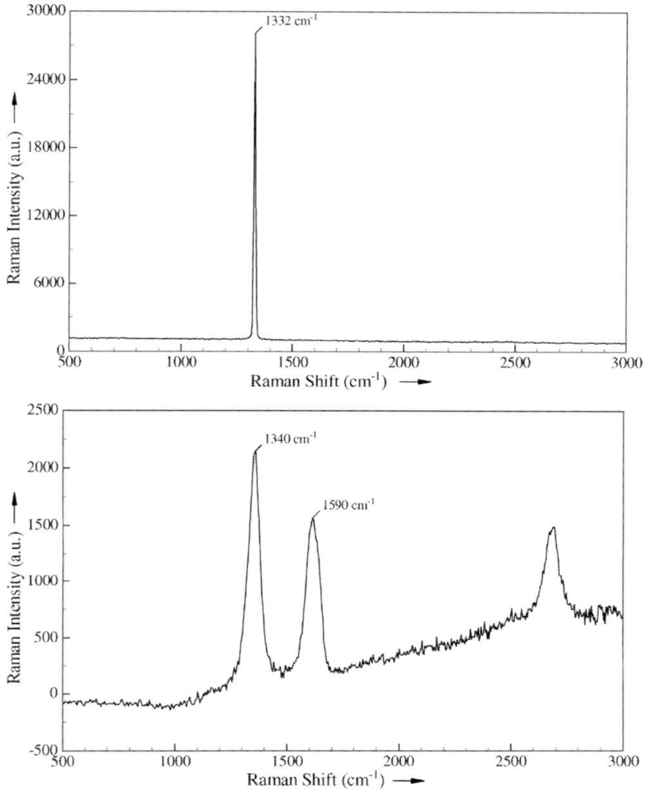

Figure 3.
(top) Micro-Raman spectroscopy spectra of a gem-quality (*tw, vsi*) diamond, and (bottom) glassy carbon.

Graphite

Another example on how Raman spectroscopy can distinguish between crystalline and amorphous order of carbon is presented by analyzing graphite. In single crystalline graphite Tuinstra and Koenig[17] only found a single line at 1575 cm^{-1} (Figure 4). This is called the G-peak (from graphite). Analysing highly ordered pyrolytic graphite (HOPG) by micro-Raman spectroscopy we can confirm the presence of this band and only this band, which suggests that the sample has very high crystallinity. In graphite of lower quality, however, one also encounters a Raman band at 1355 cm^{-1}, called D-line (from disordered graphite, Figure 4b). The ratio G-band versus D-band Raman intensity tells us about the ordered/disordered state of graphite, and thus e.g. about its conductivity.

C_{60} Buckminsterfullerene

Looking with Raman spectroscopy to Buckminsterfullerenes (C_{60} molecules), on the other hand, results in more complex Raman spectra than for diamond, amorphous carbon or graphite. C_{60} molecules are produced together with C_{70} according to the method of Krätschmer *et al.*[18] by heating graphite in an inert gas. From the soot produced the pure compound can be separated by liquid chromatography on silica. Due to the high symmetry of the molecule, the infrared spectrum shows only four lines. The Raman spectrum, as it is complementary to the IR spectrum, shows much more lines, and if, measured with excitation by visible light, is complicated by the fact that a resonance Raman spectrum is produced. Bethune *et al.*[19] measured the Raman spectrum of C_{60} with an argon ion laser and reported a line at 1470 cm^{-1} as the strongest,

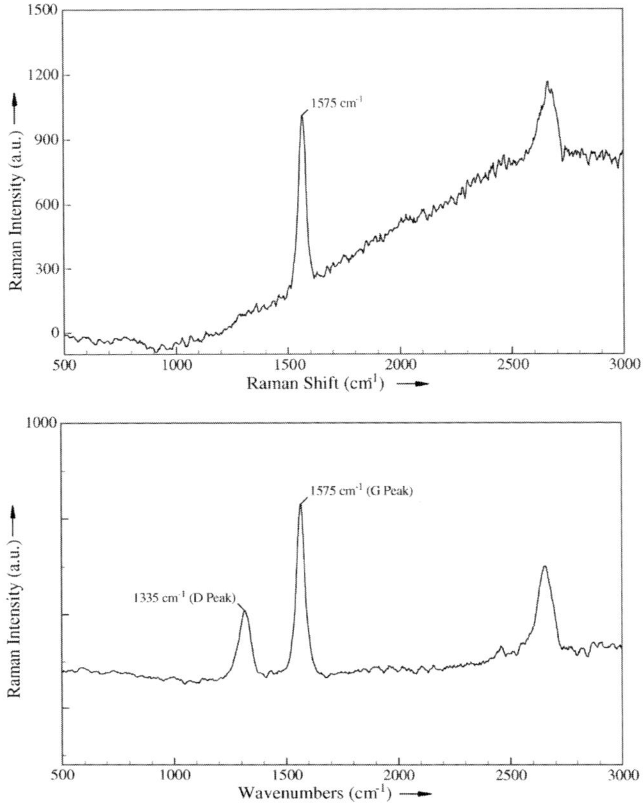

Figure 4.
Micro-Raman spectra of (top) highly ordered pyrolytic graphite, and (bottom) conventional graphite.

which corresponds well with our own measurements (Figure 5). Neugebauer *et al.*[20] calculated the Raman and IR vibrational spectrum of Buckminsterfullerenes on a high theoretical level, but without taking into account resonance effects. The visible lines in our experimental spectrum match the calculated ones in frequency but not in intensity, as

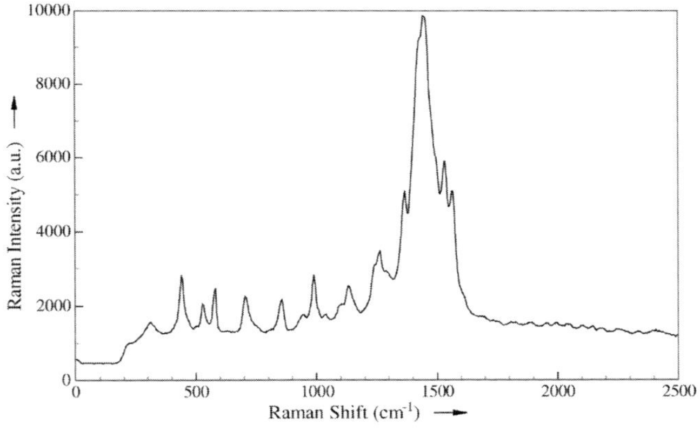

Figure 5.
Micro-Raman spectrum of a Buckminsterfullerene (C_{60}) film obtained with a laser source with wavelength of 633 nm.

our tip-enhanced Raman spectrum actually is a resonance Raman spectrum.

Single Wall Carbon Nanotubes

Single wall carbon nanotubes (SWCNTs) can be grown from the vapour phase at higher temperatures using transition metal catalysts. Formally one can build these tubes by rolling a single graphene sheet and connecting the edge carbon atoms. Similar to graphite, the Raman spectrum of SWCNTs shows as main lines the D-line at 1390 cm^{-1}, and the G^{+}-line 1594 cm^{-1}, slightly shifted when comparing with graphite, and an additional Raman band, the radial breathing mode (RBM) at 290 cm^{-1}, which can be used to calculate the diameter of the tubes. The ratio of D to G^{+}-line can be used advantageously in scans of SWCNTs to detect defects in the otherwise perfect periodically arranged atoms of the tube.[13] For the conventional Raman spectroscopy setup, a large quantity of SWCNTs are analysed at the same time and the average amount and type of defects can

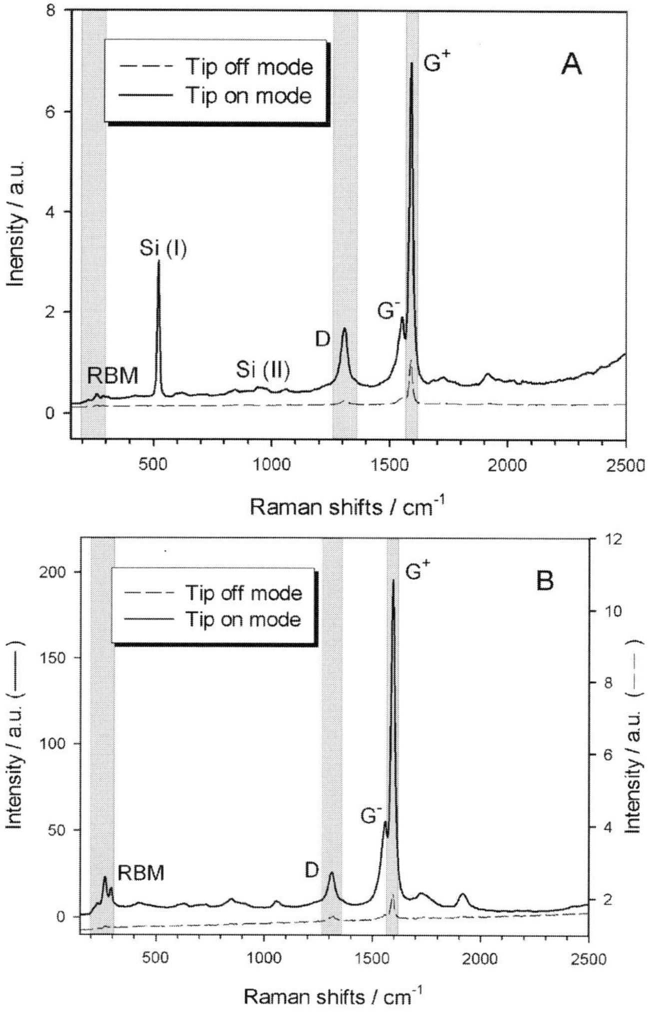

Figure 6.
Tip-enhanced Raman spectra of carbon nanotube bundles, (top) taken with a gold-coated AFM tip and (bottom) a self-made gold SFM tip. Note the much larger enhancement of the gold SFM tip (from Ref. 14).

be calculated, however, e.g. to learn more about local defect distribution at individual SWCNT to optimise procedures for functionalisation of SWCNTs, recently TERS was introduced.

Based on the concept of evanescent waves existing in the near-field (<100 nm) optical measurements beyond the diffraction limit are possible for attaining ultrahigh resolution in optical spectroscopy. A practical implementation of that has become possible by combining optical spectroscopy and scanning probe microscopy (SPM), often referred to as apertureless near-field optical microscopy, and in particular, TERS. The crucial role in TERS is played by the SPM tip as a nanoscopic scatterer and/or lighting source. In the first case, the tip disturbs a confined non-radiating electromagnetic field in the proximity of a nanometer-sized specimen and converts it to a radiating one, which can be then detected by standard diffraction-limited optics. In the second case, a tip localises and enhances the scattered optical radiation over the incident one due to the coupled excitation of free electrons and the electromagnetic field present (called localised surface plasmon) in the metal of the tip. The latter is caused by the fact that namely metals, due to their small skin effect, provide the highest enhancement and scattering efficiency. An additional contribution to the field enhancement,

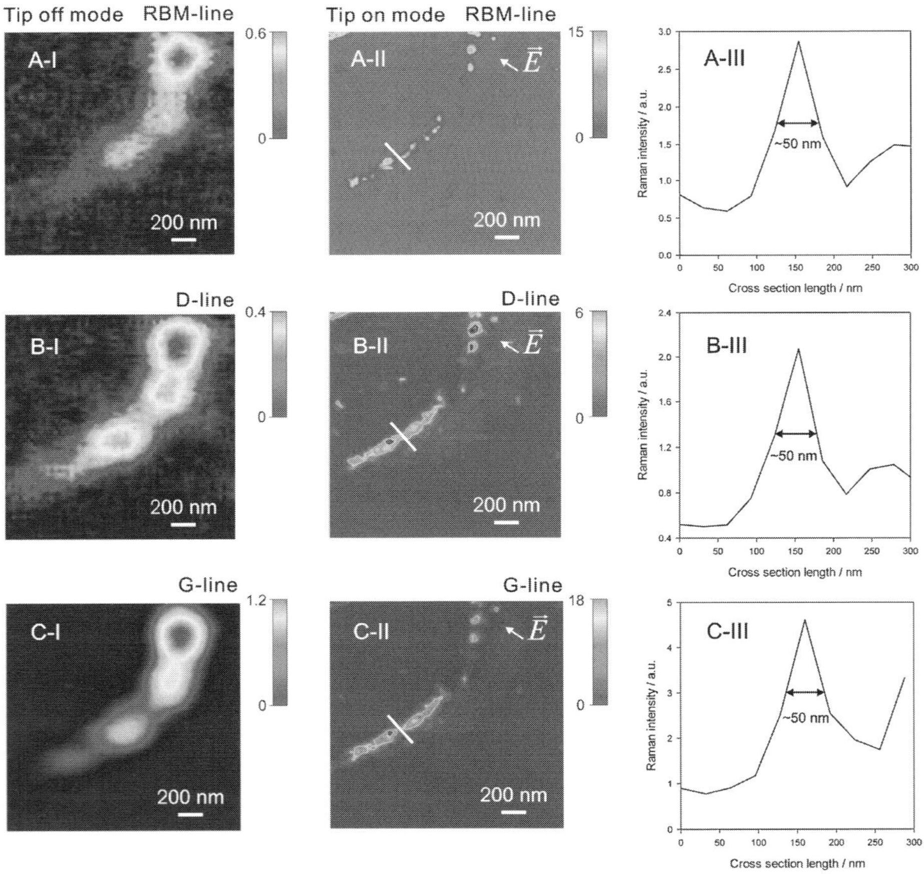

Figure 7.
Confocal (left) and tip-enhanced Raman (middle) scans on carbon nanotubes, using the radial breathing mode (RBM) line as well as the D- and G^+-lines. The cross sections indicated as a white bar in the middle scans are shown on the right (from Ref. 14).

known as quasi-static lighting rod effect, comes from a purely geometrical factor of the tip resulting in a quasi-singularity of the electromagnetic field near its apex. The material composition of the tip, its geometry and the polarization state of the incident light in the local excitation-based scheme are of the greatest importance for efficient enhancements.

Figure 6 shows tip-enhanced Raman spectra of SWCNT bundles that were taken with a gold-coated AFM tip as with a self-made gold SFM tip for tip-off (SPM tip is far away from the sample surface) and tip-on operation mode (tip is close to the sample surface and active in the nearfield). The gold SFM tip shows a much larger enhancement than the gold-coated AFM tip (256 compared to 9 for the G-line, linear enhancement, not area corrected). More details on the experiments performed can be found in reference.[14]

As the next step towards achievement of local Raman data by TERS is spectroscopic imaging. We have compared confocal and tip-enhanced Raman topographical scans on SWCNTs (Figure 7), using the radial-breathing mode (RBM) line as well as the D- and G^+-lines. Cross sections from these scans indicate a resolution of at least 50 nm (original data, 30 nm from a Gaussian fit) for the TERS scan, compared to approximately 300 nm for the confocal scan. These results demonstrate the potential of TERS imaging for local nanometer scale spectroscopic analysis of functional materials.

Graphene

Finally, we would like to introduce first TERS data obtained on graphene. It has

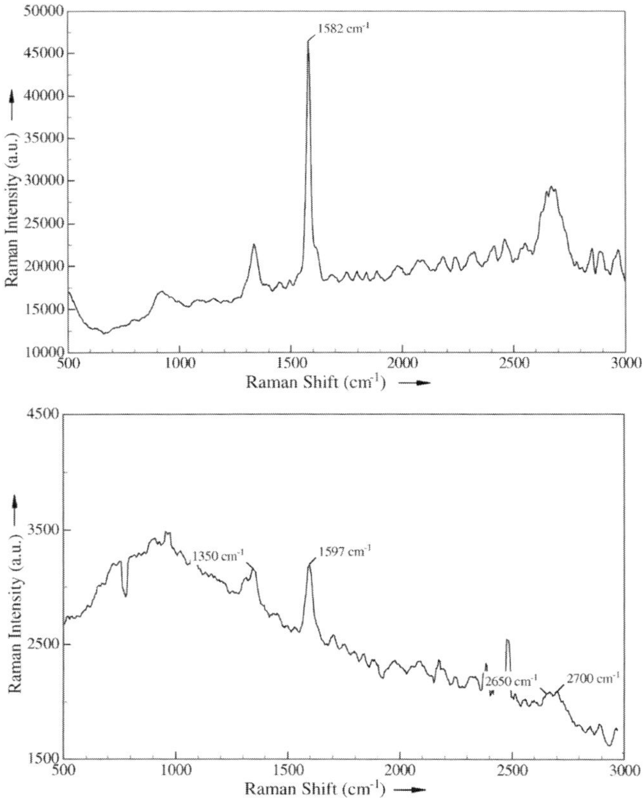

Figure 8.
Micro-Raman spectrum of (top) impure bulk grapheme, and (bottom) tip-enhanced Raman spectrum of impure graphene sheets.

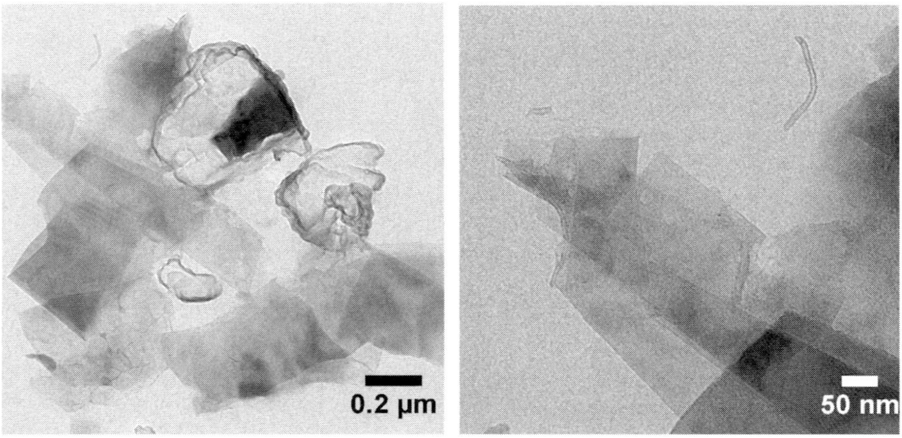

Figure 9.
TEM images of representative graphene samples.

been first described by Novoselov *et al.*[3] as monocrystalline graphitic films, which are a few atoms thick but are nonetheless stable under ambient conditions, metallic, and of remarkably high quality. The films were found by the authors to be a two-dimensional semimetal with a tiny overlap between valence and conductance bands. The conventional Raman spectrum of graphene and graphene layers has been studied in great detail by Ferrari *et al.*[21]. Graphene shows a Raman spectrum very similar to that of graphite, the differences observed mirror the missing interaction between the layers. The D peak second order changes in shape, width, and position for an increasing number of layers, reflecting the change in the electron bands via a double resonant Raman process, and the G peak slightly down-shifts.

We report here the first measured TERS spectrum of graphene. The graphene used by us has been prepared from HOPG and is still containing graphite flakes. These and the edges of the graphene flakes can be detected in the graphene bulk spectrum by the line at 1350 cm^{-1} (Figure 8). A representative transmission electron microscopy (TEM) image of a similar sample is shown in Figure 9, which shows single as well as stacked layers of graphene. The spectrum was taken from a part of the sample (graphene and flakes) where no material was detected optically indicating only very little amount of material was present. The TERS spectrum suggests that the area under inspection was at the edge of graphene flakes, as the band at 2700 cm^{-1} shows a split structure of approximately equal height, while pure graphene would only produce a single line (that at 2650 cm^{-1}). The band at 2700 cm^{-1} in the graphite Raman spectrum at higher resolution has a shoulder at lower wavenumbers, it is the second order of zone-boundary phonons, the results presented are only preliminary data and our study of graphene still is in progress.

Conclusion

Applying Raman spectroscopy helps understanding better the organisation of various carbon allotropes. Raman spectra of diamond, amorphous carbon, graphite, single wall carbon nanotubes (SWCNTs) and graphene sheets show different bands and band positions dependent on the degree of crystallinity, perfectness of their organisation or number of defects of the material under investigation. These variations can be used to determine the quality of the carbon-based materials, e.g. by comparing the intensities of the G^+- and D-bands. Ultimately, Raman spectroscopy combined with a scanning probe microscopy setup, so-called apertureless near-field optical microscopy, and in particular tip-enhanced Raman spectroscopy and imaging, allows detection of such defects with lateral resolution in the nanometer range. We have demonstrated TERS on SWCNT and graphene samples, and in both cases high enhancement factors are obtained. In case of SWCNT, TERS imaging with lateral resolution far better than 50 nm is obtained and the results obtained reflect the local heterogeneity of individual SWCNTs on the nanometer length scale. Currently, similar investigations on graphene samples are in progress

Acknowledgements: The authors are grateful to Dr. J. Yu (Donghua University, Shanghai, China), Ing. S. van Bavel, and Dr. K. Lu for their help in preparing samples, to Dr. S. S Kharintsev and Dr. A. Kodentsov for carbon and graphite samples and stimulating discussions. The authors also acknowledge technical assistance from Dr. P. Dorozhkin and Dr. I. Dushkin (NT-MDT, Russia). This research has been supported by the Ministry of Economic Affairs of the Netherlands via the Technologische Samenwerkings project QUANAP (SenterNovem TSGE3108).

[1] H. Kroto, J. R. Heath, S. C. O'Brian, R. F. Curl, R. E. Smalley, *Nature* **1985**, *318*, 162.
[2] S. Iijima, *Nature* **1991**, *354*, 56.
[3] K. S. Novoselov, A. K. Geim, S. V. Morozov, D. Jiang, Y. Zhang, S. V. Dubonos, I. V. Grigorieva, A. A. Firsov, *Science* **2004**, *306*, 666.
[4] K. I. Katsnelson, *Materials today* **2007**, *10*, 20.
[5] J. Filik, *Spectroscopy Europe* **2005**, *17*, 10.
[6] I. R. Lewis, P. R. Griffith, *Appl. Spectrosc.* **1996**, *50*, 12A.
[7] R. K. Chang, "*Surface Enhanced Raman Scattering*", T. E. Furtak, Eds., Plenum Press, New York **1982**.
[8] Y. Inue, S. Kawata, *Opt. Lett.* **1994**, *19*, 159.
[9] L. Novotny, E. J. Sanchez, X. S. Xie, *Ultramicroscopy* **1998**, *71*, 21.
[10] E. J. Sanchez, L. Novotny, X. S. Xie, *Phys. Rev. Lett.* **1999**, *82*, 4014.
[11] A. Hartschuh, E. J. Sanchez, X. S. Xie XS, L. Novotny, *Phys. Rev. Lett.* **2003**, *90*, 095503.
[12] L. Novotny, S. J. Stranick, *Annu. Rev. Phys. Chem.* **2006**, *57*, 303.
[13] N. Anderson, A. Hartschuh, S. Cronin, L. Novotny, *J. Am. Chem. Soc.* **2005**, *127*, 2533.
[14] S. S. Kharintsev, G. G. Hoffmann, P. S. Dorozhkin, G. de With, J. Loos, *Nanotechnology* **2007**, *18*, 315502.
[15] D. S. Knight, W. B. White, *J. Mater. Res.* **1990**, *5*, 385.
[16] M. I. Nathan, J. E. Smith, Jr., K. N. Tu, *Journal of Applied Physics* **1974**, *45*, 2370.
[17] F. Tuinstra, J. L. Koenig, *J. Chem Phys.* **1970**, *53*, 1126.
[18] W. Krätschmer, K. Fostiropoulos, D. R. Huffman, *Chem. Phys. Lett.* **1990**, *170*, 167.
[19] D. S. Bethune, G. Meijer, W. C. Tang, H. J. Rosen, W. G. Golden, H. Seki, C. A. Brown, M. S. de Vries, *Chem. Phys. Lett.* **1991**, *179*, 181.
[20] J. Neugebauer, M. Reiher, C. Kind, B. A. Heß, *J. Comp. Chem.* **2002**, *23*, 895.
[21] A. C. Ferrari, J. C. Meyer, V. Scardaci, C. Casiraghi, M. Lazzeri, F. Mauri, S. Piscanec, D. Jiang, K. S. Novoselov, S. Roth, A. K. Geim, *Phys. Rev. Lett.* **2006**, *97*, 187401.

 DOI: 10.1002/masy.200850502

Broadband Dielectric Relaxation Spectroscopy in Polymer Nanocomposites

Polycarpos Pissis,[*1] Daniel Fragiadakis,[1] Athanasios Kanapitsas,[2] Kostas Delides[3]*

Summary: Dielectric spectroscopy in the frequency domain and thermally stimulated depolarization currents techniques, covering together a broad frequency range ($10^{-4}-10^{9}$ Hz), were employed to investigate molecular dynamics in relation to structure and morphology in polymeric nanocomposites. Several systems were investigated, three of them with the same epoxy resin matrix and different inclusions (modified smectite clay, conducting carbon nanoparticles and diamond nanoparticles) and two with silica nanofiller (styrene-butadiene rubber/silica and polyimide/silica nanocomposites). Special attention was paid to the investigation of segmental dynamics associated with the glass transition of the polymer matrix, in combination also with differential scanning calorimetry measurements. Effects of nanoparticles on local (secondary) relaxations and on the overall dielectric behavior were, however, also investigated. Several interesting results were obtained and discussed for each of the particular systems. Two opposite effects seem to be common to the nanocomposites studied and dominate their behavior: (1) immobilization/reduction of mobility of a fraction of the chains at the interface to the inorganic nanoparticles, due to chemical or physical bonds with the particles, and (2) loosened molecular packing of the chains, due to tethering and geometrical confinement, resulting in an increase of free volume and of molecular mobility.

Keywords: dielectric spectroscopy; glass transition; polymer nanocomposites; segmental dynamics

Introduction

The mechanical and the physical properties of polymer nanocomposites, i.e. composite materials with a polymeric matrix and, typically, inorganic fillers with characteristic size in the range of 1–100 nm, are often significantly improved, as compared to those of the polymer matrix, for much smaller filler content than would be required for conventional macroscale or microscale composites.[1,2] Polymer nanocomposites also exhibit distinctive properties related to the small particle size and correspondingly small mean interparticle spacing (typically also in the nanometer range).[3]

There is yet no satisfactory theoretical explanation for the origin of improvement of the properties of polymer nanocomposites. It is generally accepted, however, that the large surface to volume ratio of the nanoscale inclusions plays a significant role. Results obtained by various experimental techniques, as well as by theory and computer simulations, indicate the presence of an interfacial polymer layer around the filler, with structure/morphology and chain dynamics modified with respect to the bulk polymer matrix.[4–9] The existence of such an interfacial layer has been postulated for conventional composites long ago and various experiments provided

[1] National Technical University of Athens, Department of Physics, Zografou Campus, 157 80 Athens, Greece
E-mail: ppissis@central.ntua.gr
[2] Technological Education Institute (TEI) of Lamia, Department of Electronics, 35100 Lamia, Greece
[3] Technological Education Institute (TEI) of West Macedonia, Laboratories of Physics and Materials Technology, Kozani, Greece

support for that.[10,11] Questions related to the existence of such an interfacial layer, its thickness and the variation of polymer properties within the layer with respect to bulk properties become crucial for nanocomposites. The reason for that is that, due to the small particle size, resulting in a large surface area presented to the polymer by the nanoparticles, the interfacial layer can represent a significant volume fraction of the polymer in nanocomposites.[3]

Broadband dielectric relaxation spectroscopy (DRS) has proved to be a powerful tool for investigation of molecular dynamics of polymers and composites.[12,13] The main advantage of DRS, as compared to other similar techniques for studying molecular dynamics, is the broad frequency range, which can be relatively easily covered[13] (10^{-4}–10^{9} Hz in the present work). This broad frequency range allows to measure on the same sample processes with very different characteristic (relaxation) times and, correspondingly, different characteristic length scales.

Guided by theory and by results obtained with model systems of geometrical confinement, we have investigated over the last few years molecular dynamics in nanostructured polymers and in polymer nanocomposites with various matrices and fillers. To that aim we employed ac dielectric spectroscopy in the frequency domain and, to a lesser extent, a second dielectric technique in the temperature domain, thermally stimulated depolarization currents – TSDC, differential scanning calorimetry (DSC) and dynamic mechanical analysis (DMA). Here we present and discuss comparatively to each other results obtained with five selected nanocomposite systems, three of them with the same epoxy resin matrix and different inclusions (modified smectite clay, conducting carbon nanoparticles and diamond nanoparticles) and two with silica nanofiller (styrene-butadiene rubber/silica and polyimide/silica nanocomposites). The preparation of the nanocomposites, the morphological characterization and details of the dynamics studies have been presented/will be presented for each particular system elsewhere. In this comparative study we focus on common features and differences in the effects of nanoparticles on the polymer matrix dynamics, as revealed by dielectric techniques. Effects on the overall dielectric behavior, on local (secondary) relaxations and, in particular, on segmental dynamics, associated with the glass transition (dynamic glass transition), are critically discussed.

Experimental Part

Details of the preparation and characterization of the materials have been given elsewhere.[14–17] The nanocomposites investigated include: (1) epoxy resin/modified smectite clay (ER/clay) nanocomposites of exfoliated structure[14]; (2) epoxy resin/nanosized carbon particle (ER/NCP) nanocomposites with a mean particle diameter of about 10 nm[15]; (3) epoxy resin/diamond (ER/diamond) nanocomposites with a mean particle diameter of about 6 nm[16]; (4) styrene-butadiene rubber/silica (SBR/silica) nanocomposites[6]; (5) polyimide-silica (PI/silica) nanocomposites prepared by sol-gel techniques.[17]

For ac dielectric spectroscopy measurements the complex dielectric function, $\varepsilon = \varepsilon'\text{-}i\varepsilon''$, was determined as a function of frequency and temperature.[12,13] In addition to ac dielectric spectroscopy measurements, the non-isothermal dielectric technique of thermally stimulated depolarization currents (TSDC) was used. TSDC consists of measuring the thermally activated release of frozen-in polarization and corresponds to measuring dielectric losses as a function of temperature at low equivalent frequencies of 10^{-2}–10^{-4} Hz.[18] Details of the measurements and of the various formalisms used for the presentation and analysis of the data have been given elsewhere.[14–17]

Results and Discussion

Overall Dielectric Behavior

Figure 1 shows results obtained with SBR/silica nanocomposites.[6] The composition

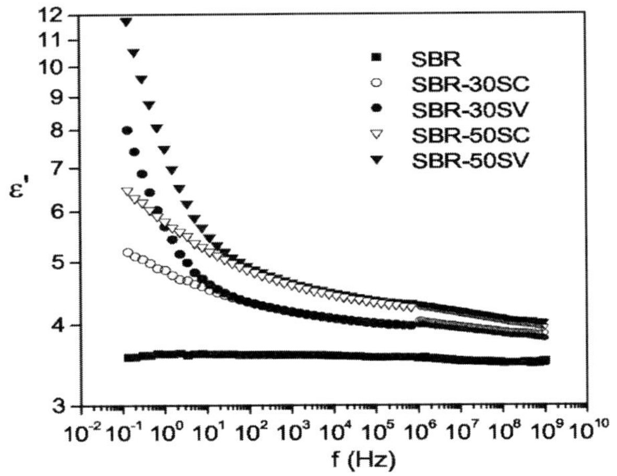

Figure 1.
Real part of dielectric function ε' against frequency f at 25 °C for the SBR/silica nanocomposites indicated on the plot.

of SBR was 23.5 wt% styrene and 76.5 wt% butadiene. The filler used (30 and 50 wt%, nominal) was a precipitated amorphous silica, non-treated (code SV) and pre-treated (code SC) to render the surface organophilic.[6] The results show that ε' increases with increasing amount of filler. This can be understood in terms of a higher dielectric constant of the filler than the matrix and effective medium formulae[17] and/or increased molecular mobility of the polymeric chains. The values of ε' in the nanocomposites of Figure 1 exceed, however, those of pure silica ($\varepsilon' = 3.8$–4.0 at 25 °C[17]), indicating that the data can not be explained solely on the basis of mixture formulae. The hypothesis of increased molecular mobility of the polymeric chains resulting from increase of free volume due to loosened molecular packing of the chains confined between the nanoparticles[19] will be further discussed later on the basis of results for the dielectric strength (magnitude) of secondary and primary relaxations.

A dielectric relaxation is observed in Figure 1 (step in $\varepsilon'(f)$) centered at 10^6–10^7 Hz. This is the segmental α relaxation associated with the glass transition of SBR to be studied in more detail in the next section. The increase in $\varepsilon'(f)$ with decreasing frequency for $f \leq 10^3$ Hz, not observed in the pure matrix, originates from space charge polarization and dc conductivity effects. The results in Figure 1 show that these effects are more pronounced in the samples with non-treated silica particles, whereas dipolar effects at higher frequencies do not depend on filler treatment. It is reasonable to assume that space charge polarization originates from the accumulation of charges in the volume of polymer trapped within agglomerates formed by the nanoparticles. The higher values of space charge polarization in the composites with non-treated filler suggest then that the degree of agglomeration is larger in these composites. These results suggest that low-frequency ac measurements are sensitive to changes in the morphology, in agreement with results for the glass transition and the α relaxation by DRS and DSC (this work) and by DMA.[6]

Figure 2 shows TSDC and Figure 3 ac results for the ER/NCP nanocomposites. The data in Figure 3 have been recorded isothermally by scanning the frequency and have been replotted here. A relatively high frequency has been chosen for the presentation, in order to eliminate conductivity effects present at lower frequencies. An

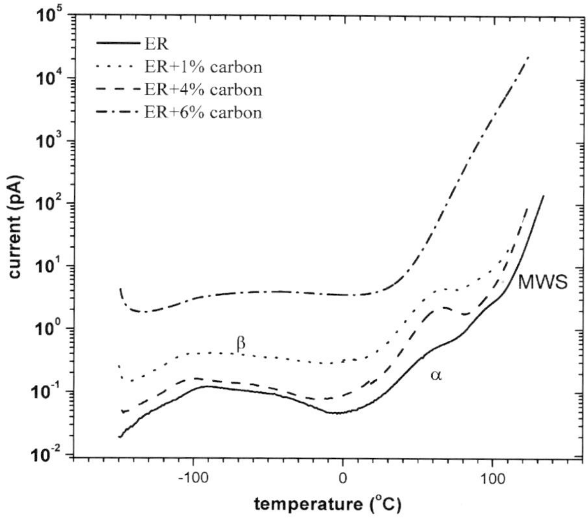

Figure 2.
TSDC thermograms for the ER/carbon nanocomposites indicated on the plot.

overall increase of molecular mobility is observed in Figure 3, in agreement with TSDC data for the same samples shown in Figure 2, in the sense that, at each temperature, ε' and ε'' increase with increasing filler content. This is to a large extent related to the formation of a percolation structure of the nanoparticles, as confirmed by the dependence of ε' (at a frequency of 1Hz and a temperature of $-50\,^{\circ}$C) on volume concentration p of NCP in the inset to Figure 3(a). The well-known equation for the dependence of ε' on p from percolation theory.[20]

$$\varepsilon'(p) = \varepsilon'_m + A|p - p_c|^{-t} \tag{1}$$

where m refers to the matrix, p_c is the percolation threshold and t the critical exponent, has been fitted to the data and the values of p_c and t determined to 7.4% and 0.69 respectively. Two relaxations, a secondary β relaxation at lower temperatures and the segmental α relaxation at

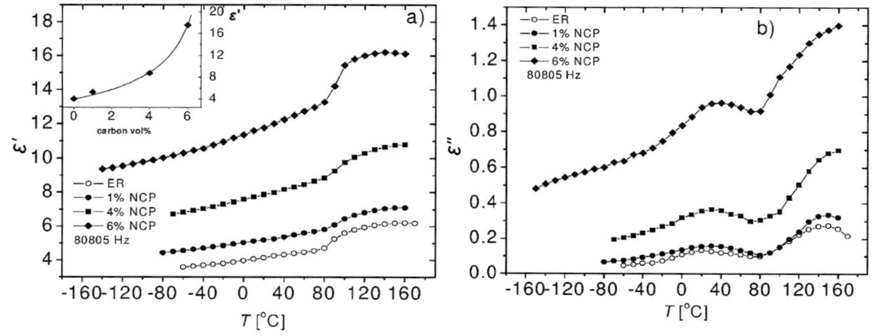

Figure 3.
Temperature dependence of the real ε' (a) and the imaginary part ε'' (b) of the dielectric function of the samples indicated on the plot at 80805 Hz. The inset shows ε' (measured at 1Hz and $-50\,^{\circ}$C) against volume concentration of NCP. The line is a fit of Equation (1) to the data.

higher temperatures, associated with the glass transition of the ER matrix, are observed in Figure 3. They will be studied in more detail in following sections. In the TSDC measurements on the same samples in Figure 2, in addition to the β and the α relaxations, an interfacial Maxwell-Wagner-Sillars (MWS) relaxation is observed in the ER matrix at higher temperatures (interestingly, however, not in the nanocomposites).

Figure 4 shows results obtained with PI/silica nanocomposites prepared by the in situ generation of crosslinked organosilicon nanophase through the sol-gel process.[17] The step at higher frequencies is due to the secondary γ relaxation of the PI matrix, to be discussed in the next section. The most interesting result in Figure 4 is the overall and monotonous decrease of ε' with increasing filler content. Moreover, the values are lower than those of bulk silica ($\varepsilon' = 3.8$–4.0), suggesting a looser molecular packing of PI chain fragments adjacent to the filler particles and/or a loose inner structure of the spatial aggregates of the organosilicon nanophase. By assuming a constant value of ε' for the PI matrix (the measured one, $\varepsilon'_m = 3.18$) and by using various effective medium theory formulae for the calculation of the dielectric function of a composite material[17,20] we obtained for the organosilicon nanophase ε'_i values between 2.47 and 1.58, depending on the composition and the specific formula used. The ε'_i values show, however, the same trend with composition, independently of the formula used. These results can be rationalized assuming that the organosilicon nanophase is made up of nanoparticles of silica ($\varepsilon'_m = 3.8$–4.0) fused together into loose spatial aggregates with a considerable fraction of empty inner pockets ($\varepsilon'_i = 1$). Effective medium theory calculations for this silica-air composite give for the volume fraction of air values in the range 0.40–0.65.[17]

Secondary Relaxations

The step in $\varepsilon'(f)$ in Figure 4 at frequencies higher than about 10^4 Hz is due to the local γ relaxation of the PI matrix, tentatively attributed to non-cooperative motions of the imide groups and/or adsorbed water.[17] Figure 5 shows results for the same γ relaxation in another series of PI/silica nanocomposites, prepared also by sol-gel techniques.[19] The magnitude of the relaxation increases with increasing silica content, without any change of the time scale. Similar results were obtained by TSDC measurements (also in the PI/silica nanocomposites of Fig. 4). Measurements on the same PI/silica samples at different water contents by ac

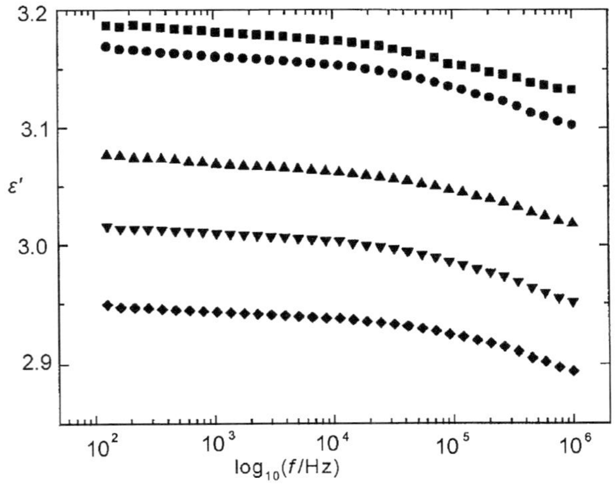

Figure 4.
Frequency dependence of the real part of the dielectric function ε' at 25 °C for PI/silica nanocomposites. The weight fraction of silica is from the uppermost to the lowermost curve 0, 8.6, 22.4, 31.7, 35.6.

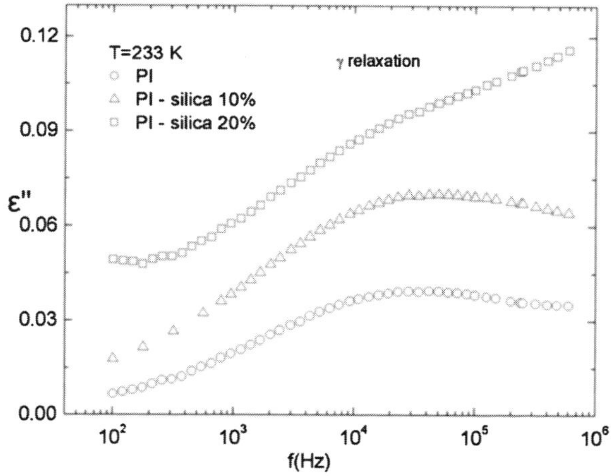

Figure 5.
Frequency dependence of dielectric losses ε'' of the PI/silica nanocomposites indicated on the plot in the region of the γ relaxation.

dielectric spectroscopy and by TSDC indicate that adsorption of water has the same effect on the γ relaxation as the increase of silica content. The effects of water in PIs are commonly explained in terms of plasticization and increase of free volume.[19] Thus, our results suggest that the same explanation may apply for the effects of silica inclusions on the local-scale dynamics in PIs.

Analysis of the various relaxation mechanisms in the nanocomposites under investigation was done by fitting an appropriate model function to the experimental data, typically the Havriliak-Negami (HN) function,[12,13] and analysis in terms of time scale, relaxation strength and shape of the response. Analysis of the data for the γ relaxation of the PI/silica nanocomposites of Figure 4 in terms of the time scale (loss peak frequency f_{max}) shows that the relaxation becomes faster in the nanocomposites. The Arrhenius equation [12,13]

$$f_{\max} = f_0 \exp(-E_{\text{act}}/kT) \qquad (2)$$

where E_{act} is the apparent activation energy, f_0 the pre-exponential frequency factor and k Boltzmann's constant, was fitted to the data for the temperature dependence of f_{max} (Arrhenius plot) and E_{act} and f_0 determined for each composition. Both E_{act} and f_0 tended to decrease with increasing filler content, from $E_{act} = 49$ kJ/mol and $log f_0 = 14.1$ in pure PI to $E_{act} = 36$ kJ/mol and $log f_0 = 12.2$ in the composite with 35.6 wt-% organosilicon nanophase. These results are in agreement with the hypothesis of increased free volume in the nanocomposites due to loosened molecular packing of the chains close to the nanoparticles. Calculations have indicated a decrease of polymer density around a sphere, in particular for short chains.[21] Measurements have shown that the self-diffusion constant of pentane in a polymer characterized by high permeability increases on addition of nanoparticles, and this result has been explained in terms of increased free volume.[22]

We discuss now the results for the β relaxation of ER in the ER/NCP nanocomposites of Figures 2 and 3. The β relaxation in ER has been associated with motions of the hydroxypropylether group.[23] The magnitude of the relaxation increases in the nanocomposites, whereas the time scale (temperature position) of the response does not change appreciably with the composition, similarly to the results for the PI/silica nanocomposites of Figure 4. The Arrhenius plot provides more details on the dynamics of the β relaxation. Similarly to the results in

PI/silica nanocomposites, the apparent activation energy in the Arrhenius Equation (2) decreases in the nanocomposites (69 kJ/mol in ER, 51, 59, 63 kJ/mol in the nanocomposites with 1, 4, 6% NCP respectively), as does the corresponding frequency factor.

Segmental α Relaxation and Glass Transition

Figure 6 refers to the SBR/silica nanocomposites of Figure 1 and shows isochronal $\varepsilon''(T)$ plots in the region of the primary α relaxation associated with the glass transition. For SBR and SBR-30SC a HN expression was fitted to the data following an evaluation method proposed by Schoenhals and coworkers.[23] The results in Figure 6 suggest that the relaxation shifts slightly to lower temperatures in the nanocomposites, whereas they are less conclusive with respect to the magnitude of the relaxation. Analysis by HN fittings suggests no significant broadening of the response in the nanocomposites. DSC measurements on the same samples show, in consistency with the dielectric data, that the glass transition temperature T_g decreases slightly, whereas the heat capacity jump ΔC_p at the glass transition decreases more than additivity would predict in the nanocomposites with respect to the pure matrix. The shift of T_g is systematically larger for the pretreated than for the non-treated samples. These results can be understood in terms of two opposite effects[19]: immobilization of a fraction of chains at the interfaces, because of interaction with the particles, which causes a deficit in ΔC_p, and loosened molecular packing of the chains, because of tethering and of geometrical confinement, which leads to increase of free volume and increased chain mobility. Interestingly, the shift of T_g is larger for the pretreated samples, in agreement with a lower degree of aggregation of nanoparticles in these samples, as indicated by the results in Figure 1.

No second glass transition was observed in the SBR/silica nanocomposites, neither by DSC nor by DRS. In poly(dimethylsiloxane)/silica (PDMS/silica) nanocomposites, however, with a fine dispersion of silica nanoparticles of about 10 nm diameter and hydrogen bonding polymer-filler interactions two α relaxations (dynamic glass transitions) were observed by DRS, the slower one corresponding to the polymer in an interfacial layer with a thickness of 2–3 nm around the nanoparticles.[9,24] Interestingly, only a drop of ΔC_p at T_g but no second glass transition were observed in these nanocomposites by DSC.[9]

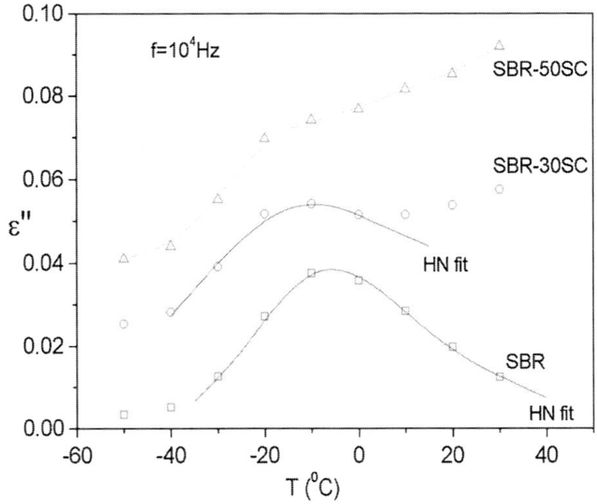

Figure 6.
Isochronal (constant frequency, $f = 10^4$Hz) plot of dielectric loss ε'' against temperature T in the region of the α relaxation for three SBR/silica samples indicated on the plot.

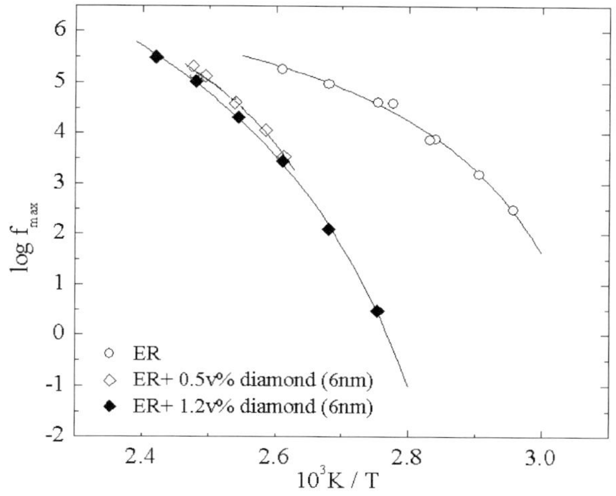

Figure 7.
Arrhenius plot for the α relaxation of the epoxy/diamond nanomposites indicated on the plot. The lines are fits of the VTF Equation (3).

Different results than those in SBR/silica and PDMS/silica, namely a shift of T_g and of the α relaxation to higher temperatures in the nanocomposites as compared to the pure polymer matrix, were obtained for ER/clay[14] and ER/diamond nanocomposites.[16] Despite strong contribution of conductivity and space charge polarization to the measured $\varepsilon'(f)$ and $\varepsilon''(f)$ data in ER/diamond nanocomposites, the dynamics of the α relaxation could be analyzed by calculating, at selected temperatures, $\varepsilon''(f)$ by a derivative method from the measured $\varepsilon'(f)$, where dc conductivity makes no contribution.[25] The frequency of the maximum of the dielectric loss f_{max} for the α relaxation was obtained from the calculated spectra at each temperature and is plotted in the Arrhenius diagram (activation diagram) of Figure 7. The temperature dependence of f_{max} is well described by the Vogel-Tammann-Fulcher (VTF) Equation,[12,13]

$$f_{max} = A \exp[-B/(T - T_0)] \qquad (3)$$

where A, B and T_0 (Vogel temperature) are temperature independent empirical constants. The α relaxation is significantly slower in the nanocomposites as compared to the matrix; note, however, that doubling the volume fraction of the filler from 0.5% to 1.2% seems to only slightly affect the dynamics.

Conclusions

The results presented above for several nanocomposites indicate that dielectric techniques are very powerful for investigating molecular dynamics in relation to morphology in nanocomposites. Several interesting results were obtained with each of the five systems studied and discussed in relation to results obtained by using other techniques. We would like to stress here two observations, which may apply for more systems. (i) Chemical treatment of the nanoparticles, polymer-filler interactions and details of the method of preparing the nanocomposites are reflected in the characteristics of the cooperative segmental relaxation. (ii) The frequency/temperature position of local relaxations of the polymer matrix is not significantly altered by the presence of the nanoparticles. However, in two of the systems studied (PI/silica and ER/NCP) significant changes of the dynamics were observed, with both the apparent activation energy and the pre-exponential factor in the Arrhenius equation decreasing significantly in the nanocomposites. Further work is needed to fully understand these changes at the molecular level.

Acknowledgements: The project is co-funded by the European Social Fund (75%) and National Resources (25%) – EPEAEK II – Archimedes II.

[1] M. Alexandre, P. Dubois, *Mater. Sci. Eng.* **2000**, *28*, 1.
[2] S. S. Ray, M. Okamoto, *Prog. Polym. Sci.* **2003**, *28*, 1539.
[3] M. Z. Rong, M. Q. Zhang, Y. X. Zheng, H. M. Zeng, K. Friedrich, *Polymer* **2001**, *42*, 3301.
[4] G. Tsagaropoulos, A. Eisenberg, *Macromolecules* **1995**, *28*, 6067.
[5] J. Berriot, H. Montes, F. Lequeux, D. Long, P. Sotta, *Macromolecules* **2002**, *35*, 9756.
[6] V. Arrighi, I. J. McEwen, H. Qian, M. B. Serrano Prieto, *Polymer* **2003**, *44*, 6259.
[7] F. W. Starr, T. B. Schroeder, S. C. Glotzer, *Macromolecules* **2002**, *35*, 4481.
[8] G. J. Papakonstantopoulos, K. Yoshimoto, M. Doxastakis, P. F. Nealey, J. J. de Pablo, *Phys. Rev. E* **2005**, *72*, 031801.
[9] D. Fragiadakis, P. Pissis, L. Bokobza, *Polymer* **2005**, *46*, 6001.
[10] G. C. Papanicolaou, S. A. Paipetis, P. S. Theocharis, *Colloid Polym. Sci.* **1978**, *7*, 625.
[11] G. C. Papanicolaou, N. K. Anifantis, L. K. Keppas, Th.V. Kosmidou, *Compos. Interf.* **2007**, *14*, 131.
[12] J. P. Runt, J. J. Fitzgerald, Eds, "*Dielectric Spectroscopy of Polymeric Materials*", American Chemical Society, Washington, DC 1997.
[13] F. Kremer, A. Schoenhals, Eds, "*Broadband Dielectric Spectroscopy*", Springer, Berlin 2003.
[14] A. Kanapitsas, P. Pissis, R. Kotsilkova, *J. Non-Cryst. Solids* **2002**, *305*, 204.
[15] R. Kotsilkova, D. Fragiadakis, P. Pissis, *J. Polym. Sci. Part B Polym. Phys* **2005**, *43*, 522.
[16] P. Pissis, D. Fragiadakis, *J. Macromol. Sci. Part B Phys.* **2007**, *46*, 119.
[17] D. Fragiadakis, E. Logakis, P. Pissis, V. Yu. Kramarenko, T. A. Shantalii, I. L. Karpova, K. S. Dragan, E. G. Privalko, A. A. Usenko, V. P. Privalko, *J. Phys. Conf. Ser.* **2005**, *10*, 139.
[18] J. vanTurnhout, " *Thermally Stimulated Discharge of Electrets*", in: "*Electrets. Topics in Applied Physics*", Vol. 33, G. Sessler, Ed., Springer, Berlin **1980**, pp. 81–215.
[19] V. A. Bershtein, L. M. Egorova, P. N. Yakushev, P. Pissis, P. Sysel, L. Brozova, *J. Polym. Sci. Part B Polym. Phys.* **2002**, *40*, 1056.
[20] R. Pelster, U. Simon, *Colloid Polym. Sci.* **1999**, *277*, 2.
[21] R. Tuinier, H. N. W. Lekkerkerker, *Macromolecules* **2002**, *35*, 3312.
[22] J. Zhong, W. Y. Wen, A. A. Jones, *Macromolecules* **2003**, *36*, 6430.
[23] E. Schlosser, A. Schoenhals, H.-E. Carius, H. Goering, *Macromolecules* **1993**, *26*, 6027.
[24] D. Fragiadakis, P. Pissis, L. Bokobza, *J. Non-Cryst. S. olids* **2006**, *352*, 4969.
[25] J. van Turnhout, M. Wuebbenhorst, *J. Non-Cryst. S. olids* **2002**, *305*, 50.

Planar Array Transient Infrared Spectroscopy: A New Tool for the Time-Resolved Analysis of Polymers

Christian Pellerin

Summary: Planar array infrared spectroscopy (PA-IR) was used for the first time to record transient infrared (TIRS) spectra. In proof-of-concept experiments, it was demonstrated that PA-TIRS can record high quality spectra of common polymeric samples. It was shown the this new technique allows acquiring spectra in as little as 17 ms, opening the door to high speed, real-time monitoring applications.

Keywords: planar array infrared spectroscopy; transient infrared spectroscopy; polymers; time-resolved studies

Introduction

Infrared (IR) spectroscopy is a powerful technique for the qualitative and quantitative characterization of polymers because it provides information about their chemical nature, concentration, structure, etc.[1] In particular, IR spectroscopy possesses excellent chemical selectivity because most functional groups absorb IR radiation at different frequencies. This feature is often used to identify materials non-destructively, but it also enables advanced studies such as the simultaneous characterization of the components of a copolymer,[2] the phases of a semi-crystalline polymer,[3] etc. Continuous-scanning Fourier transform IR (FT-IR) spectroscopy is by far the most widely used technique to record static IR spectra and to follow dynamic phenomena with time constants on the order of a second or more. However, its time resolution is adversely affected when long scanning times are necessary to provide an appropriate signal-to-noise ratio, for instance when measuring the reflectance spectrum of an ultrathin film.

The time resolution of FT-IR can also be a limiting factor for measurements in the emission mode because the emitted radiation is usually very weak, unless the species of interest are highly excited (chemically, thermally, or otherwise). Because absorption and emission of IR light involve the same energy levels, both techniques can, in principle, yield the same molecular information about a sample. According to Kirchhoff's law, the emittance (ε) of a sample at a given wavelength, ν, is equal to its absorptance (α) and is related to its transmittance (τ) and reflectance (ρ) through $\varepsilon_{(\nu)} + \tau_{(\nu)} + \rho_{(\nu)} = 1$.[4] A thick sample with a transmittance close to zero yields a featureless (saturated) spectrum in both the absorption and emission modes.

McClelland et al. have developed a technique that circumvents this problem, the so-called transient IR spectroscopy (TIRS).[5,6] In TIRS, a hot stream of air is directed on a moving sample in order to create a thin surface layer with a higher temperature than the bulk of the sample. This temperature gradient allows recording the emission spectrum of a transient "thin sample" that does not saturate even if the bulk sample is thick. Such surface measurements are often performed in the absorption mode, in particular trough the use of attenuated total reflectance (ATR) spectroscopy. The penetration depth in ATR can be easily controlled through selection of the crystal (index of refraction) and the incidence angle of the incoming radiation. In contrast, the thickness probed in a TIRS experiment depends on the time between

Département de chimie, Université de Montréal, Montréal, QC, H3C 3J7, Canada
E-mail: c.pellerin@umontreal.ca

the exposure of the sample to the hot stream and its passage in the optical path of the spectrometer, as well as on the coefficient of thermal diffusion of the sample. For this reason, the use of ATR is much more convenient under normal sampling conditions. Nevertheless, the TIRS approach becomes advantageous for continuous, real-time, applications for which contact with the sample and/or immobilization are not acceptable. The interest of TIRS has been demonstrated both in the laboratory and in various industrial settings, for instance for the in-line characterization of polymer extrudates, wood chips, etc.[5–7] Time resolution is critical in TIRS because heat diffusion tends to homogenize the sample temperature, leading over time to a featureless grey body emission spectrum. The best option is often to circulate the sample very rapidly, but the spectrum then corresponds to an average over long distances or even to multiple samples.

In the last few years, we have coupled fast and sensitive focal plane array (FPA) detectors to IR spectrographs.[8,9] Such planar array IR (PA-IR) spectrographs allow recording broadband spectra with a time resolution typically between 17 ms and 100 μs, as limited by the frame rate of the camera. In a recent article, we have demonstrated for the first time that PA-IR spectroscopy can be used to record emission spectra of polymers in as little as 17 ms.[10] In this paper, we present our first attempt to perform TIRS measurements using a PA-IR spectrograph (PA-TIRS).

Experimental Part

The PA-IR spectrometer and TIRS setup used in this work are shown schematically in Figure 1. The main instrumental difference between the usual absorption configuration and the emission configuration is the absence of an infrared source, which is replaced by the emitting sample. The spectrometer was custom-built around a Czerny-Turner monochromator equipped with a 300 μm slit and a 50 grooves/mm planar reflection grating. The diffracted radiation was dispersed on a 256 × 256 mercury-cadmium-telluride (MCT) FPA (Santa Barbara Focalplane, Goleta, CA) equipped with a Ninox compound lens

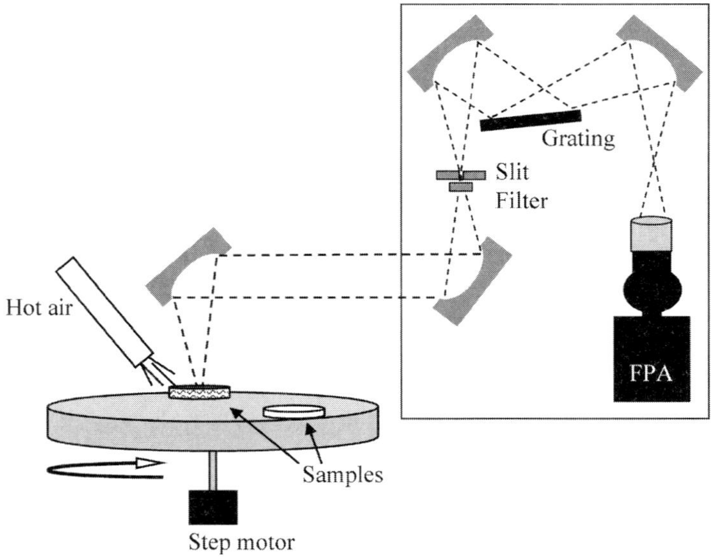

Figure 1.
Schematic representation of the planar array infrared spectrometer and transient infrared spectroscopy sampling accessory.

(Janos Technology, Keene, NH). Under such configuration, the spectral resolution varies from 7 cm^{-1} at 1100 cm-1 to approx. 14 cm^{-1} at 1600 cm^{-1}. The integration time and frame rate were fixed at 99 μs and 114 Hz, respectively. A 1950 cm^{-1} low-pass filter was used to block eventual unwanted high-order diffraction of the grating. PA-TIRS measurements were performed using a custom-built device to simulate on-line monitoring conditions. Polymer samples were deposited on an aluminum-covered stage rotating at a controlled speed using a stepper motor. A stream of hot air was funnelled onto the sample, upstream but as close as possible to the field of view of the spectrometer (radius ~6 mm), using a commercial temperature-controlled heat gun. A rheostat was used to control the flow rate and limit the stream temperature to approx. 150 °C. The emitted IR radiation was collected using a downward-looking off-axis parabolic mirror and focused on the entrance slit of the monochromator. The spectrometer was not purged.

TIRS emittance spectra were calculated by normalizing the luminance of the sample, L_S, by that of a blackbody source, L_B, at an elevated temperature.

$$\varepsilon(\upsilon) = \frac{L_S(\upsilon) - D_S(\upsilon)}{L_B(\upsilon) - D_B(\upsilon)}$$

Blackbody spectra were approximated by recording the emission from a metal plate covered by a thick layer of flat black paint. The blackbody plate was heated to different temperatures in a static mode, and the most appropriate blackbody spectrum was selected to match the luminance of the sample using the procedure described by McClelland et al.[6] All spectra were corrected by subtracting a dark background spectrum, $D(\nu)$, recorded with a closed monochromator entrance slit. This spectrum eliminates the contribution of stray light and corrects for residual pixel-to-pixel intensity variations. The absorption spectrum of a thin polystyrene film was recorded in transmission, using the blackbody plate as a source, to calibrate the frequency axis of the TIRS spectra. The procedure described by Pelletier et al. was finally used to correct the spectral images for slit curvature.[9]

Results and Discussion

Figure 2 shows the spectral images recorded for the blackbody source and for an inch-thick slab of expanded polystyrene. The horizontal axis corresponds to the frequency of the emitted radiation, while the vertical axis corresponds to the height along the field of view of the spectrometer. The useful frequency range, for which significant radiation can be observed in Figure 2, ranged from approx. 1850 to 1050 cm^{-1}. The spectral images were corrected for the dark background emission. Such correction is of utmost importance for PA-TIRS or emission measurements because the FPA is a DC-coupled detector that is sensitive to non-modulated radiation such as the IR emission from the environment. In fact, the maximum intensity for the dark background-corrected images in Figure 2 is 250 counts, as compared to a raw signal on the order of 5000 counts. Dark background spectra were thus recorded frequently to compensate for any drift originating from a slight variation in the laboratory temperature. This can easily be performed using a shutter inserted behind the slit in the monochromator.

The spectral image of the blackbody source (Figure 2a) shows a signal envelope typical of that measured for the background in an absorption experiment.[8] The multiple vertical dark stripes are due to absorption by water vapour, as the spectrometer was not purged. The image of the polystyrene sample also shows a broad intensity envelope and water absorptions but is dominated by polystyrene emission bands. For instance, the most intense band appears around pixel #93 (1450 cm^{-1}) and is attributed to a phenyl semi-circle stretching and CH_2 deformation.[11]

Even if the image curvature appears limited in Figure 2, the spectral images were straightened before averaging 60 pixel rows

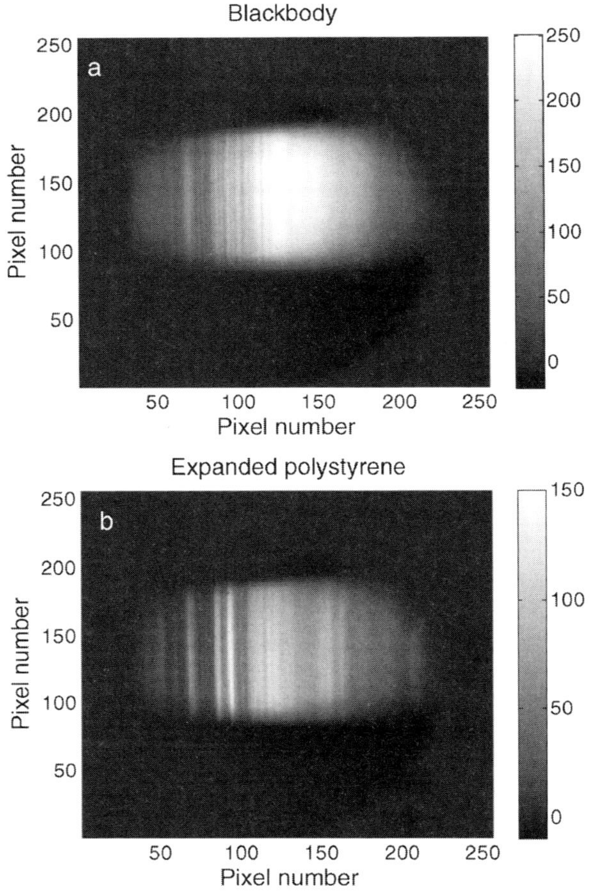

Figure 2.
Spectral images recorded for a) the blackbody source and b) the expanded polystyrene sample. The intensity scale is in counts.

to improve the signal-to-noise ratio.[9] Figure 3 shows the resulting spectra for the expanded polystyrene sample and for the blackbody source at two temperatures, as obtained by exposing the metal plate to the hot stream for different times. The shape of the blackbody spectra is due to the superimposition of a Planck's law distribution with the response function of the spectrometer, including the diffraction efficiency of the grating and the wavelength-dependent sensitivity of the FPA detector. The FPA cut-off precludes measurements below 950 cm^{-1}, while noise level considerations limit studies above 1800 cm^{-1} unless the sample can be heated to a high temperature without degradation. The blackbody spectra of Figure 3 also show water vapour absorption bands above 1300 cm^{-1} as down-going peaks.

In contrast, the sample spectrum shows multiple, chemically-specific, bands that are superimposed on a blackbody-type spectrum. The latter is due to emission from the greatest observable depth of the sample.[6] The heated surface layer of the sample can, in principle, produce bands with a maximum intensity determined by the radiance of a perfect blackbody emitter at the same temperature, as is the case in standard emission measurements.[4] The sample spectrum is thus sandwiched between two blackbody-type spectra, with the intense (weak) bands appearing close to the high

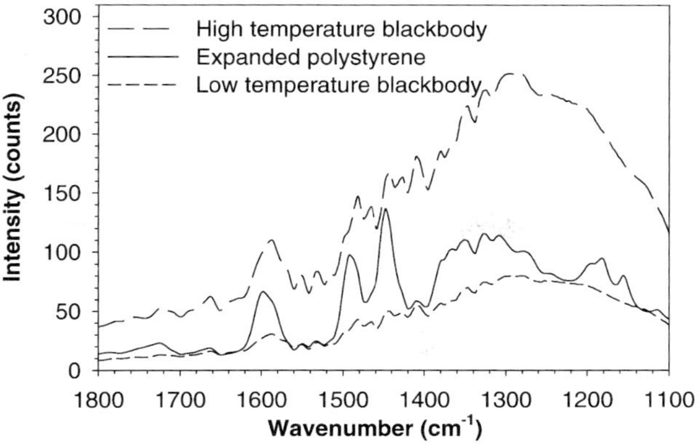

Figure 3.
TIRS radiance spectrum of a thick expanded polystyrene sample and of the blackbody emitter at low and high temperatures.

(low) temperature blackbody spectrum. However, because the sample temperature is not uniform with depth (and even laterally because heating is not perfectly homogeneous), it is not possible to define a single temperature for the surface layer. As a consequence, the temperature of the blackbody to be used for calculating the TIRS emittance is not strictly defined either. The usual TIRS practice is to select a normalizing blackbody spectrum that only slightly exceeds the most intense feature of the sample spectrum.[6] Several blackbody curves were thus recorded at different temperatures in order to match the intensity of the sample curve and calculate the final TIRS spectra.

Figure 4 shows the spectra calculated using for the expanded polystyrene as well as for two other common packaging materials: an air cellular cushioning material (bubble wrap) and a second type of foam. These everyday products are meant to represent polymeric materials that could be found on a recycling unit. The spectra were not baseline corrected but were shifted for clarity. As expected, the three spectra show clear differences that allow easy distinction. The spectrum of the expanded polystyrene shows all the expected bands, in addition to a carbonyl band most likely due to additives. The second packaging foam shows the characteristic bands of polyolefins and is likely composed of polypropylene. The

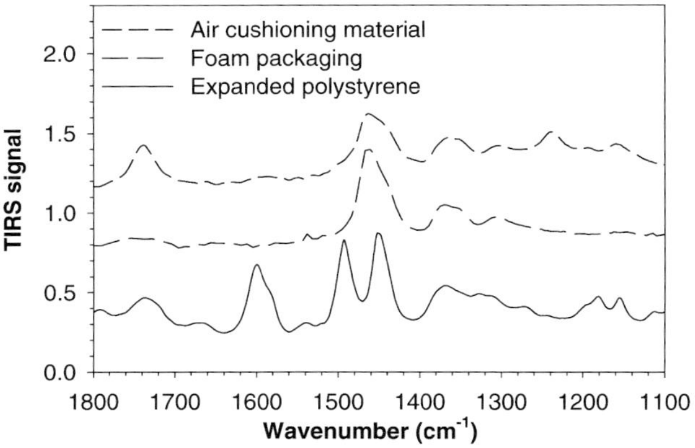

Figure 4.
TIRS spectra of various polymer samples.

spectrum of the cushioning material is more complex and also appears to contain a large fraction of polypropylene. Additional bands, which could help identifying the other component contained in this material, are visible at 1735 and 1235 cm^{-1}. The expanded polystyrene TIRS spectrum is qualitatively similar to an ATR spectrum (not shown) of the same material. Small differences in band shape exist because the spectral resolution varies as a function of wavenumber in spectra recorded with a dispersive spectrometer. The band intensities are somewhat different from those in conventional ATR or transmission spectra, with the weaker bands showing larger relative intensities in the TIRS spectra. This is due to the fact that the emitting layer was relatively thick in these measurements, leading to equivalent absorbance values beyond the linear (Beer law) range. It was shown that TIRS spectra can be used for quantitative analysis using bands of appropriate intensity.[7]

The results of Figure 4 demonstrate that good quality TIRS spectra can be recorded using a no moving part PA-IR spectrometer. This feature could be advantageous in the context of on-line monitoring, where the interferometer of an FT-IR spectrometer can be affected by strong vibrations. A rather large bandwidth was recorded in these experiments by using a grating with a low groove density, but at the expense of a spectral resolution limited to 7–14 cm^{-1}. Alternatively, a higher resolution could be obtained at the expense of bandwidth. This compromise does not have to be made when using FT-IR spectrometers, for which the two parameters are independent. Nevertheless, a potential major advantage of PA-IR spectroscopy over FT-IR spectroscopy for TIRS measurements is the possibility of performing faster time-resolved acquisition, as was demonstrated in absorption experiments.[8,12]

Figure 5 shows spectra of the expanded polystyrene sample recorded by averaging between 300 and 2 camera frames, corresponding to acquisition times of 2.6 s, 435 ms, 87 ms and 17 ms. The four spectra are almost identical, albeit with slightly different baselines. The signal-to-noise level is very good, even for the spectrum recorded in a mere 17 ms. Similar results were obtained with the air cushion sample, while the second foam became noisier in 87 ms acquisitions. In that case, however, the intensity of the most intense band was only 46 counts, as compared to 133 counts for the expanded polystyrene in Figure 3.

Such fast acquisition times could greatly improve the performance of TIRS in terms of sample resolution. Indeed, the samples are often circulated at a speed around 40 cm/s but which can be as high as several m/s.[6] In either case, recording an FT-IR spectrum in as little as 1 s would still mean averaging a sample (or several samples) over a long

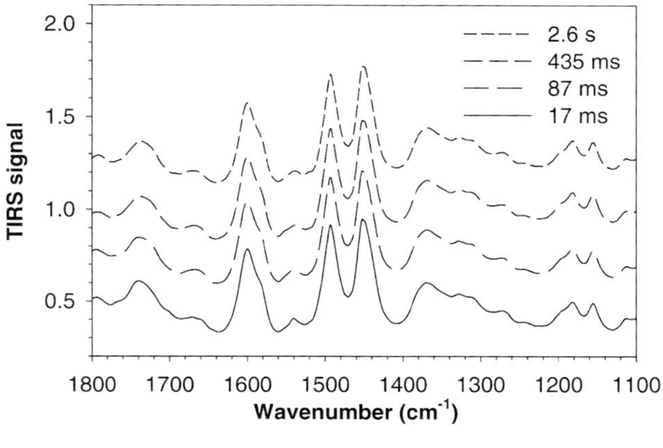

Figure 5.
TIRS spectra of expanded polystyrene recorded using different acquisition times.

distance. In contrast, a PA-TIRS setup with a time resolution of 25 ms could in principle provide a spectrum per cm, opening the door to true real-time TIRS monitoring.

An important limitation of our demonstration setup should be acknowledged. Because the PA-IR spectrograph was built on an optical table without any protection for the optical elements, it was necessary to keep the temperature of the hot stream relatively low, and the linear velocity of the samples on the rotating stage was restricted to 10 cm/s. Both aspects are detrimental because 1) they reduce the temperature gradient between the surface and bulk of the sample, leading to a decreased TIRS signal, and 2) the longer heat diffusion time creates a thicker surface layer that can lead to saturation of the signal. This was indeed a problem for dense samples with better heat diffusion properties, leading to ill-structured or even blackbody-type spectra. It should nevertheless be stressed that these limitations are of an engineering nature and that, under improved sampling conditions, the performance of the PA-TIRS technique is expected to match that of Figure 4 and 5 for such samples.

Acknowledgements: The financial support of the Natural Sciences and Engineering Research Council of Canada (NSERC) and the Canada Foundation for Innovation (CFI) is acknowledged. The author also thanks J.F. Rabolt and D.B. Chase for the use of equipment at the University of Delaware, and I. Pelletier for useful discussions.

[1] P. R. Griffiths, J. A. De Haseth, *Fourier Transform Infrared Spectrometry*, 2nd ed., Wiley-Interscience, Hoboken, NJ 2007.
[2] Y. Wang, C. Pellerin, C. G. Bazuin, M. Pézolet, *Macromolecules* **2005**, *38*, 4377.
[3] K. C. Cole, C. Depecker, M. Jutigny, J. M. Lefebvre, P. Krawczak, *Polym. Eng. Sci.* **2004**, *44*, 231.
[4] J. Mink, Infrared Emission Spectroscopy. In *Handbook of Vibrational Spectroscopy*, Vol. 2, P. R. Griffiths, J. M. Chalmers, Eds., John Wiley & Sons, Chichester **2002**, pp. 1193–1214.
[5] R. W. Jones, J. F. McClelland, *Anal. Chem.* **1989**, *61*, 650.
[6] R. W. Jones, J. F. McClelland, S. J. Bajic, Transient Infrared Spectroscopy. In *Handbook of Vibrational Spectroscopy*, Vol. 2, P. R. Griffiths, J. M. Chalmers, Eds., John Wiley & Sons, Chichester **2002**, pp. 1215–1230.
[7] R. W. Jones, R. R. Meglen, B. R. Hames, J. F. McClelland, *Anal. Chem.* **2002**, *74*, 453.
[8] C. Pellerin, C. M. Snively, D. B. Chase, J. F. Rabolt, *Appl. Spectrosc.* **2004**, *58*, 639.
[9] I. Pelletier, C. Pellerin, D. B. Chase, J. F. Rabolt, *Appl. Spectrosc.* **2005**, *59*, 156.
[10] C. Pellerin, I. Pelletier, J. F. Rabolt, D. B. Chase, *Anal. Chem.* **2007**, *79*, 2037.
[11] S. Krimm, *Adv. Polym. Sci.* **1960**, *2*, 51.
[12] C. Pellerin, S. Frisk, J. F. Rabolt, D. B. Chase, *Appl. Spectrosc.* **2004**, *58*, 799.

 DOI: 10.1002/masy.200850504

Spectroscopical Investigation of Ski Base Materials

Jörg Fischer,[*1] Gernot M. Wallner,[2] Alois Pieber[3]

Summary: Raman spectroscopy was applied to perform a comprehensive morphological analysis of polyethylene (PE) ski base materials at different processing levels. The morphological characterization included determination and evaluation of Raman spectra and examination of the crystallinity values by differential scanning calorimetry (DSC). A good agreement between Raman and DSC crystallinity fractions was obtained, thus corroborating the Raman spectroscopy approach. While for the PE grade with the lowest average molar mass no significant morphological changes due to processing from the raw material via the extruded film to the post-treated film was found, higher molar mass PE grades exhibited a decrease of crystallinity, but an increase of the amorphous fraction along the process chain.

Keywords: DSC; morphology; polyethylene; Raman spectroscopy; ski base

Introduction

From a market point of view polyethylene (PE) is the most important commercially used polymer with overall market shares of about 30%.[1] While commodity PE grades are dominating in various industrial sectors, such as the packaging industry, engineering PE grades have also been developed for technical applications, requiring special functional properties. An outstanding example is ultra-high-molecular-weight polyethylene (UHMWPE) with advanced tribological behaviour. In the literature many investigations on UHMWPE are described focusing on the application as bearing material in orthopaedics.[2–6] However, PE grades with special tribological properties are also used for other applications which attract less scientific attention. Regarding the leisure and sports industry PE materials are of special relevance for running surfaces (i.e., bases), which are a key element of alpine and nordic skis determining their glide characteristics. For this application currently polyethylene extrusion grades with molar mass between 300,000 and 600,000 g/mol (high density polyethylene (HDPE)) and sinter grades with molar mass between 3,000,000 and 12,000,000 g/mol (UHMWPE) are used. The process from the raw material to a finished ski base surface consists of various steps, including the formulation and compounding of the material, the extrusion or sintering of semi-finished films, post-treatment (e.g., flaming) of the film, and production of the ski. To improve the base performance advanced material formulations and surface structuring techniques have been developed, implemented and used. While relationships between the additive content of ski base materials, the topographical structure of ski base surfaces and the gliding characteristics were established, up to now no systematic investigation on the effect of material morphology on the performance of ski base surfaces has been carried out. Hence, the overall objective of our research work is to study the morphology of ski base materials on different processing levels (raw material (compound), semi-finished film and post-treated film) by various techniques.

[1] Polymer Competence Center Leoben GmbH, Parkstrasse 11, Leoben, 8700, Austria
Fax: (+43) 3842 402 2102;
E-mail: jfischer@pccl.at
[2] Institute of Material Science and Testing of Plastics, University of Leoben, Franz-Josef-Strasse 18, Leoben, 8700, Austria
[3] Fischer GmbH, Fischerstrasse 8, Ried/Innkreis, 4910, Austria

HDPE and UHMWPE are linear forms of polyethylene with semi-crystalline morphology. Important variable parameters of semi-crystalline polymers are the degree of crystallinity and the semi-crystalline morphology, which is considered as a superposition of three phases, the crystalline component (often lamellae), the melt-like amorphous component and the disordered component of anisotropic nature (interphase).[7]

Relevant parameters of polyethylene are commonly derived by density measurement, differential scanning calorimetry (DSC), infrared (IR) and Raman spectroscopy, nuclear magnetic resonance spectroscopy (NMR), and wide and small angle X-ray scattering (WAXS and SAXS) techniques. While for most of these methods an adequate sample preparation (i.e., the destruction of the sample) is required, Raman spectroscopy has proved to be an appropriate tool to investigate morphological parameters of PE without any kind of damage. Furthermore, Raman spectroscopy allows for a rather comprehensive description of morphological parameters including the full characterization of the three phase components. Although there are some controversies in the literature concerning the feasibility of Raman spectroscopy for the determination of the disordered phase[8–10], various recent studies[6,11] have used the three phase model to describe morphology changes due to processing or loading and obtained reasonable results from a polymer physical point of view.

The sliding characteristic of ski bases is mainly affected by structural and morphological parameters of the surface layer, which is in contact with the snow counterpart. Therefore, the primary purpose of this paper is to describe morphological parameters at the surface of PE ski base grades as a function of different processing levels. In addition to the various advantages described above, Raman spectroscopy is a sensitive method with high lateral and vertical resolution to determine the morphology of materials surfaces. Therefore, Raman spectroscopy was applied in this study as the prime method. To corroborate the crystallinity values determined by Raman spectroscopy DSC measurements were also carried out. Regarding DSC special attention was given to the sample preparation. An appropriate technique to cut the sample from the surface layer was implemented and used.

Experimental Part

Materials

Within this work five polyethylene grades differing in their average molar mass and distribution were selected, compounded with additives but without modifiers (e.g., carbon black) and processed to semi-finished products (ski base film) (Table 1). The compounds and semi-finished products were supplied by Fischer GmbH (Ried, Austria). Depending on the material grade two different techniques, extrusion (200–220 °C) and sintering (200–230 °C, 250 bar, 12 h, heating/cooling rate: 30 °C/h) were applied to manufacture the ski base film, which was further treated (post-treated film) by flaming with an oxygen rich (yellow) flame (200–300 °C, 0.5–1 s). These processing steps are common before the production of the ski.

Table 1.
Characteristics of the polyethylenes used in this work.

Designation in this work	Material	Processing method	Density (g/cm^3)	$M_W \times 10^{-3}$ (g/mol)
M2	HDPE	extrusion	0.914	250
M4	HDPE	extrusion	0.936	400
H4	UHMWPE	sintering	0.911	4,000
H9	UHMWPE	sintering	0.918	9,000
HA	UHMWPE	sintering	0.918	10,500

Due to secrecy reasons further processes after the ski production are not considered within this paper.

Raman Spectroscopy

Raman spectra were determined by using a LabRam confocal-Raman spectrometer (HORIBA Jobin Yvon GmbH, Bensheim, D). Excitation of Raman bands was obtained upon irradiating the sample with a frequency doubled Nd-YAG laser of 100 mW operating at 532 nm. All measurements were performed in a backscattering geometry, using a 100× long working distance (3.4 mm) microscope objective with a numerical aperture value of 0.80, providing a lateral resolution of about 800 nm. To obtain a spectral resolution of about 3 cm^{-1} a diffraction grating of 1800 grooves/mm was used. Single point spectra were derived from 5 scans with an accumulation time of 5 s. By measuring on three different positions of a specimen mean and standard deviation values were evaluated.

The Raman spectra were manipulated by smoothing and baseline correction. To determine the areas under the peaks of interest, which was used as the measure for the intensity values (I), a curve-fitting procedure was carried out with the software package PeakFit v4.12 (SeaSolve Software Inc.). Appropriate fitting results were obtained by peak fitting with freely floating ratios of Gaussian and Lorentzian amounts automatically optimized by the software (Figure 1). The reproducibility of the fit was corroborated by fitting the peaks multiple times.

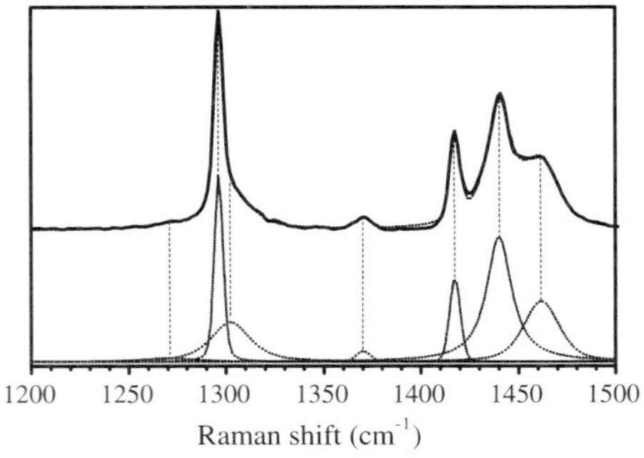

Figure 1.
Raman spectra of PE with typical curve fitting results.

To evaluate the degree of crystallinity (α_c) the method devised by Strobl and Hagedorn[7] was used (see Eq. (1)). Because ski bases are produced by at least partial melting and crystallization orthorhombic crystalline structures were assumed.

$$\alpha_c = \frac{I_c}{0.45 \cdot I_T} \cdot 100\% \tag{1}$$

where I_c is the integrated area of the Raman band representing the orthorhombic crystalline phase, I_T is the integrated area of the CH_2 twisting vibration region (1200–1350 cm^{-1}), which is not affected by the morphology and therefore used as an internal intensity reference and 0.45 is a normalization coefficient, which was found through experiments on 100% crystalline PE. [12,13]

Furthermore the fraction of the amorphous phase (α_a) was calculated by using Eq. (2)[7]:

$$\alpha_a = \frac{I_a}{I_T} \cdot 100\% \tag{2}$$

where I_a is the integrated area of the Raman band representing the amorphous phase.

Finally, the fraction of the interphase (α_b) is given by Eq. (3)[7]:

$$\alpha_b = 100\% - \alpha_a - \alpha_c \tag{3}$$

Differential Scanning Calorimetry

The DSC measurements were conducted on a Mettler Toledo DSC 822^e instrument (Mettler Toledo GmbH, Schwerzenbach, CH). Before testing, temperature and heat flow signals from the apparatus were calibrated using indium and zinc. Thermal analyses were carried out in air with PE specimens whose weight ranged between 2 and 3 mg. Samples with a thickness of 0.2 mm and a diameter of 4 mm were prepared from the semi-finished and post-treated films by planning a layer with the defined thickness and by cutting with the defined cross-section (diameter). Each DSC curve was recorded from −30 to 180 °C applying a heating rate of 10 °C/min. The degree of crystallinity was determined as the ratio between the heat of fusion of the specimen and the heat of fusion of a perfect (100%) crystalline PE (293.6 J/g).[14] To check possible superimposed melting and crystallization effects due to a low heating rate, various heating rates ranging from 10 to 50 °C/min were applied in preliminary investigations. Because no effect of heating rate on the DSC based crystallinity values was found, the standardized rate of 10 °C/min was used. By measuring on three different specimens mean and standard deviation values were evaluated.

Results and Discussion

Raman Spectroscopy

As discussed in various references[2,4–7,11–13,15–20], in Figure 2 a representative Raman spectrum of PE material H9 is exhibited along with the assignment of the Raman bands to vibration modes. Attributions of Raman bands to morphological phases and to conformational states of the PE chains are well established.[2,4–7,11–13,15–20] In the spectral region from 1000 to 1550 cm^{-1} the bands between 1400 and 1480 cm^{-1} correspond to methylene bending vibrations ($\delta(CH_2)$). The crystalline phase is represented by the peak at 1416 cm^{-1} (crystallinity band)[7]. This band is in conjunction with a crystal field splitting phenomenon, which is resulting from an interaction between the two chains of the orthorhombic unit cell in all-trans

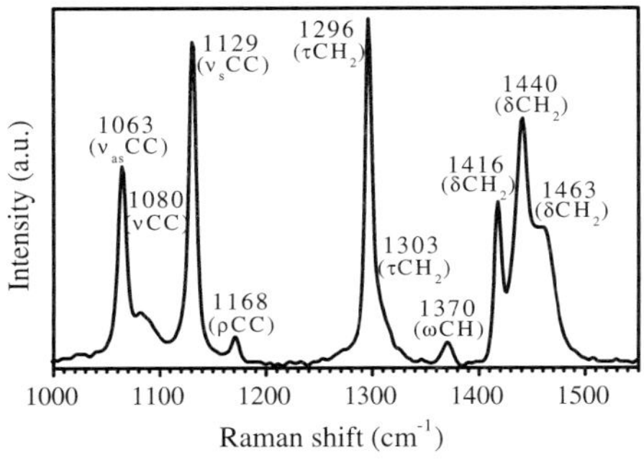

Figure 2.
Raman spectrum of PE together with indication of peak wavenumbers and vibration modes.

conformation.[16] The crystallinity band at 1416 cm^{-1} has been used to determine the degree of orthorhombic crystallinity of PE samples.[2–13,15,19] For the monoclinic crystalline structure, which is a further lattice structure observed in PE, high chain folding energy is required. Hence, a direct crystallization from the melt into the monoclinic form is impossible. [5,21] A prerequisite for recrystallization from the orthorhombic to the monoclinic structure is a high energy input (i.e., shear or compression deformation).[22–24] For melt crystallized PE samples without significant energy input the orthorhombic crystallinity represents the overall crystalline fraction.

The bands at 1080 and 1303 cm^{-1} are assigned to the amorphous phase with PE chains in gauche conformation. Both bands have been used to determine the amount of the amorphous fraction. However, the values of the amorphous content derived from the band at 1303 cm^{-1} (amorphous band) are reported to be more reliable.[15] The Raman peaks at 1063 and 1123 cm^{-1} are attributed to the symmetric and asymmetric C–C stretching vibrations, respectively, occurring in conformational trans sequences longer than 10 all-trans bonds.[16,25] Because of the highly localized character of the methylene twisting vibrations ($\tau(CH_2)$), the peaks at about 1300 cm^{-1} are almost independent of conformation and temperature. Therefore, the integrated intensity of the methylene twisting bands is used as an internal intensity standard.

In Figure 3 Raman spectra of the investigated PE samples at the various processing levels are displayed. A qualitative interpretation of the crystalline and amorphous fractions is done by analysis of the spectral regions from 1400 to 1550 cm^{-1} (bending vibrations) and 1250 to 1350 cm^{-1} (twisting vibrations), respectively. For the semi-finished and post-treated films higher intensities of the amorphous band at 1303 cm^{-1} are observable.

Hence, due to extrusion or sintering of the raw material an increase of the amorphous phase is established. Concerning the crystallinity band no distinct trends

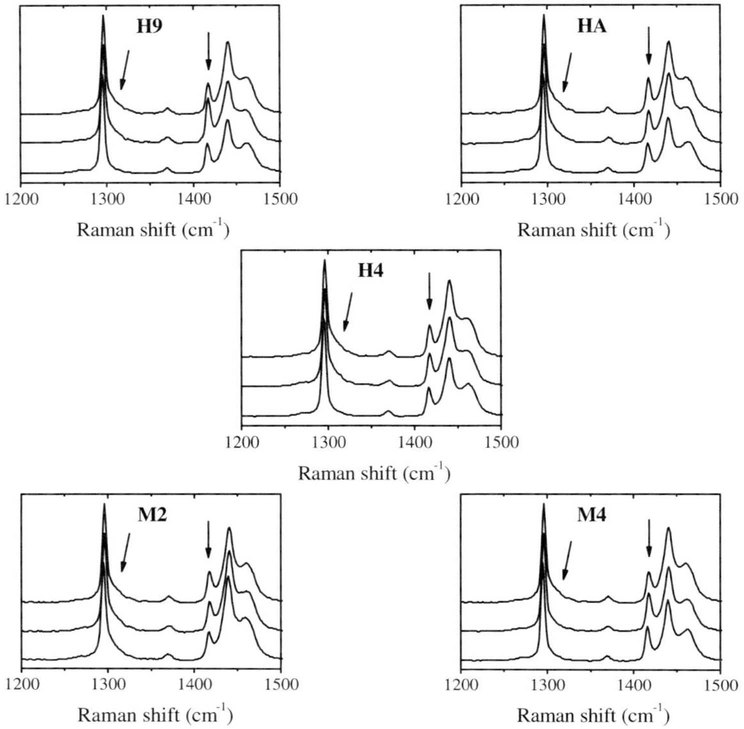

Figure 3.
Raman spectra of the investigated PE grades (M2, M4, H4, H9 and HA) at raw material, semi-finished film and post-treated film level (from bottom to top). The spectra are normalized to the internal intensity standard band at 1296 cm^{-1}.

along the process chain are ascertainable. Therefore, a comprehensive analysis of the three phases was carried out. Quantitative data of the crystalline, amorphous and interfacial contents for the investigated PE materials are revealed in Figure 4 and summarized in Table 2. For many of the investigated PE grades the highest crystallinity values but lowest amorphous fractions were obtained for the raw material (exception: M2 grade with the lowest average molar mass value). As an effect of film extrusion, sintering and post-treatment the crystallinity of the samples M4, H4, H9 and HA decreased whereas the content of the amorphous phase increased. Hence, the progresses of the crystallinity and amorphous content for polymers with an average molar mass higher than about 400,000 g/mol are following the same trend. Again, M2 with the lowest average molar mass exhibited a different trend with less significant changes in the crystalline and amorphous fractions. Regarding the interphase the highest values were established for the raw materials. Due to processing a tendency for a decrease of the interfacial content is observable.

As to the effect of average molar mass on the morphology, for the various raw materials the highest crystallinity value was obtained for the extrusion grade M4. For the UHMWPE grades slightly lower crystallinities, but higher interfacial fractions and thus the lowest amorphous contents were identified. The lowest crystalline fraction determined for the extrusion grade M2 with an average molar mass of only 250,000 g/mol is presumably related to a higher branching density.

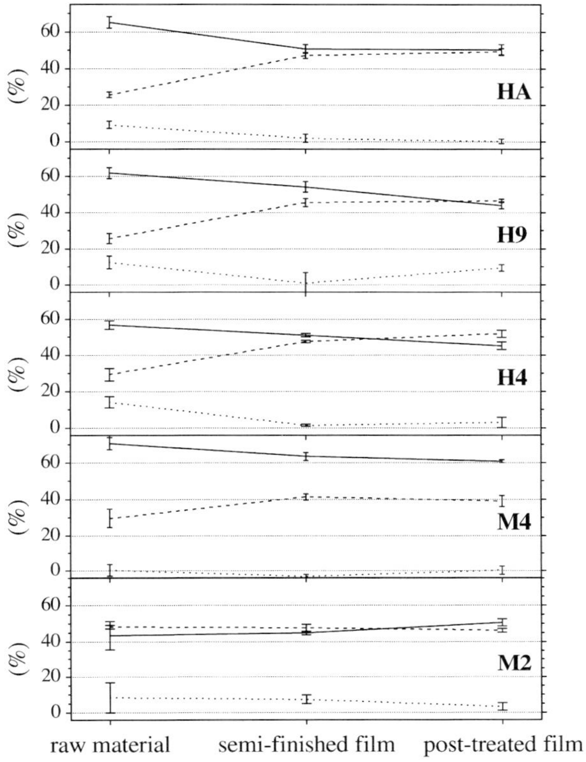

Figure 4.
Crystalline —, amorphous --- and interfacial ········ fractions as a function of processing level.

At semi-finished film level it is remarkable that the various UHMWPE grades investigated exhibit comparable phase fractions. Thus, it can be concluded that the morphology after sintering is not affected by the average molar mass values but by the sinter process. In contrast, after post-treatment different phase fractions were identified for the UHMWPE samples, which can be related to the differences in the molecular structure.

Differential Scanning Calorimetry

In Figure 5 the DSC traces of the investigated PE grades are shown for the various process levels. While the extrusion grade M2 exhibits the lowest melting peak temperatures of about 120 °C, the values of the second extrusion grade M4 are comparable to the melting peak temperatures of the investigated UHMWPE polymers. For all grades extrusion or sintering is associated by a significant shift of the melting

Table 2.
Mean and standard deviation values (%) of the three phase fractions of the investigated materials.

Sample	Raw material			Semi-finished film			Post-treated film		
	α_a	α_b	α_c	α_a	α_b	α_c	α_a	α_b	α_c
M2	48 ± 1	8 ± 8	43 ± 8	48 ± 2	7 ± 2	45 ± 1	46 ± 1	3 ± 2	50 ± 2
M4	30 ± 5	0 ± 3	71 ± 3	41 ± 2	-3 ± 1	63 ± 2	39 ± 3	0 ± 2	61 ± 1
H4	29 ± 3	14 ± 3	57 ± 2	48 ± 1	2 ± 1	51 ± 1	52 ± 2	3 ± 3	45 ± 2
H9	26 ± 3	13 ± 3	62 ± 3	45 ± 2	1 ± 6	54 ± 3	46 ± 1	9 ± 2	44 ± 2
HA	26 ± 2	9 ± 2	65 ± 3	47 ± 2	2 ± 2	51 ± 2	49 ± 2	0 ± 1	50 ± 3

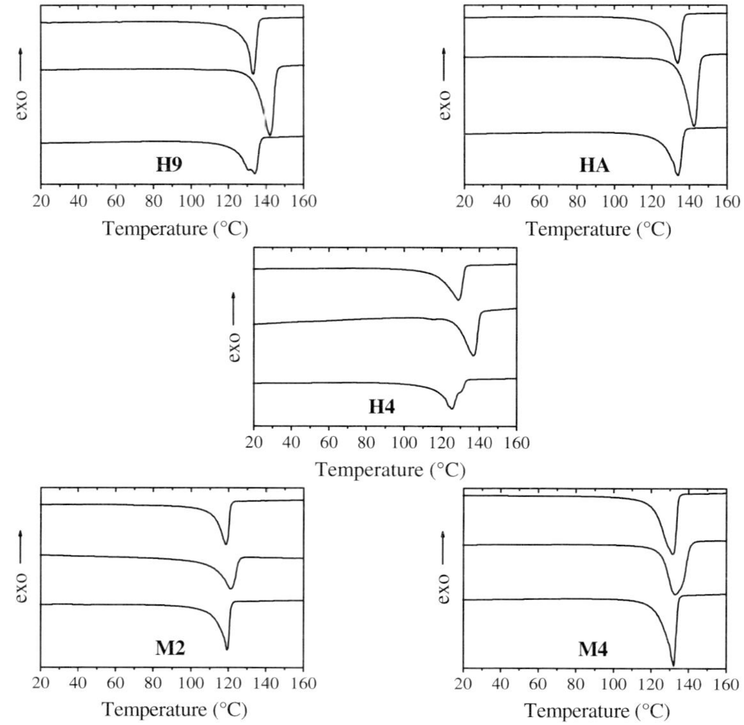

Figure 5.
DSC melting endotherms of the investigated PE grades (M2, M4, H4, H9 and HA) at raw material, semi-finished film and post-treated film level (from bottom to top).

peak to higher temperatures. This shift is most likely attributed to thicker crystal lamellae. However, because of the lower crystalline and interfacial fractions derived from Raman spectroscopy presumably less crystal lamellae are formed associated with thicker amorphous layers, which can be attributed to the different thermal and process histories during polymerization and film production.

For the post-treated films, compared to the semi-finished films, a shift of the melting peak temperatures to lower values is ascertained. Due to flaming the samples were exposed to temperatures above the melting temperature, thus leading to a melting and recrystallization of the film surface. This negative shift is most likely in conjunction with the formation of thinner crystal lamellae. Because of the lower crystalline fractions derived from Raman spectroscopy for many of the investigated PE grades (M4, H4, H9 and HA), a comparable amount of crystal lamellae, but with lower thickness is evolved. The different morphology is probably related to a higher cooling rate during post-treatment.

A comparison of the crystallinity values determined by DSC and Raman spectroscopy is revealed in Figure 6 and summarized in Table 3. In general, a good agreement between the DSC and Raman crystallinities was found, thus corroborating the results obtained by Raman spectroscopy and the assumption of an orthorhombic crystalline structure. With exception of the M2 grade the correlation coefficients are equal to 0.90 or higher. The lower correlation coefficient for the M2 grade is related to the fact that the crystallinity values are within a small range. The tendency for lower Raman crystallinity

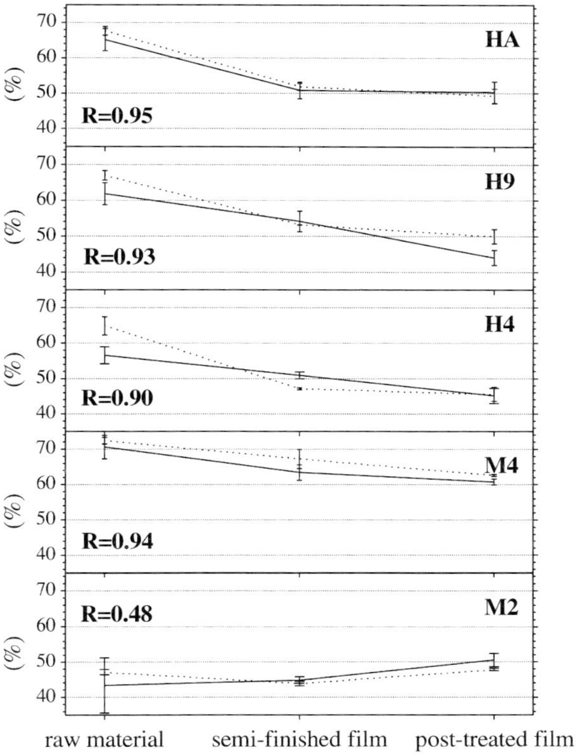

Figure 6.
Crystallinity as a function of processing level of the investigated materials determined by Raman spectroscopy — and DSC ········; R represents the correlation coefficient between Raman and DSC results.

Table 3.
Mean and standard deviation of the crystallinity values (%) of the investigated materials obtained by DSC and Raman spectroscopy.

Sample	Raw material		Semi-finished film		Post-treated film	
	DSC	Raman	DSC	Raman	DSC	Raman
M2	47 ± 1	43 ± 8	44 ± 1	45 ± 1	48 ± 0	50 ± 2
M4	72 ± 1	71 ± 3	67 ± 3	63 ± 2	63 ± 0	61 ± 1
H4	65 ± 3	57 ± 2	47 ± 0	51 ± 1	46 ± 2	45 ± 2
H9	67 ± 1	62 ± 3	53 ± 0	54 ± 3	50 ± 2	44 ± 2
HA	68 ± 1	65 ± 3	52 ± 1	51 ± 2	49 ± 2	50 ± 3

values compared to DSC reported in the literature[5,11,19] is confirmed to some extent. As discussed in section Raman spectroscopy the crystallinity values are decreasing significantly for the samples M4, H4, H9 and HA along the process chain. In contrast the crystallinity of the M2 grade is nearly unaffected by the various process steps investigated.

Conclusions

For various PE ski base materials it has been demonstrated that Raman spectroscopy allows for a comprehensive morphological analysis obtaining fractions for the crystalline, amorphous and interfacial phase. Crystallinity values derived by Raman spectroscopy are in good agreement with

DSC based data, thus corroborating the implemented and used Raman technique. Along the process chain from the raw material to the post-treated extruded or sintered film for many of the investigated PE grades a decrease of the crystalline and interfacial fraction associated by an increase of the amorphous content was detected. UHMWPE grades with different average molar mass values exhibited morphological differences at raw material and post-treated film level. However, after sintering the phase fractions are comparable and thus independent of the average molar mass.

Acknowledgements: The research work of this paper was performed at the Polymer Competence Center Leoben GmbH (PCCL, Austria) within the framework of the Kplus-program of the Austrian Ministry of Traffic, Innovation and Technology with contributions by the Institute of Materials and Testing of Plastics at the University of Leoben (Austria) and the Fischer GmbH (Austria). The PCCL is funded by the Austrian Government and the State Governments of Styria and Upper Austria. The authors wish to express their acknowledgement to Ao.Univ.-Prof. Dr. Ronald J. Bakker of the Institute of Mineralogy and Petrology at the University of Leoben (Austria) for his cooperation in course of this study.

[1] Association of Plastic Manufacturers in Europe [APME], *"The Compelling Facts About Plastics"*, 2007.

[2] A. Bertoluzza, C. Fagnano, M. Rossi, A. Tinti, G. L. Cacciari, *Journal of Molecular Structure* **2000**, *521*, 89.

[3] S. Affatato, G. Bersaglia, D. Emiliani, I. Foltran, P. Taddei, M. Reggiani, P. Ferrieri, A. Toni, *Biomaterials* **2003**, *24*, 4045.

[4] M. Visentin, S. Stea, M. de Clerico, M. Reggiani, C. Fagnano, S. Squarzoni, A. Toni, *J. Biomater. Appl.* **2006**, *21*(2), 131.

[5] M. Reggiani, A. Tinti, P. Taddei, M. Visentin, S. Stea, M. de Clerico, C. Fagnano, *Journal of Molecular Structure* **2006**, *785*, 98.

[6] M. Kyomoto, Y. Miwa, G. Pezzotti, *J. Biomater. Sci. Polymer Edn.* **2007**, *18*, 165.

[7] G. R. Strobl, W. Hagedorn, *J. Polymer Sci.* **1978**, *16*, 1181.

[8] D. Dothée, M. Berjot, W. Marx, *J. Polymer Degradation and Stability* **1988**, *20*, 149.

[9] F. Rull, A. C. Prieto, J. M. Casado, F. Sobron, H. G. M. Edwards, *J. Raman. Spectrosc.* **1993**, *24*, 545.

[10] C. C. Naylor, R. J. Meier, B. J. Kip, K. P. J. Williams, S. M. Mason, N. Conroy, D. L. Gerrrard, *Macromolecules* **1995**, *28*, 2969.

[11] W. Lin, M. Cossar, V. Dang, J. Teh, *Polymer Testing* **2007**, *26*, 814.

[12] L. Mandelkern, R. G. Alamo, M. A. Kennedy, *Macromolecules* **1990**, *23*, 4721.

[13] M. A. Rodríguez-Pérez, R. A. Campo-Arnáiz, R. F. Aroca, J. A. de Saja, *Polymer* **2005**, *46*, 12093.

[14] M. Pyda, *ATHAS DataBank* **2007**, http://athas.prz.rzeszow.pl.

[15] G. Keresztury, E. Földes, *Polymer Testing* **1990**, *9*, 329.

[16] J. C. Rodriguez-Cabello, J. Martin-Monge, J. M. Lagaron, J. M. Pastor, Macromol, *Chem. Phys.* **1998**, *199*, 2767.

[17] C. Fagnano, M. Rossi, R. S. Porter, S. Ottani, *Polymer* **2001**, *42*, 5871.

[18] J. M. Lagaron, *J. Materials Sci.* **2002**, *37*, 4101.

[19] R. P. Paradkar, S. S. Sakhalkar, X. He, M. S. Ellison, *J. Appl. Polymer Sci.* **2003**, *88*, 545.

[20] M. Zheng, W. Du, *Vibrational Spectroscopy* **2006**, *40*, 219.

[21] T. Yemni, R. L. McCullogh, *J. Polymer Sci.* **1973**, *11*, 1385.

[22] K. M. Furuheim, D. E. Axelson, H. W. Anthonsen, T. Helle, *J. Appl. Polymer Sci.* **2004**, *91*, 218.

[23] L. Kurelec, S. Rastogi, R. J. Meier, P. J. Lemstra, *Macromolecules* **2000**, *33*, 5593.

[24] J. M. Lagaron, N. M. Dixon, W. Reed, J. M. Pastor, B. J. Kip, *Polymer* **1999**, *40*, 2569.

[25] R. J. Meier, *Polymer* **2002**, *43*, 517.

Temperature Effect on Drying and Swelling of Kappa Carrageenan Gels: A Steady State Fluorescence Study

Özlem Tarı,[*1] Önder Pekcan*[2]

Summary: A novel technique based on in situ steady state fluorescence (SSF) measurements is introduced for studying drying and swelling of κ- carrageenan (kappa carrageenan) gels at various temperatures. κ- carrageenan gels were completely dried and then swelled in water vapor. Pyranine was embedded in κ- carrageenan and used as a fluorescence probe. Scattered light intensities, I_{sc} and fluorescence intensities, I were monitored during the drying and swelling of κ- carrageenan gels. It was observed that the fluorescence intensity decreased linearly as drying time was increased. A simple model consisting of Case II diffusion was used to quantify the drying processes of the κ- carrageenan gels. This moving boundary model provided packing constant, k_0. During swelling, fluorescence intensity increased exponentially as time is increased. The increase in I, was modeled using Li-Tanaka equation from which swelling time constants, τ_c and cooperative diffusion coefficients, D_c were determined. It was observed that swelling time constants, τ_c decreased and diffusion coefficients, D_c increased as the swelling temperature was increased. Activation energies for drying and swelling were also obtained and found to be 53.9 and 47.2 kJ mol^{-1}, respectively.

Keywords: cooperative diffusion coefficients; drying; fluorescence; κ- carrageenan; swelling

Introduction

Two network phases of different degree of swelling can exist and the transition from one state of network to the other is called volume phase transition.[1] Volume phase transitions in gels may occur from dry to swollen states, either continuously, or by sudden jumps between them.[2,3] The swelling, shrinking and drying kinetics of physical gels are important in many technological applications. Especially in pharmaceutical industries in designing controlled release of drugs and in using cosmetic ingredients, understanding the kinetics of gels is highly desirable. The knowledge of the gel kinetics is an important requirement for producing storable foods in agricultural industry and developing artificial organs in medical applications. In general the elastic and swelling properties or permanent networks can be understood by considering two opposing effects: osmotic pressure and restraining force.[4–6] Usually the total free energy of a chemically crosslinked network can be separated into two terms: bulk and shear energies. In a swollen network the characteristic quantity of the bulk free energy is the osmotic bulk modulus, *K*. The shear energy as the other important energy, keeps the gel in shape by minimizing the non-isotropic deformation. The characteristic coefficient of these forces is the shear modulus, *G* which can be most directly evaluated by stress–strain measurements.[7]

The theory of kinetics of swelling for a spherical chemical gel was first developed by Tanaka and Fillmore.[4] They assumed that the shear modulus *G* is negligible compared to the osmotic bulk modulus. However,

[1] Department of Physics, Istanbul Technical University, Maslak, 34469 Istanbul, Turkey
Fax: (+90) 2122856386;
E-mail: otari@itu.edu.tr
[2] Department of Physics, Işık University, Kumbaba, Şile, 34980 Istanbul, Turkey

several studies have shown that the shear modulus is the same order of magnitude as the osmotic bulk modulus.[8,9] Later, Peters and Candau have derived a model for the kinetics of swelling of spheres, cylinders and disks made of polymer gels by assuming non-negligible shear modulus.[10] Li and Tanaka have developed a model where the shear modulus plays an important role that keeps the gel in shape due to coupling of any change in different directions.[11] This model predicts that the geometry of the gel is an important factor, and swelling is not a pure diffusion process.

Several experimental techniques have been employed to study the kinetics of swelling, shrinking and drying of chemical and physical gels, e.g. neutron scattering[12], quasielastic light-scattering[10], macroscopic experiments[13] and in situ interferometric[14] measurements. Hawlader et al.[15] used a one-dimensional diffusion model to describe the heat and mass transfer in the wet and dry regions of materials undergoing shrinkage during drying. Coumans[16] has provided an excellent tutorial overview of the uses of the diffusion equation to analyze drying characteristic of slabs, including lumped diffusion models, retreating front models, and the characteristic drying curve model. The method given by Coumans relates to porous and nonporous materials. Some of these models also enable the evaluation of moisture dependent diffusivities from experimental drying curves of slabs. Steady state and time-resolved fluorescence techniques were applied to drying process of selected silane gels in oxygen free atmosphere. A kinetic model of drying was suggested and drying rate constants were determined.[17] The steady-state fluorescence technique was performed for studying drying and swelling kinetics in disc shape gels[18–21]. Recently, fast transient fluorescence (FTRF) technique was used in our laboratory to study gel swelling[22,23] and drying[24,25] processes.

In this work, we studied drying and swelling of κ- carrageenan gels at various temperatures by using steady-state fluorescence technique. κ- carrageenan gels doped with pyranine were completely dried and then swelled in water vapour. Drying of these gels were quantified by employing moving boundary model from which linear relaxation constants, k_0 were determined. Li-Tanaka equation was used to determine the swelling time constants, τ_c and cooperative diffusion coefficients, D_c for the swelling processes. It was observed that swelling time constant, τ_c decreased and cooperative diffusion coefficients, D_c increased as the swelling temperature increased.

Theoretical Considerations

Kinetics of Drying

The linear transport mechanism is characterized by the following steps. As the water molecules desorp from the gel, that is, as the gel starts drying, a moving boundary forms. This boundary proceeds with a constant velocity.

Now, consider a cross section of a gel with thickness d, under going Case II Diffusion[26] as in Figure 1, where L is the position of the advancing desorption front, C_0 is the initial molecule concentration and k_0 (mg/cm^{2}min) is defined as the packing constant. In fact, here k_0 represents the parameter for the packing of helices during drying of the gel. The kinetic expression for the desorption in the slab of an area A is given by

$$\frac{dM_t}{dt} = -k_0 A \tag{1}$$

where the amount of water molecules, M_t at time t is given by

$$Mt = -\int_0^t k_0 A dt + M_0 \tag{2}$$

here $M_0 = C_0 A d$ is the initial amount of water molecules trapped in the swollen gel at time zero. The amount of desorbed molecules at time t, can be written as

$$(M_0 - M_t) = k_0 A t \tag{3}$$

Since $M_t = C_0 A L$, then Equation (3) provides

$$C_0 A(d - L) = k_0 A t \tag{4}$$

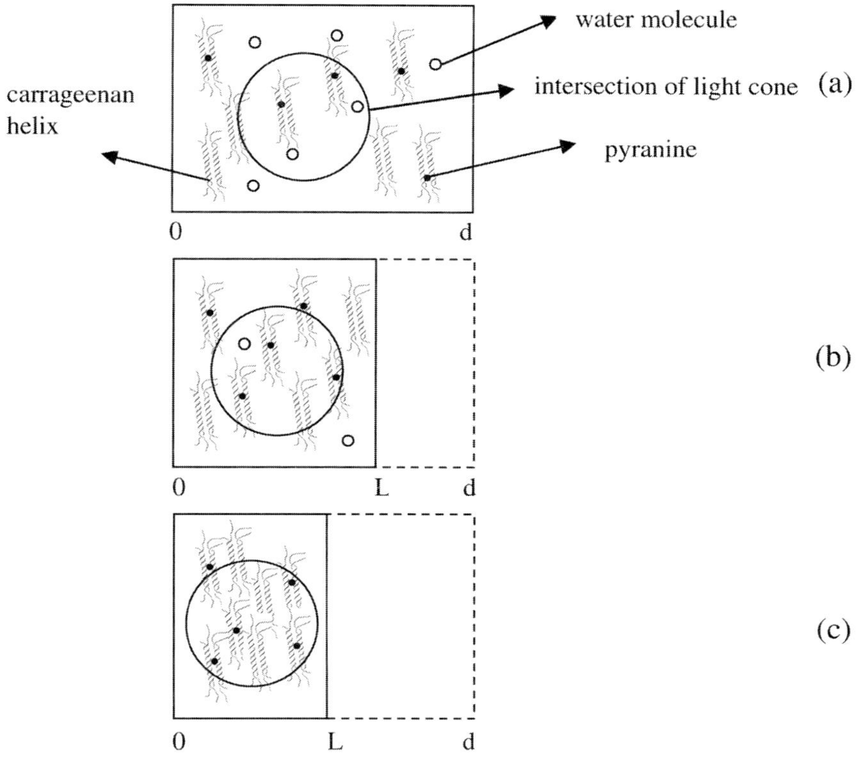

Figure 1.
A schematic representation of drying process (a) swollen gel, (b) drying gel and (c) dried gel.

The time derivative of Equation (4) produces the following relation

$$\frac{dL}{dt} = -\frac{k_0}{C_0} \tag{5}$$

Equation (5) can predict that the packing front, position at L, moves toward the origin with a constant velocity, k_0/C_0. The algebraic relation for L as a function of time, t is then described by Equation (6)

$$L = -\frac{k_0}{C_0}t + d \tag{6}$$

Kinetics of Swelling

Li and Tanaka[11] showed that the kinetics of swelling and shrinking of a polymer network or gel obey the following relation,

$$\frac{W(t)}{W_\infty} = 1 - \sum_{n=1}^{\infty} B_n e^{-t/\tau_n} \tag{7}$$

where W(t) and W_∞ are the degree of swelling or solvent uptake at time t and at infinite equilibrium, respectively. Here B_n represents a constant related to the ratio of the shear modulus, G and the longitudinal osmotic modulus, M is defined by the combination of shear and osmotic bulk modulus as $M = (K + 4G/3)$ and τ_n is the swelling rate constant. In the limit of large t or if τ_1 is much larger than the rest of τ_n, all higher terms ($n \geq 2$) in Equation (7) can be neglected, then Equation (7) becomes

$$\frac{W(t)}{W_\infty} = 1 - B_1 e^{-t/\tau_c} \tag{8}$$

Here B_1 is given by the following relation[11]:

$$B_1 = \frac{2(3-4R)}{\alpha_1^2 - (4R-1)(3-4R)} \tag{9}$$

where $R = G/M$ and α_1 is given as a function of R, i.e.

$$R = \frac{1}{4}\left[1 + \frac{\alpha_1 J_0(\alpha_1)}{J_1(\alpha_1)}\right] \tag{10}$$

where J_0 and J_1 present Bessel functions. In Equation (8), τ_c is related to the collective cooperative diffusion coefficient D_c of a gel disk at the surface and given by the relation

$$D_c = \frac{3a_\infty^2}{\tau_c \alpha_1^2} \tag{11}$$

Here, a_∞ is the half thickness of the gel in the final equilibrium state. Once the quantities τ_c and B_1 are obtained, R, α_1, and D_c can be calculated.

Experimental Part

Carrageenan (Sigma) at 3% (wt) concentration and pyranine were dissolved in distilled water (pH 6.5) by heating. The pyranine concentration were kept at 4×10^{-4} M, which is low enough to ensure that any excitation transfer effects are negligible. The heated carrageenan solution was held at 80 °C and was continuously stirred by a magnetic stirrer. Then this solution was cooled down to room temperature. After the gels were formed, we cut the gel samples into disc shape gels to use in drying. Disc-shaped gel samples were placed on the wall of 1 × 1 quartz cell for the drying and swelling experiments. In fluorescence spectrometer the position of the gel and the incident light beam, I_o for the fluorescence measurements are shown in Figure 2a during drying in air. In Figure 2b the position of the gel during swelling in water vapor is presented. These gels were first dried at 30, 40, 50 and 60 °C respectively before the swelling measurements started. The drying and swelling experiments of κ- carrageenan gels were performed at temperatures of 30, 40, 50 and 60 °C respectively. The fluorescence intensity measurements were carried out using the Model LS-50 spectrometer of Perkin-Elmer, equipped with temperature controller.

All measurements were made at 90° position and slit widths were kept at 5 nm. Pyranine was excited at 460 nm during *in situ* experiments and emission intensities of the pyranine were monitored at 515 nm as a function of drying and swelling time. Typical spectra of pyranine at various drying and swelling times are presented in Figure 3a and Figure 3b, respectively.

Results and Discussion

Drying

Scattered light, I_{sc} and fluorescence intensities, I during drying of κ- carrageenan gels for the various temperatures are presented in Figure 4 and Figure 5 respectively.

It is seen in Figure 4 and Figure 5 that the fluorescence intensities decreased as the scattered light intensity increased, predicted that gel becomes turbid during drying. Figure 1 shows that as water molecules

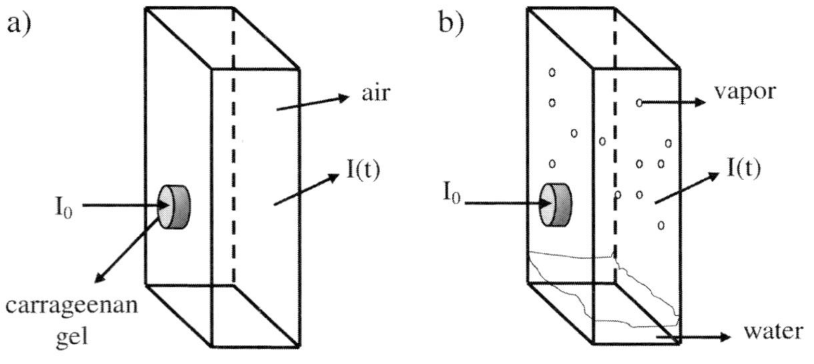

Figure 2.
The position of κ- carrageenan gel in the fluorescence cell (a) during drying, (b)during swelling in water vapour. I_o is the excitation and $I(t)$ is the emission intensities at 460 nm and 515 nm, respectively.

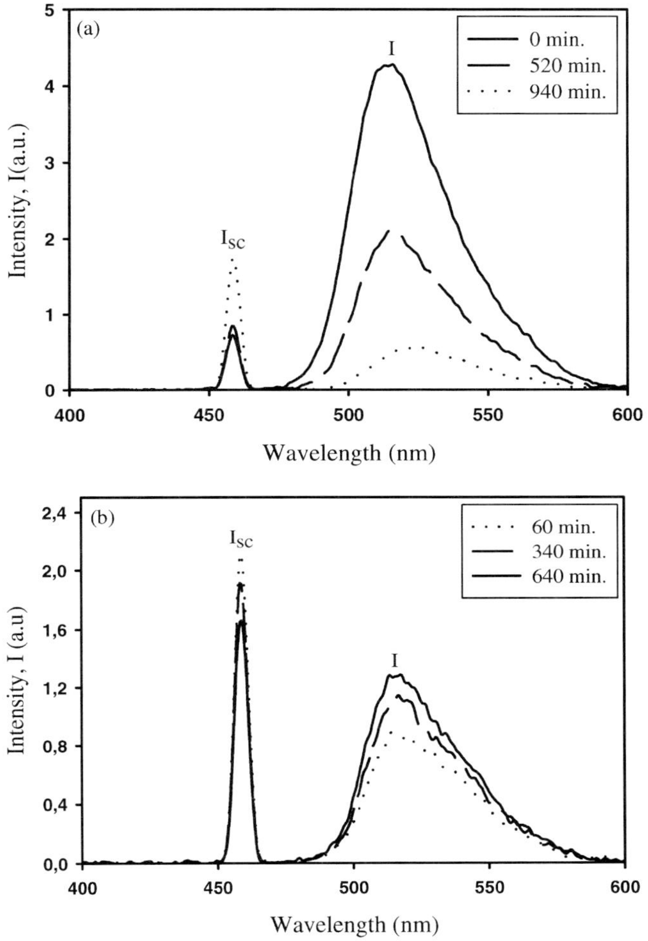

Figure 3.
Fluorescence spectra of pyranine during (a) drying, (b) swelling.

desorp from the drying gel double helices pack and crowd into the light cone (incident light I_0) intersection. Crowding helices prevent the incident light beam to penetrate into the gel sample. As a result less pyranine molecules can be excited, which cause a decrease in the fluorescence light intensity. Since drying occurs in the gel state of κ- carrageenan, no quenching of pyranine molecules are expected. In other words pyranine molecules are assumed to be buried in the double helices.

This behavior of fluorescence intensity, I during drying can be modeled by using Equation (3), where M_0 and M values are assumed to be proportional to I_0 and I values at time zero and at time t. Then, Equation (3) becomes

$$\frac{I_0 - I}{I_0} = \frac{k_0}{C_0 d} t \tag{12}$$

Organizing Equation (12) provides us with a very useful relation

$$\frac{I}{I_0} = 1 - \frac{k_0}{C_0 d} t \tag{13}$$

Equation (13) predict that fluorescence intensity decreases linearly as the drying time is increased, due to packing of double helices i.e. due to increasing turbidity of carrageenan gel which scatters the incident light. Fitting Equation (13) to the data in

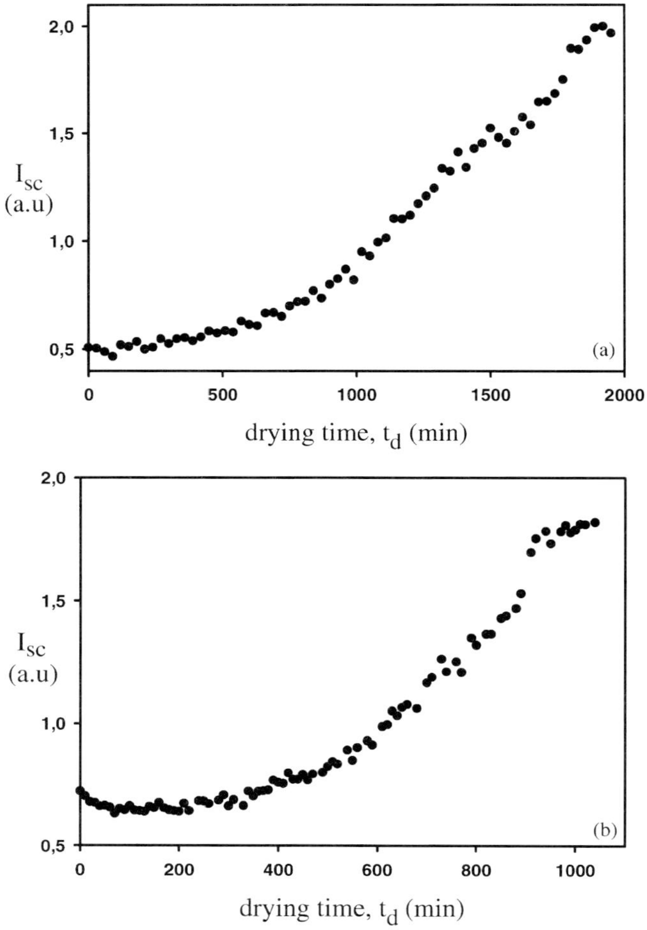

Figure 4.
Scattered light intensities of pyranine at (a) 30 and (b) 40 °C during drying time

Figure 5 produces k_0 values which are listed in Table 1 together with the other measured parameters of the gel samples where d is the diameter, a_i and a_∞ are the thickness, m_i and m_∞ are the weights of the gel before and after the drying process. It is seen that k_0 value increases as the temperature is increased, as expected, helices can be packed faster at higher temperatures.

The temperature-dependence of the packing constant, k_0, for the helices can be treated by the well-known Arrhenius relation given below

$$k_0 = k_{00} e^{-\Delta E/kT} \tag{14}$$

where ΔE is the energy for packing process of carrageenan gel, k is the Boltzmann constant, T is the temperature and k_{00} is the pre exponential factor. Figure 6 present the fitting of Equation (14) to the k_0 data in Table 1. The activation energy, ΔE measured from the fluorescence intensity is found to be 53.9 kJ mol^{-1}.

Swelling

The plots of fluorescence, I and scattered light intensities versus time during swelling of κ- carrageenan gels at various temperatures are presented in Figure 7 and Figure 8, respectively. It is seen that fluorescence intensity increased, however, scattered light

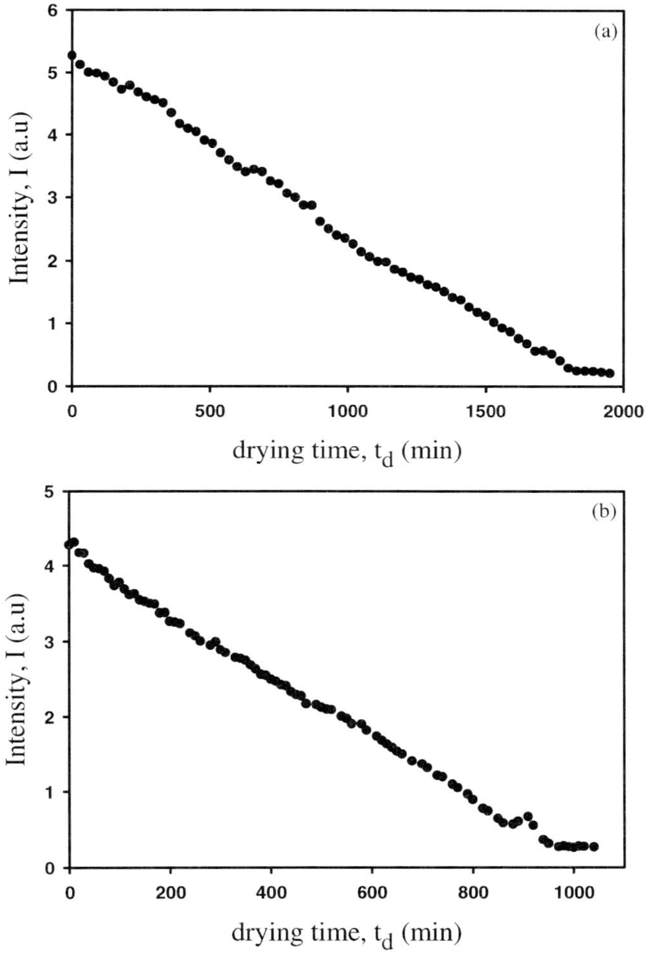

Figure 5.
Fluorescence intensities of pyranine at (a) 30 and (b) 40 °C during drying time.

intensity decreased during swelling. Since the transmitted light intensity, $I_{tr} = 1 - I_{sc}$ increases, the gel became transparent, as a result I increases.

Here since swelling occurs in gel state of carrageenan then one has to assume that pyranine molecules are embedded in the helices, so that no quenching can take place.

Table 1.
Experimentally obtained drying parameters.

Gel properties	Temperature			
	30 °C	40 °C	50 °C	60 °C
a_i (mm)	3.3	3.2	3.35	3.2
a_∞ (mm)	0.45	0.7	0.8	0.90
m_i (g)	0.201	0.208	0.2055	0.1825
m_∞ (g)	0.0108	0.0104	0.0102	0.0085
d(mm)	8.85	8.7	8.8	8.8
$k_o \times 10^{-8}$ ($mm^2g^{-1}s^{-1}$)	3.37	5.29	14.3	20.6

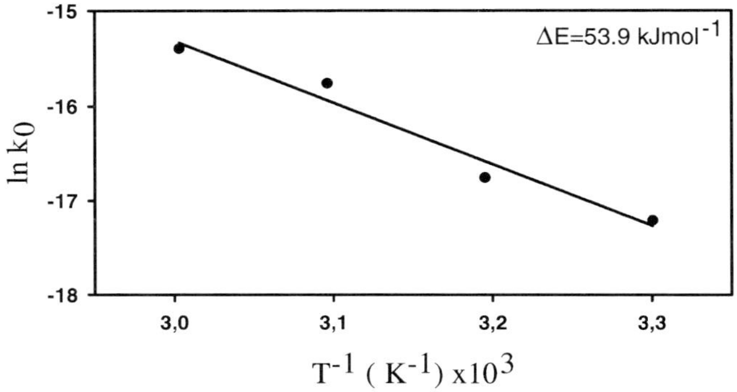

Figure 6.
The logarithmic plot of k_0 values versus temperature T^{-1} according to Equation. (14). The slope of the linear relation produces the activation energy, ΔE for drying process.

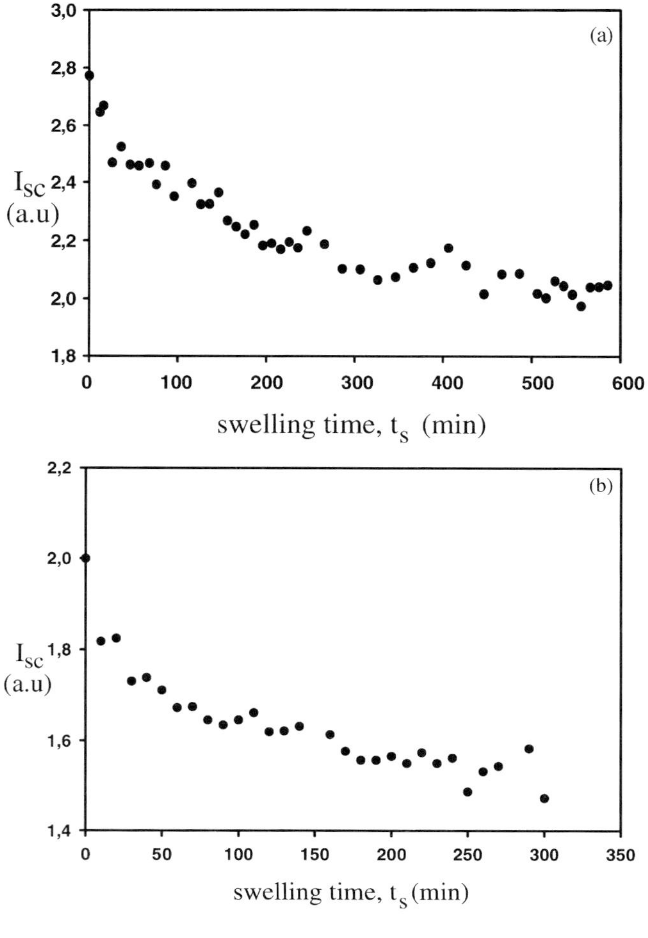

Figure 7.
Scattered light intensities of pyranine at (a) 30 and (b) 50 °C during swelling time

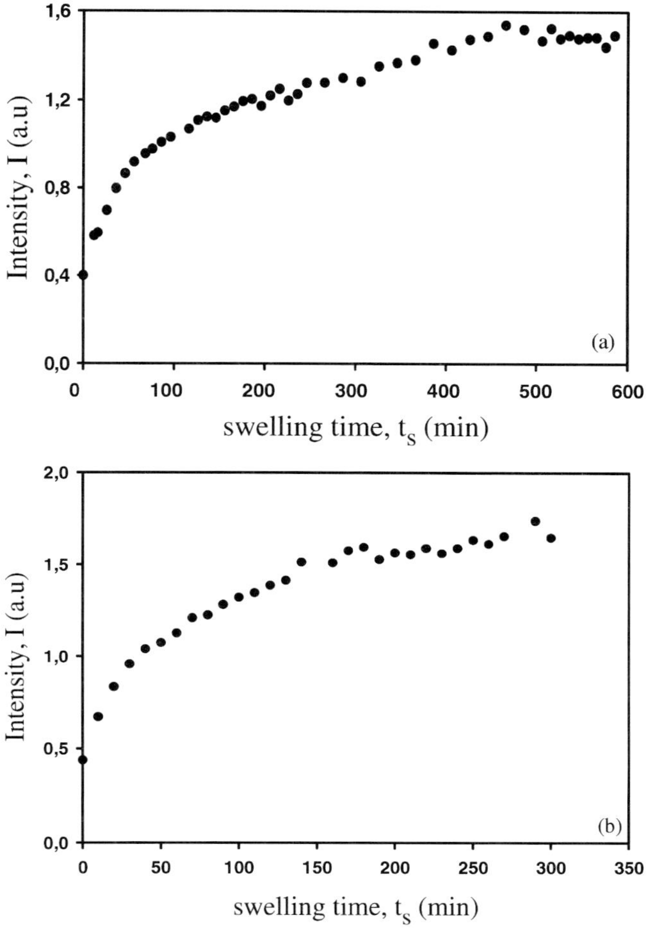

Figure 8.
Fluorescence intensities of pyranine at (a) 30 and (b) 50 °C during swelling time.

At the equilibrium state of swelling, the fluorescence intensity reaches I_∞, where the vapor uptake is W_∞. The relation between the vapor uptake W and the fluorescence intensity, I is then given by

$$\frac{W}{W_\infty} = \frac{I}{I_\infty} \tag{15}$$

This relation predicts that as W increases, I increases. Combining Equation (15) with Equation (8) and calculating the logarithm of them, the following relation can be obtained

$$\ln\left(1 - \frac{I}{I_\infty}\right) = \ln B_1 - \frac{t_s}{\tau_c} \tag{16}$$

where $t = t_s$ is taken in Equation (8) to present the swelling time in Equation (16). Logarithmic plots of $(1 - \frac{I}{I_\infty})$ are presented in Figure 9. Linear regression of curves in Figure 8 provides us with the B_1 and τ_c values from Equation (16).

Taking into account the dependence of B_1 on R, one obtains R values and from α_1-R dependence, α_1 values were produced. Then using Equation (11) cooperative diffusion coefficients D_c were determined for these disc-shaped carrageenan gels. Experimentally obtained τ_c and D_c values for kappa carrageenan gels at various temperatures are presented in Table 2. It is seen in Table 2 that as temperature is increased, time constant, τ_c presented a

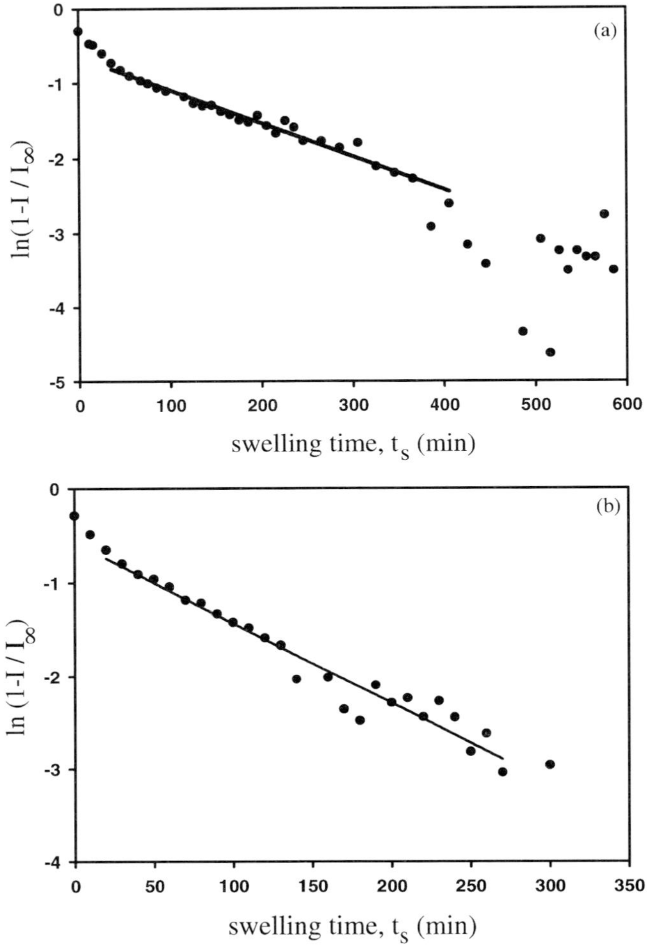

Figure 9.
Logaritmic plots of I, according to Equation (16) for κ- carrageenan gels in water vapour at (a) 30 and (b) 50 °C during swelling time.

decrease as expected, i.e. the time for the network homogenization decreased as the temperature is increased. Here we note that decrease in I_{sc} (increase in I_{tr}) may correspond to homogenization of the gel network, which produce high transparency.

The behavior of D_c versus temperature predicts that gel segments (helices) move

Table 2.
Experimentally obtained swelling parameters.

Gel properties	Temperature			
	30 °C	40 °C	50 °C	60 °C
a_i (mm)	0.45	0.7	0.8	0.90
a_∞ (mm)	1.6	1.6	1.7	1.9
m_i (g)	0.0108	0.0104	0.0102	0.0085
m_∞ (g)	0.03	0.0327	0.0332	0.0056
τ_c (min)	298	196	109	87
$D_c \times 10^{-7}$ (cm^2s^{-1})	2.24	3.43	7.22	11.4

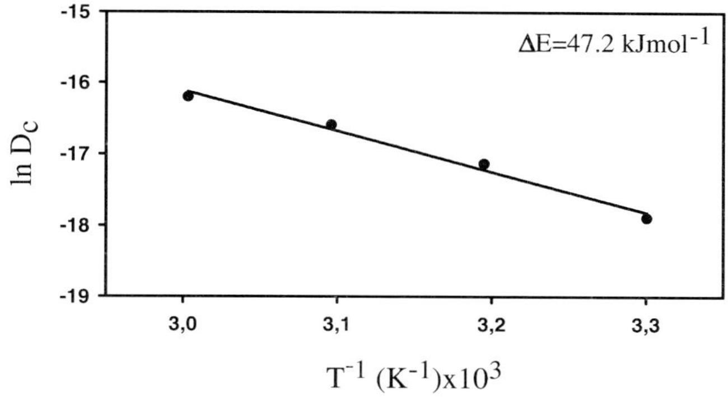

Figure 10.
The logarithmic plot of D_c values versus temperature T^{-1} according to Eq. (17). The slope of the linear relation produces the activation energy, ΔE for swelling process.

much faster at higher temperatures during vapor penetration. As seen in Table 2, D_c values increased as the temperature is increased, predicts that the D_c-T relation may obey the following Arrhenius law.

$$D_c = D_{c0}\exp(-\Delta E/kT) \qquad (17)$$

where the ΔE is named as the activation energy for swelling, k is the Boltzmann's constant and D_{co} is the cooperative diffusion coefficient at $T = \infty$. The logarithmic form of the D_c is plotted versus T^{-1} in Figure 10, where the slope of the linear relation produces the activation energy, ΔE for the swelling gel as 47.2 kJ mol^{-1}.

Conclusion

In summary, this paper presents a novel method for studying drying and swelling kinetics of κ- carrageenan gels at various temperatures. Case II diffusion model was used to measure the packing constants, k_0 during drying and the activation energy, ΔE for drying process. Also the swelling time constants, τ_c and the cooperative diffusion coefficients, D_c were measured during the swelling of κ- carrageenan gels by using Li-Tanaka model. It was observed that swelling time constant, τ_c decreased and cooperative diffusion coefficients, D_c increased as the swelling temperature increased. D_c values were used to obtain the swelling activation energy. It is interesting to note that the swelling activation energy was found to be smaller than the drying activation energy, indicating that the energy need for packing of helices is higher than unpacking of them. In other words contribution of hydrogen bonding to the swelling process was found to be much less than during drying process.

[1] M. Shibayama, T. Tanaka, *Adv. Polym. Sci.* **1993**, *109*, 1.
[2] K. Dusek, D. Paterson, *J. Polym. Sci.* **1968**, *A2*, 1209.
[3] T. Tanaka, *Phys. Rev. Lett.* **1980**, *45*, 1636.
[4] T. Tanaka, D. Filmore, *J. Chem. Phys.* **1979**, *70*, 1214.
[5] A. Peters, S. J. Candau, *Macromolecules* **1986**, *19*, 1952.
[6] E. Geissler, A. M. Hecht, *Macromolecules* **1980**, *13*, 1276.
[7] S. Candau, J. Bastide, M. Delsanti, *Adv. Polym. Sci.* **1982**, *44*, 27.
[8] M. Zrinyi, F. Horkay, *J. Polym. Sci. Polym. Phys. Ed.* **1982**, *20*, 815.
[9] E. Geissler, A. M. Hecht, *Macromolecules* **1981**, *14*, 185.
[10] A. Peters, S. J. Candau, *Macromolecules* **1988**, *21*, 2278.
[11] Y. Li, T. Tanaka, *J. Chem. Phys.* **1990**, *92*, 1365.
[12] J. Bastide, R. Duoplessix, C. Picot, S. Candau, *Macromolecules* **1984**, *17*, 83.
[13] M. Zrinyi, J. Rosta, F. Horkay, *Macromolecules* **1993**, *26*, 3097.
[14] C. Wu, C. Y. Yan, *Macromolecules* **1994**, *27*, 4516.
[15] M. N. A. Hawlader, J. C. Ho, Z. Qing, *Dry. Technol.* **1999**, *17*, 27.
[16] W. J. Coumans, *Chem. Eng. Process.* **2000**, *39*, 53.
[17] E. Miller, *J. Photoch. Photobio.* **2002**, *A152*, 249.
[18] Ö. Pekcan, Y. Yılmaz, *Progr. Colloid Polym. Sci.* **1996**, *102*, 89.

[19] Ö. Pekcan, Y. Yılmaz, *Polymer* **1998**, *39*, 5351.
[20] M. Erdogan, Ö. Pekcan, *J. Plym. Sci. Polym. Phys.* **2000**, *38*, 739.
[21] Ö. Tarı, Ö. Pekcan, *J. Appl. Polym. Sci.* **2007**, *106*, 4164.
[22] Ö. Pekcan, D. Kaya, M. Erdoğan, *Polymer* **2000**, *41*, 4915.
[23] Ö. Pekcan, D. Kaya, M. Erdoğan, *J. Appl. Polym. Sci.* **2000**, *76*, 1494.
[24] M. Erdogan, Ö. Pekcan, *Polymer* **2003**, *44*, 2129.
[25] Ö. Pekcan, M. Erdogan, *Compos. Interfaces* **2003**, *10*(6), 547.
[26] Ö. Pekcan, S. Kara, *J. Appl. Polym. Sci.* **2001**, *82*, 894.

Spectroscopic Investigations of Phase-Separated Thermotropic Layers Based on UV Cured Acrylate Resins

*Katharina Resch,**[1] *Gernot M. Wallner,*[2] *Reinhold W. Lang*[1,2]

Summary: In this paper spectroscopic techniques were adapted and applied to characterize the optical and morphological properties of thermotropic resins. Thermotropic films were prepared by variation of resin base and thermotropic additives. By UV/Vis/NIR spectroscopy the solar optical properties, the switching temperature, the switching process and the residual transmittance in the opaque state were determined. To control the sample temperature from ambient to 100 °C a conventional UV/Vis/NIR spectrophotometer equipped with an Ulbricht-sphere was adapted by a heating stage. Additive content, distribution and scattering domain size were characterized by Differential Scanning Calorimetry, Attenuated Total Reflectance spectroscopy and microscopic techniques. For the investigated film types hemispheric solar transmittance values ranging from 80 to 87% and 75 to 85% were obtained in the clear and the scattering state, respectively. The diffuse solar transmittance values increased significantly from 14 to 40% below the switching temperature to 36 to 70% at elevated temperatures. The thermotropic resins exhibit a steep and rapid switching process with switching temperatures between 45 and 70 °C. The materials with the best switching properties are characterized by distinct differences in refractive index between the components, a uniform additive distribution and scattering domains with diameters below 1000 nm.

Keywords: infrared spectroscopy; overheating protection; refractive index; thermotropic resins; UV-Vis spectroscopy

Introduction

Polymeric materials are playing a key-role in the future development of solar energy systems, as they offer significant potential for cost savings in solar thermal collectors and may thus benefit a broader utilization of solar energy, specifically for low-temperature heating purposes. Especially for domestic hot water generation and space heating applications, where the required maximum operating temperature of the absorber is about 80 °C, cost-efficient plastics will be applied. However, conventional solar thermal collectors reach stagnation temperatures up to 200 °C, which exceed the maximum operating temperatures of cost-efficient plastics (~80–140 °C). Absorbers made of such plastics tend towards irreversible deformation and/or degradation when exposed to elevated temperatures of 200 °C for prolonged periods. Therefore an appropriate overheating protection is required. The energy flux in an all-polymeric flat-plate collector can be controlled by thermotropic layers.[1,2] Thermotropic materials change their light transmission behavior from highly transparent to light diffusing upon reaching a certain threshold temperature.[3,4] Thus such actively switchable layers permit the light and energy flux to be adapted dynamically to climatic

[1] Polymer Competence Center Leoben GmbH, Roseggerstrasse 12, 8700 Leoben, Austria
Fax: (+43) 3842 402 2102;
E-mail: resch@pccl.at

[2] Institute of Materials Science and Testing of Plastics, University of Leoben, Franz-Josef-Strasse 18, 8700 Leoben, Austria

demands. By theoretical modeling of an all-polymeric flat plate collector it was found that for adequate overheating protection (i.e. stagnation temperatures ranging from 80 to 140 °C) switching temperatures between 55 and 60 °C as well as a residual hemispheric solar transmittance of 25 to 60% in the opaque state are effectual.[1]

In thermotropic materials that undergo a transition from transparent to light diffusing, the light is scattered from particles, which exhibit an index of refraction, which is different to that of the matrix.[3,4] In the past various thermotropic systems for active daylight control in transparent facades have been developed and investigated.[3–10] Whereas in thermotropic polymer blends and hydrogels the switching is involved in a phase separation, in thermotropic resins, the scattering domains are embedded statically in a matrix material.[3] To the best of our knowledge, comprehensive and systematic development and research on thermotropic materials for solar thermal systems has not yet been performed. Hence, the overall objective of our research is to perform a comprehensive development and characterization of thermotropic materials for overheating protection of solar collectors and to establish structure-property relationships. Various thermotropic resins were manufactured by systematic variation of base resin formulation as well as of additive type. The materials are characterized as to their optical properties and switching characteristics on a UV/Vis/NIR spectrophotometer equipped with an Ulbricht-sphere to differentiate between diffuse and direct transmitted light. Additionally the indices of refraction are considered. Scattering domain size, additive content and distribution of the thermotropic resins are characterized applying Differential Scanning Calorimetry, light microscopy and Attenuated Total Reflectance spectroscopy.

Material Preparation

Thermotropic layers were prepared by dissolving the thermotropic additives in a photo-crosslinkable matrix solution, which consists of oligomers, a reactive diluent as well as a photoinitiator. For the present study 2 different oligomer systems (oligomers A and B) serving as the main component of the matrix material were selected. In order to meet the requirements as to overheating protection of solar thermal systems, 3 thermotropic additives exhibiting switching temperatures of 47, 54 and 60 °C were selected (additive types 1, 2 and 3).[1] The additive concentration chosen for the preparation of the films was 5 wt%. The dissolution of the thermotropic additive in the resins was filled into an intervening space located between two glass panes, which were sealed around the edge. Afterwards the mixture was cured by UV-radiation (Ultra Vitalux, Osram GmbH, München, GER) for 5 minutes. The film thickness was 800 μm. As to the nomenclature A-1 indicates a film based on resin A containing additive type 1, for example.

Characterization of Optical Properties

The refractive indices of the additives and the resins were determined on a m-line apparatus (Metricon model 2010 (Metricon Corp., Pennington, USA)) based on the prism coupling technique. All measurements were performed at temperatures from 25 to 100 °C using a He-Ne laser with a wavelength of 633 nm.

The thermotropic materials were analyzed as to solar transmittance values in clear and opaque state, switching temperature and switching process applying UV/Vis/NIR spectroscopy. Hemispheric and diffuse transmittance spectra were recorded at normal incidence within a wavelength range between 250 and 2500 nm on a conventional UV/Vis/NIR spectrophotometer (Lambda 950, Perkin Elmer Instruments GmbH, Überlingen, GER) equipped with an Ulbricht-sphere (diameter 150 mm). For the given measurement apparatus the radiation passing through the specimen outside a cone of approximately 5° relative to the incident beam direction is defined as diffuse (scattered) component.

The measured spectral data were weighted in steps of 5 nm by the AM 1.5 global solar irradiance source function, given in Bird and Hulstrom[11] and representing the spectral distribution of solar radiation energy, when the sun is at approximately 42° above the horizon at sea level where the solar radiation has traveled 1.5 atmospheres. By this way integral solar transmittance values were determined.

To investigate the switching behavior (i.e. switching characteristic and temperature) the spectrophotometer was adapted by a heating stage allowing for controlling the sample temperature from ambient to 100 °C. For the determination of films thermal transitions the samples were mounted in a heating-chamber within the standard sample chamber of the spectrophotometer using a purpose-built clamp. This clamp has been additionally adapted by sensors to acquire the temperatures of the chamber and the sample. Due to this setup the sample is not positioned directly at the entrance of the integrating sphere. Especially for highly diffuse scattering samples this causes significant chamber recess error, which in turn results in considerable deviations between the recorded values and the effective radiation transmitted by the sample. To avoid these radiative losses on the one hand the inner boundaries of the light transmitting recess in the heating chamber were covered with a highly reflective mirror film which exhibits direct-direct and direct-diffuse solar reflectance values of 80 and 5%, respectively. On the other hand, the dimension of the incident light beam striking the sample has been reduced by a pinhole that was mounted at the front side of the heating chamber. With this configuration even for highly diffuse scattering samples more accurate transmittance values were measured. However, slight differences between the detected values and the effective radiation transport were obtained, which can be attributed to absorption of the mirror film (merely 85% solar reflectance). Thus a calibration of the recorded data is required. For this purpose hemispheric and diffuse transmittance spectra of numerous weak and highly scattering samples were recorded both by positioning the samples direct within the portholes of the Ulbricht-sphere and by mounting the materials in the adapted heating stage. In Figure 1 hemispheric and diffuse solar transmittance values, recorded at the different sample positions are compared. Excellent linear correlations with regression coefficients of 0.99 and 0.96 for hemispheric and diffuse solar transmittance values, respectively,

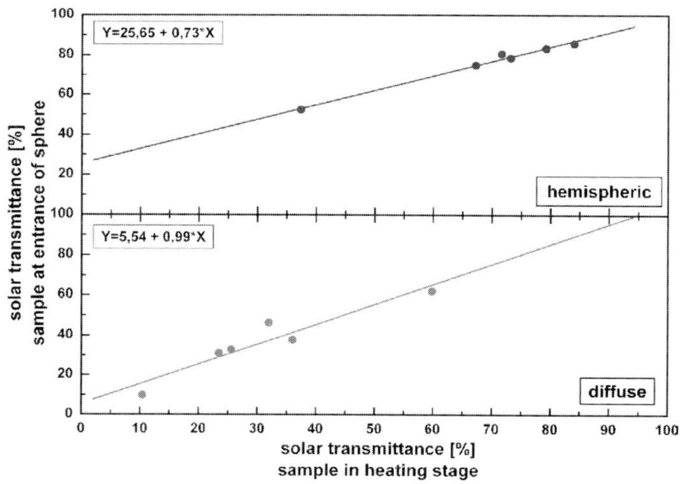

Figure 1.
Calibration for hemispheric and diffuse transmittance.

were obtained. All data presented in the following have been corrected by the corresponding equations of the linear regressions given in Figure 1.

Transmittance measurements were performed in steps of 5 K. Prior to the measurements the sample was maintained at the selected temperature for 10 minutes.

Characterization of Morphology

The additive contents of the produced thermotropic layers were determined by Differential Scanning Calorimetry operating in Temperature Modulation mode (TM-DSC). TM-DSC was performed under static air from 0 °C to 130 °C at a heating rate of 1 K/min, a modulation amplitude of ±0.5 °C and a modulation period of 75 s on a Mettler Toledo DSC822e (Schwarzenbach, CH). Weighting the transition enthalpy of the thermotropic layer by the heat of fusion of the pure additive allowed for the evaluation of the films additive content.

To obtain information as to film composition and additive distribution on the surface of the thermotropic materials Fourier Transform Infrared (FT-IR)-spectroscopy was performed in Attenuated Total Reflection mode (ATR) within a wavenumber range from 700 to 4000 cm^{-1} at a resolution of 1 cm^{-1} using a diamond crystal (Spectrum One, Perkin Elmer Instruments GmbH, Überlingen, GER). The recorded spectra were corrected for the wavelength-dependent penetration depth of the evanescent wave as well as for the baseline by the *Spectrum v. 5.0* software (Perkin Elmer Instruments GmbH, Überlingen, GER).

Additive particle sizes were characterized applying optical microscopy. Transmitted light micrographs were recorded on an Olympus BX51 (Olympus GmbH, Hamburg, GER) microscope at a magnification of 50×.

Results and Discussion

In Figure 2 a representative thermotropic resin (layer A-1) in transparent and light diffusing state is depicted. The thermotropic material is perfectly clear below the switching temperature (left side). At elevated temperatures the light scattering of the sample increases, which causes the material to turn opaque.

Figure 2.
Thermotropic layer A-1 in clear (left) and scattering state (right).

In Figure 3 to 6 hemispheric and diffuse transmittance spectra of the different thermotropic layers based on the resins A and B in the clear and scattering state are depicted, respectively. In the transparent state (solid line) all layers show a hemispheric transmittance of about 80% in the visible and the near-infrared region (Figure 5 and 7). No significant differences between the various materials are apparent. Interactions between the films and the solar radiation which reduce transmission occur below 380 and above 1200 nm. Below 380 nm the transmittance decreases due to absorption of incorporated ultraviolet absorbers. Absorption bands at approximately 1200, 1400 and 1700 nm and above 2225 nm can be attributed to vibrations of carbon-hydrogen single bonds (C–H). The peak at 2150 nm appears due to stretching of carbon-oxygen double bonds in the polymers (C=O). The absorption peak at about 1900 nm refers to water molecules absorbed by the material. In the scattering state (dotted line) material B-1 displays slightly higher transmittance values compared to the clear state. For the other layers

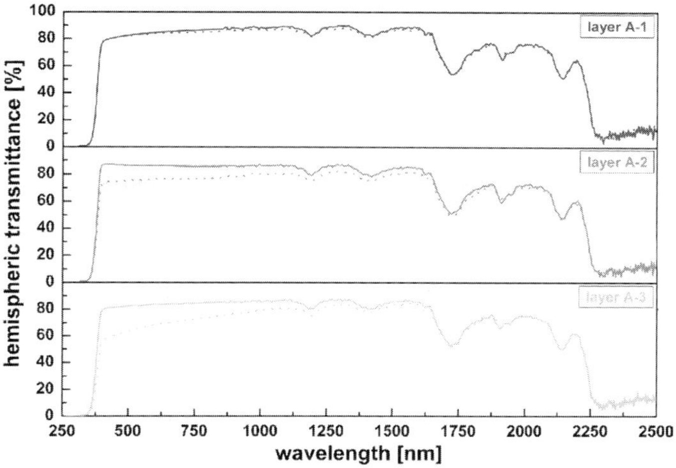

Figure 3.
Hemispheric transmittance spectra of thermotropic layers based on resin A in clear and scattering state; top: layer A-1; mid: layer A-2 and bottom: layer A-3 (solid line: clear state; dotted line: scattering state).

a decrease of the hemispheric transmission is discernible at elevated temperatures. In general these layers show a stronger decrease of transmission in the region between 300 and 1500 nm than at longer wavelengths. The differences in transmission between the clear and the scattering state increase with decreasing wavelength. The reduction of transmittance is more pronounced for materials A-2, A-3 and B-3. This indicates that these layers exhibit better light shielding properties than film types A-1, B-1 and B-2.

The diffuse transmittance spectra of the investigated thermotropic materials shown in Figure 4 and 6 give information as to their scattering properties. The spectra reveal significant differences in the film's scattering behavior. All film types exhibit an increase of diffuse transmittance with

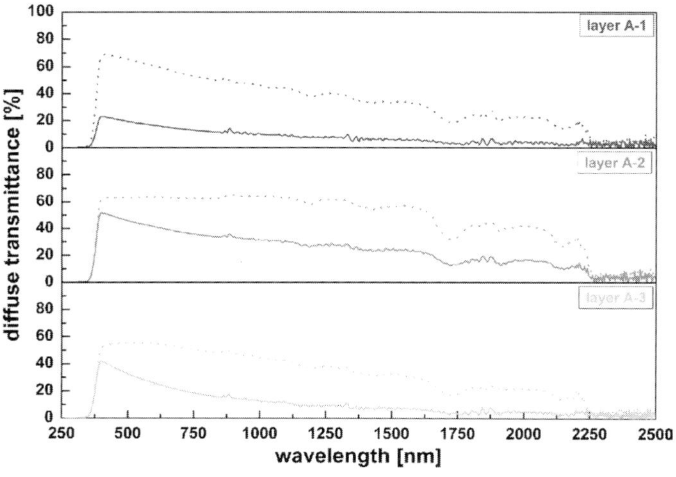

Figure 4.
Diffuse transmittance spectra of thermotropic layers based on resin A in clear and scattering state; top: layer A-1; mid: layer A-2 and bottom: layer A-3 (solid line: clear state; dotted line: scattering state).

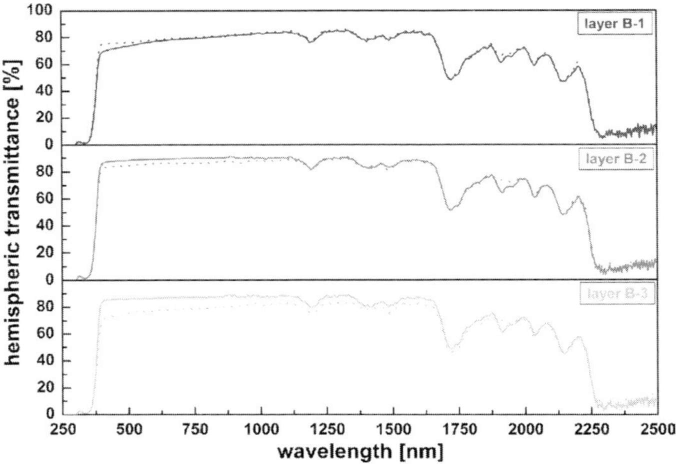

Figure 5.
Hemispheric transmittance spectra of thermotropic layers based on resin B in clear and scattering state; top: layer B-1; mid: layer B-2 and bottom: layer B-3 (solid line: clear state; dotted line: scattering state).

decreasing wavelength. Whereas additive type 1 yields lower diffuse scattering for layers based on resin A, additives types 2 and 3 reduce the diffuse scattering power for layers produced on the basis of resin B in the clear state (solid line). For the films A-1, B-2 and B-3 the diffuse transmittance values were found to remain below 30% over the whole wavelength range only slightly increasing with decreasing wavelength. For the layers A-2, A-3 and B-1 a more pronounced increase of the diffuse transmittance values in the short wavelength range was observed associated with also higher diffuse transmittance values in the near infrared wavelength range. In the scattering state (dotted line) a significant increase of diffuse transmittance is discernible in the wavelength range between 380 and 2250 nm. The most pronounced rise

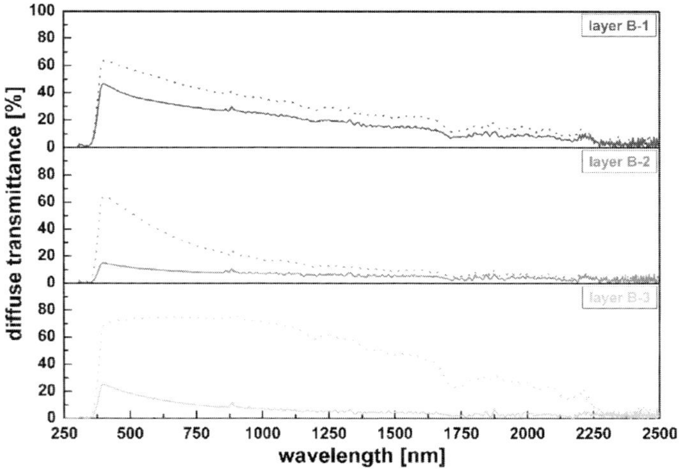

Figure 6.
Diffuse transmittance spectra of thermotropic layers based on resin B in clear and scattering state; top: layer B-1; mid: layer B-2 and bottom: layer B-3 (solid line: clear state; dotted line: scattering state).

of diffuse scattering power was found for the layers A-1 and B-3. Layer B-1 exhibits the less distinct changes. The differences in diffuse scattering below and above the switching temperature increase with decreasing wavelength for layers A-1, B-1 and B-2. For the layers showing the most significant reduction of hemispheric transmittance (layers A-2, A-3, B-3) this tendency is not observable.

The hemispheric and diffuse transmittance spectra of the samples clearly indicate that the thermotropic layers undergo a transition from a clear to a scattering state. Accurate relationships between material formulation and the scattering behavior may be emphasized by integral solar optical properties determined over the solar spectral intensity AM 1.5. Thus in Table 1 integral solar optical properties (hemispheric and diffuse transmittance) of the investigated thermotropic layers in the clear and the opaque state are summarized. Below the switching temperature all samples exhibit an excellent hemispheric transmittance above 80%. For the diffuse transmittance considerable differences between the various material formulations are observable. The values range from 14 to 40%. Whereas for layers based on resin A additive type 1 yields lower diffuse scattering, materials with resin B matrix show better optical properties with the additive types 2 and 3. In the scattering state at elevated temperatures the diffuse transmittance increases significantly to values exceeding 36%. The most distinct changes are observable for the layers B-3 and A-1, where the forward scattered light increases by 53 and 37%. The diffuse transmittance of the other layers increases by 13 to 26%. Already a change of the diffuse transmittance of 13% is associated by a loss of view-through behavior. However, the overheating protection properties are determined by the reduction of hemispheric transmittance. The hemispheric transmission of the films is decreased above the switching temperature to a smaller extent. The most significant changes were found for thermotropic layers being formulated with additive type 3. By dispersing this additive type in the resins A and B the hemispheric transmittance is reduced by 6 and 9%, respectively. A reasonable decrease of the hemispheric transmission was also obtained for material A-2 (6%). For material B-1 an increase of transmittance from 80 to 81% is discernible. This layer exhibits a significant increase of transmittance to values of 85% at temperatures between 30 and 45 °C.

To describe the switching characteristics of the materials in Figure 7 and 8 the diffuse solar transmittance is plotted as a function of temperature for thermotropic materials based on resin A and B, respectively. For thermotropic layers being formulated with additives 1 and 2 the switching gradient is reasonably steep, so that the diffuse solar transmittance changes within a small temperature range of 10 to 15 K. The temperature at which this transition occurs (switching temperature) ranges between 40 and 50 °C for these film types. Whereas for the thermotropic layer produced from

Table 1.
Hemispheric and diffuse solar transmittance values of the thermotropic layers in clear and scattering state.

Resin	Additive type	Solar transmittance			
		Clear state		Scattering state	
		Hemispheric	Diffuse	Hemispheric	Diffuse
		[%]	[%]	[%]	[%]
A	1	85	19	84	56
	2	85	40	79	64
	3	84	25	75	51
B	1	80	34	81	47
	2	87	14	85	36
	3	86	17	80	70

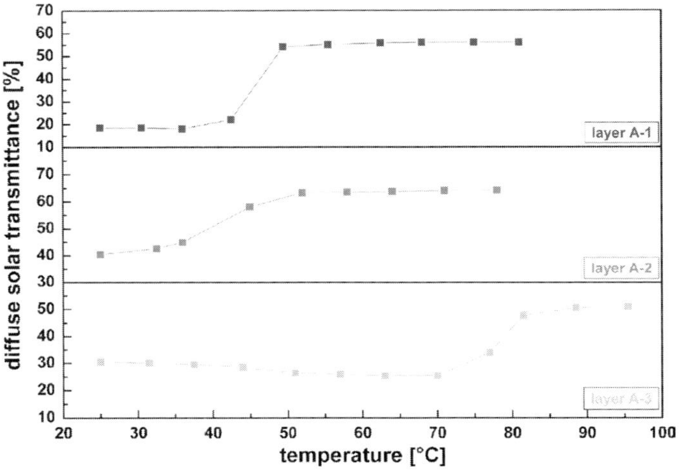

Figure 7.
Diffuse solar transmittance of thermotropic layers based on resin A as a function of temperature; top: layer A-1; mid: layer A-2 and bottom: layer A-3.

additive type 3 and resin B the switching again proceeds at a temperature of about 45 °C, the transition from the clear to the scattering state is more gradual. In contrast the thermotropic layer produced from resin A and additive type 3 shows a very steep switching process. However, the switching temperature of this film is shifted to above 70 °C.

Both, the scattering spectra and the solar optical transmittance data presented in Table 1 reveal significant differences in the scattering behavior between the various thermotropic layers. In general the optical properties and the switching behavior of the thermotropic materials are determined both, by the differences in refractive index between the additives and the matrix

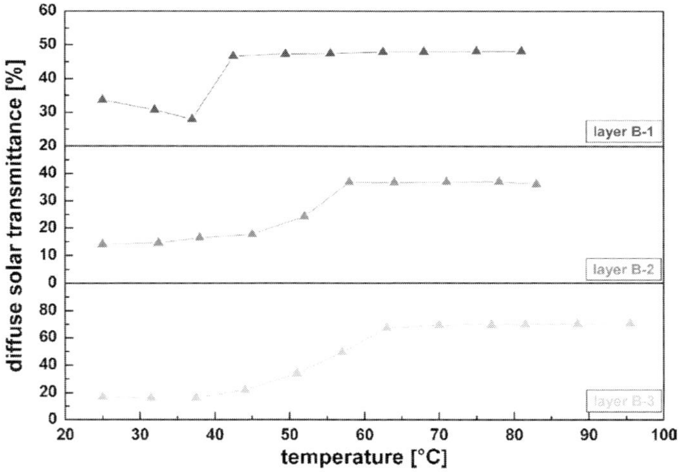

Figure 8.
Diffuse solar transmittance of thermotropic layers based on resin B as a function of temperature; top: layer B-1; mid: layer B-2 and bottom: layer B-3.

material below and above the switching temperature, and by the scattering domain size and number of scattering particles, which is correlated to additive content.[12] The consideration of these parameters is of significant importance for the establishment of a fundamental understanding of the layers switching characteristics and a systematic material development and optimization.

In Figure 9 the refractive indices of the additives and the matrix materials are compared at different temperatures. For the resins A and B refractive indices of 1.52 and 1.50 were determined at low temperatures, respectively. The insignificant decrease of 1.1 (resin A) and 1.3% (resin B) in the temperature range between 25 and 95 °C indicates that the refractive index of the resins is not affected by changes in temperature. The refractive index of additive type 1 switches steeply from 1.52 to 1.45 within 40 and 50 °C. For additive type 2 the refractive index decreases from 1.50 to 1.42 in the temperature range from 25 to 50 °C. For additive type 3 the decrease of the refractive index is more gradual. It changes from 1.52 to 1.43 in a temperature range between 30 and 90 °C.

In general below the switching temperature a reasonable match of the refractive indices of the resins and the thermotropic additives with differences between 0 and 0.02 was detected. Whereas hemispheric transmittance values were found to be insignificantly affected by the differences in the refractive index of the components, the diffuse transmittance varies between the different film types. An increasing diffuse transmittance with increasing difference in refractive index is observable. Figure 9 shows that the differences in the refractive indices of the components at elevated temperatures are more pronounced for the additive types 2 and 3 compared to additive type 1. This may be attributed to a more distinct change in the refractive index of the additive types 2 and 3. However, at elevated temperatures no correlations between the differences in refractive index and the scattering properties of the samples are observable. This indicates that the optical properties of the considered thermotropic layers are further dependent on the additive concentration and distribution as well as on scattering domain size.

As to additive content significant deviations from the theoretical concentration were ascertained, especially for the resins formulated with additive types 2 and 3. For additive type 2 dispersed in the resins A and

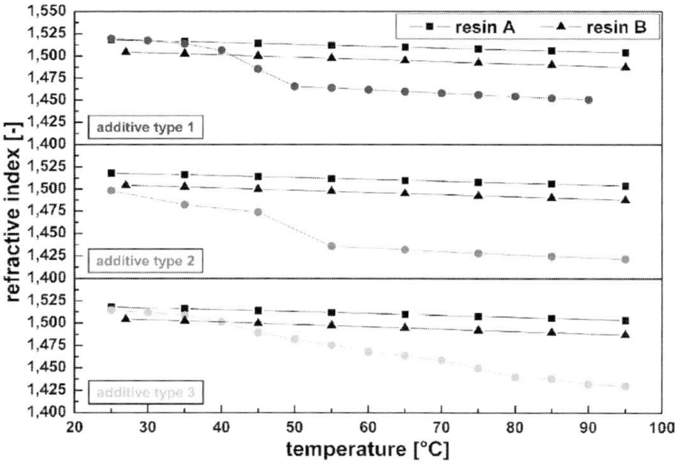

Figure 9.
Refractive indices of resins A and B and the thermotropic additives at a wavelength of 633 nm as a function of temperature; top: resins vs. additive type 1, mid: resins vs. additive type 2 and bottom: resins vs. additive type 3.

B the additive content amounts to 1.33 and 1.60 wt.%, respectively. Scattering domain concentration in films formulated with additive type 3 is even lower with 0.95 and 1.09 wt.% for layers based on resin A and B. The highest additive content was found for thermotropic materials containing additive type 1 with values of 3.87 and 4.80 wt.% in layers A-1 and B-1.

In Figure 10 and 11 ATR spectra in the wavenumber range from 3050 to 2800 cm^{-1} of the various thermotropic layers and the neat resins are depicted. For the thermotropic materials spectra are shown both, for the front side, i.e. the side faced to the UV light source during the curing and for the opposite side, i.e. the back side of the films. Information as to additive concentration on the surface can be deduced from the decrease of transmission at wavenumbers of 2848 and 2915 cm^{-1}, which can be attributed to symmetric and asymmetric stretching vibrations of methylene groups ($–CH_2–$), respectively. In general the absorption bands are more pronounced for thermotropic layers produced from additive type 1 than for materials formulated with the additive types 2 and 3. This can be ascribed to the higher additive concentration of layers containing additive type 1. Whereas for thermotropic materials based on the additive types 2 and 3 no differences between the two film sides are observable, additive concentration is higher on the front surfaces of layers formulated with additive type 1. It is observable that the absorption peaks are more distinct for thermotropic materials based on resin B. On the one hand this can be attributed to the slightly higher additive content of these materials. On the other hand it may be an indication for a less uniform additive distribution across the film thickness.

The scattering domain sizes of the thermotropic layers were determined applying light microscopy. For materials based on resin A and formulated with the additive types 1, 2 and 3 average particle diameters of 1.7, 1.0 and 0.8 μm were obtained, respectively. Thermotropic layers based on resin B and formulated with additive types 1, 2 and 3 exhibit scattering domains with diameters of 1.4, 2.2 and 1.1 μm, respectively. In

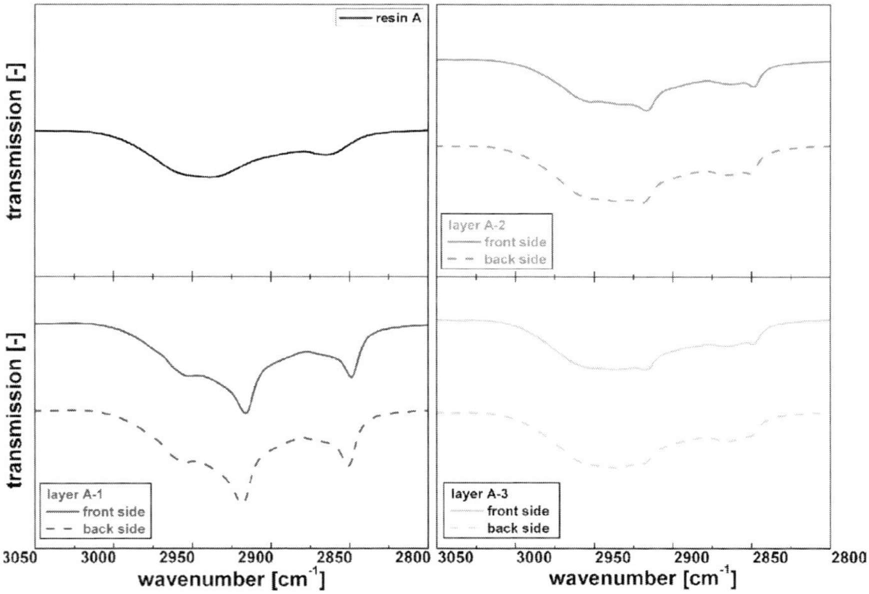

Figure 10.
ATR spectra of resin A and the thermotropic layers based on resin A in the wavenumber range between 3050 and 2800 cm^{-1}; top left: resin A, bottom left: layer A-1, top right: layer A-2 and bottom right: layer A-3.

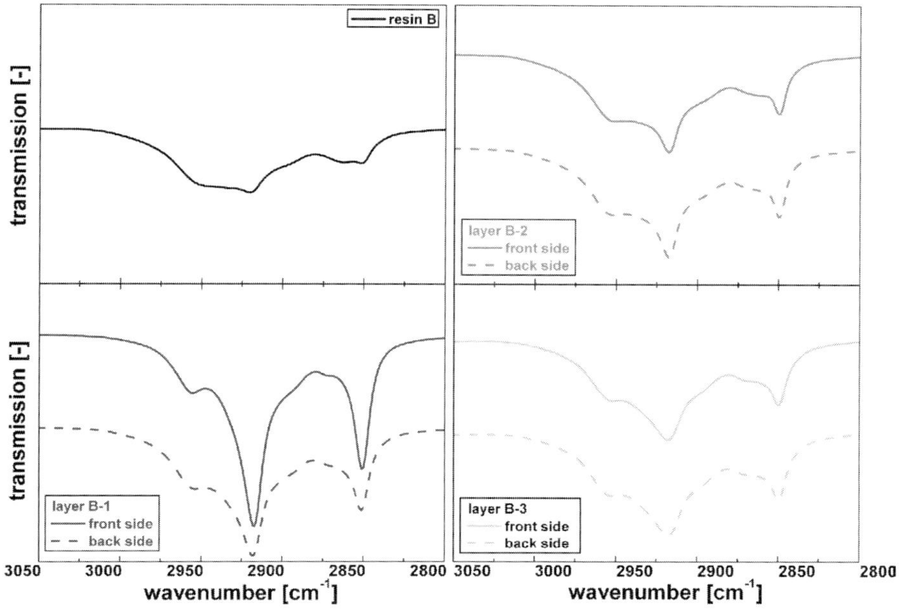

Figure 11.
ATR spectra of resin B and the thermotropic layers based on resin B in the wavenumber range between 3050 and 2800 cm^{-1}; top left: resin B, bottom left: layer B-1, top right: layer B-2 and bottom right: layer B-3.

general a tendency was observed that with decreasing particle sizes the light shielding properties of the thermotropic layers are enhanced. Nevertheless, for maximum reflectance efficiency in the solar wavelength range scattering radii of 100 to 200 nm are required.[5,6,12,13] The scattering domain sizes of the thermotropic materials produced within the present study are significantly larger, which results in less light-shielding capability.

Furthermore the micrographs of the materials depicted a higher aggregation of additives for materials based on resin B. This is convenient with ATR spectra, revealing higher additive concentrations on the surface for these materials. For layer B-3 the additives are less pronounced on the surfaces. Because light microscopy does not allow for a detailed analysis of scattering domain size distribution further investigations are currently carried out using Atomic Force Microscopy (AFM).

As to the influence of morphological parameters on the optical properties of the films it was found that although additive concentration is higher for thermotropic layers based on resin B, both the forward and the back scattering power above the switching temperature is higher for materials produced from resin A. This may be ascribed to the more uniform additive distribution across the film thickness for these materials. The decreased hemispheric solar transmittance of the layers A-1, A-2 and B-3 above the switching temperature can be attributed to the smaller scattering domain size of these films. In general a tendency is observable that decreasing scattering domain size increases the diffuse scattering power. Furthermore a more uniform distribution of the additives across the film thickness yields better light-shielding properties, and thus higher opacity in the scattering state.

Summary and Conclusion

In this paper the optical and morphological properties of thermotropic resin materials are described and discussed. In the clear state the investigated films exhibit direct-hemispheric solar transmittance values of 80 to 87%, with diffuse fractions ranging from 14 to 40%. The materials undergo a transition from the clear to the scattering state at temperatures between 40 and 75 °C. The investigated thermotropic resins exhibit a steep and rapid switching process. Above the switching temperatures the hemispheric solar transmission decreases to values between 75 to 85%, with diffuse fractions ranging from 36 to 70%. The materials show a stronger decrease of transmittance in the visible than in the near infrared region.

The optical properties of the films in the clear and the scattering state were found to be significantly dependent on the differences in the refractive indices of the components, additive content and distribution as well as on scattering domain size. The light-shielding properties of the thermotropic layers and thus the opacity in the opaque state are enhanced by more pronounced differences in refractive index between matrix and additive (>0.04), a uniform additive distribution across the film thickness and scattering particles with dimensions below 1000 nm.

Acknowledgements: The research work of this paper was performed at the Polymer Competence Center Leoben GmbH within the framework of the K_{plus} Program of the Austrian Ministry of Traffic, Innovation and Technology with the contributions by the University of Leoben, Graz University of Technology, Johannes Kepler University Linz, JOANNEUM RESEARCH Forschungsgesellschaft mbH and Upper Austrian Research GmbH. The PCCL is funded by the Austrian Government and the State Governments of Styria and Upper Austria.

This work is funded by the State Government of Styria, Department *Zukunftsfonds Steiermark*.

ZUKUNFTSfonds STEIERMARK   Das Land Steiermark

The authors wish to express their acknowledgements to Cytec Surface Specialties (Drogenbos, BEL), Sasol Germany GmbH (Hamburg, GER) and Chemson Polymer Additive AG (Arnoldstein, AUT), for providing the materials.

[1] K. Resch, R. Hauner, G. M. Wallner, in *"Proceedings of ISES SWC 2007"*, September 18–21 2007, in press.
[2] G. M. Wallner, K. Resch, R. Hausner, in *"Proceedings of Gleisdorf Solar 2006"*, September 6–8 2006, p. 105ff.
[3] A. Seeboth, J. Schneider, A. Patzak, *Solar Energy Materials and Solar Cells* **2000**, *60*, 263.
[4] P. Nitz, H. Hartwig, *Solar Energy* **2005**, *79*, 573.
[5] A. Beck, T. Hoffmann, W. Körner, J. Fricke, *Solar Energy* **1993**, *50*, 407.
[6] A. Beck, W. Körner, H. Scheller, J. Fricke, W. J. Platzer, V. Wittwer, *Solar Energy Materials and Solar Cells* **1995**, *36*, 339.
[7] H. Watanabe, *Solar Energy Materials and Solar Cells* **1998**, *54*, 204.
[8] A. Georg, W. Graf, D. Schweiger, V. Wittwer, P. Nitz, H. R. Wilson, *Solar Energy* **1998**, *62*, 215.
[9] J. Schneider, A. Seeboth, *Materialwissenschaft und Werkstofftechnik* **2001**, *32*, 231.
[10] A. Raicu, H. R. Wilson, P. Nitz, W. Platzer, V. Wittwer, E. Jahns, *Solar Energy* **2002**, *72*, 31.
[11] R. E. Bird, R. L. Hulstrom, *Solar Energy* **1983**, *30*, 563.
[12] P. Nitz, *"Optische Modellierung und Vermessung thermotroper Systeme"*, 1999, Ph.D. Thesis, Albert- Ludwigs-Universität in Freiburg im Breisgau, Germany.
[13] C. G. Granqvist, V. Wittwer, *Solar Energy Materials and Solar Cells* **1998**, *54*, 39.

Investigation of Lignins by FTIR Spectroscopy

Olga Derkacheva, Dmitry Sukhov*

Summary: Applying special computer mathematical treatments to increase resolution of experimental spectra there were established a set of stable characteristic bands for isolated softwood lignins. In the 740–1620 cm^{-1} spectral range the band maximum positions did not change but values of bandwidths and peak intensities were varied in limits 15% and 32%. After analysis of the infrared spectra of 30 investigated samples a softwood lignin spectral model was constituted. This model allowed to clear discrepancies in the bands parameters of different mild isolated (lignins of Bjorkman, Pepper and Freudenberg), dioxane and technical lignin spectra. It was helpful for studying lignin structure changes during degradation procedures.

Keywords: degradation; FTIR spectroscopy; mild isolated and technical lignins; softwood lignin; spectral model; structure

Introduction

Lignins are polymeric aromatic constituents in plant cell walls. They are traditionally considered to be dehydrogenative polymers from three monolignols: p-coumaryl alcohol, coniferyl alcohol and sinapyl alcohol (see Figure 1). After structure lignins have been grouped into several types, characteristic of hardwoods, softwoods and grasses. The hardwood (angiosperm) lignins are built from the guaiacyl propane units (the abbreviation is G) and the syringyl propane units (the abbreviation is S). Lignins obtained from straws are built from the guaiacyl propane units, the syringyl propane units and 4-hydroxylphenyl propane units (the abbreviation is H).[1,2] The softwood (gymnosperm) lignins are built from the guaiacyl propane units (92–95%) with small amount of the syringyl propane and 4-hydroxylphenyl propane units (3–8%).[3] These aromatic building units are linked with a variety of ether and carbon-carbon bonds. The predominant linkage is the so-called β-O-4 llinkage. About 40–60% of all interunit linkages in lignin is via this ether bond. Different types of linkages between phenyl propane units form three-dimensional net structure and makes it difficult to completely degrade non-regular lignin macromolecules.

Because of structural complexity there are not uniformic knowledges about chemical and physical structure of this polymer and it is possible to consider only chemical lignin models. These models are based on analysis of degradation products from softwood and hardwood lignins. Such structural models with different amount of phenyl propane units having good agreement with the various analytical data in the lignin chemistry were proposed.[4–6] The fragment of one such softwood lignin model[6] is presented in Figure 2.

Recent investigations[7–8] pointed out on existence of some organization of lignin macromolecules in cell wall. Raman microprobe studies of secondary walls in black spruce revealed that phenyl rings of lignin are aligned preferentially in the plane of the cell walls[7] and this observation was supported by dynamic FTIR spectroscopy investigations.[8]

Considerable part of articles is devoted to studying so named technical lignins

St.Petersburg State Technological University of Plant Polymers, Ivana Chernuch 4, St.Petersburg, 198095, Russia
E-mail: derkachevaou@rambler.ru

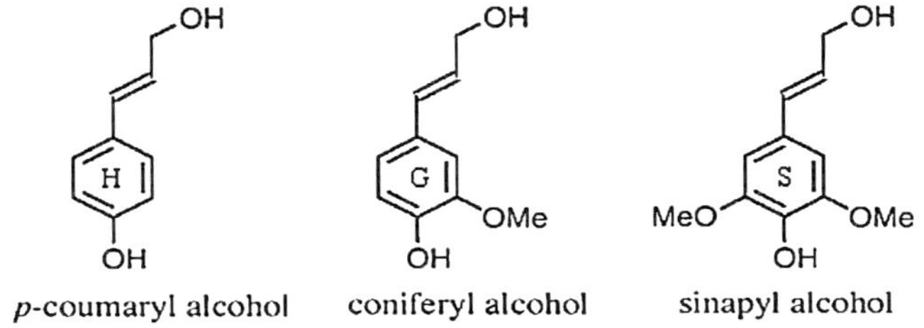

Figure 1.
The monolignols.[1]

that are ones extracted in severe or industrial conditions during delignification of wood.

At studying native (in situ) lignins two main problems occur. The one is searching isolation methods which do not change strongly structure of lignin. Lignins isolated under mild conditions are believed to be close to structure of native lignin after chemical data.[1]

The second problem is connected with selecting experimental method of analysis. A number of chemical and physical methods used for lignin structure characterization are destructive.

We assume that FTIR spectroscopy is non-destructive and most informative method of lignin investigation. The method opens perspective to find structural discrepancies in lignins isolated by different methods. Molecular spectra see differences in chemical structure and physical organization of different lignins that it is invisible for other analytical methods.

There are many articles devoting to FTIR spectra of lignins but full band interpretations are under the discussion.

In the work we used a new computer program method for enhancing lignin spectral information. It allowed revealing a complete lignin spectral band set in the diapason 740–1840 cm^{-1} and find some dependencies in behavior of spectral and structural parameters.

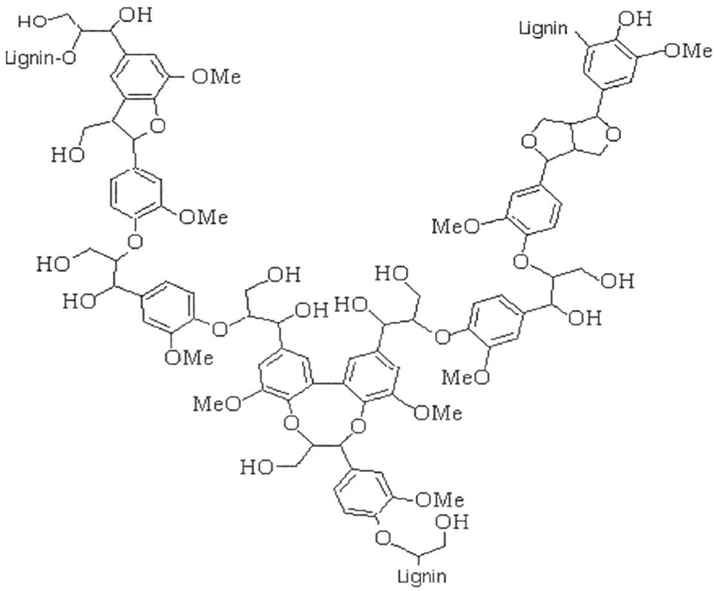

Figure 2.
Fragment of the softwood lignin model.[6]

Experimental Part

In the present work some softwoods mild isolated lignins and technical (kraft) lignins were investigated. There were three samples of mild extracted lignins (lignins of Bjorkman, Pepper and Freudenberg), dioxane lignin and 20 laboratory cook lignins extracted under different conditions. Three mild extracted wood lignins and dioxane lignin were extracted from spruce by standard methods.[1]

Technical lignins were isolated from black liquors of laboratory kraft cooks of spruce and fir woods.[9]

Lignin Isolation and Characterization

Bjorkman's lignin or milled wood lignin (MWL) was prepared in accordance with the method of Bjorkman.[10]

Milled wood was stirred for two hours with dioxane-water (9:1 by volume) that dissolved crude milled wood lignin. After removal of solvent from the filtrate, the crude preparation was purified by precipitation of a solution in 90% acetic acid into water. The product was further purified by dissolving into 1,2 dichloroethane-ethanol mixture (2:1 by volume) and precipitation into ethyl ether. The purified milled wood lignin required extensive drying under vacuum to bring it to a water-free condition.[1]

Pepper's lignin and dioxane lignin used in the present work were extracted from spruce wood by means of Peper's method.[11]

Peper's lignin was isolated by treatment of wood at elevated temperatures by a 9:1 mixture of dioxane and dilute hydrochloric acid (about 1 hour) in Argon atmosphere.

Dioxane lignin was isolated by treatment of wood at elevated temperatures by a 9:1 mixture of dioxane and dilute hydrochloric acid (more than 1 hour) in air atmosphere.

Freudenberg's lignin or cuoxam lignin was isolated by alternating treatments with boiling 1% sulfuric acid (1–2 h.) and extractions with copperammonia (12 h.). Four to five hydrolysis-extraction treatments were necessary to obtain reasonably carbohydrate-free preparations.[1]

Technical Lignins

Technical lignins were isolated from black liquors of kraft cooks of spruce and fir chips of different duration (cook duration was: 45, 75, 105, 120, 135, 150, 165, 180 minutes). The active alkali charge for the kraft cook was 17% (as Na_2O), the sulfidity was 25%.

All laboratory cooks of spruce and fir chips were carried out in a series of rotating autoclaves. Every 30 min from the beginning of the cook at 80 °C (every 15 minutes at 170 °C) a autoclave was withdrawn from the glycerin bath and was quickly cooled. The temperature was increased from 80 °C to 170 °C for 120 min., and the temperature was held at 170 °C for 60 min.

Dissolved lignins were isolated by a process of acidification of the black liquors with diluted sulfuric acid to pH 2.0, followed by the samples being centrifuged, washed repeatedly with water and dried by P_2O_5 in vacuum.

Reduction with Sodium Borohydride

Procedure of lignin reduction with sodium borohydride $NaBH_4$ was described in.[12]

The Functional Groups

The determination of content of phenolic hydroxyl groups in the investigated samples was performed by aminolysis according to.[27,28]

Content of the inter-unit alkyl – aryl ether bonds (α-O-4 and β-O-4 linkages) was calculated by the special developed method.[13]

The FTIR Spectroscopy

The FTIR absorption data were obtained by using a Bruker IFS-113v Fourier Spectrometer in spectral range 740–1840 cm^{-1} (resolution 2 cm^{-1}, number of scans 150). The lignin powder samples of 1–2 mg in weight were pressed with 300 mg of KBr powder as disks. For all spectra base line was constructed as linear line going through two points with minimal absorption around wavenumbers 740 and 1840 cm^{-1}. Then such corrected spectroscopic data were normalized on integral absorption over the considered spectral range.

Fitting Procedure

After our practice the most suitable mathematical model used for the non-linear least-squares fitting of the FTIR lignin spectra includes superposition of two line types, Lorentzian and Gaussian (equation (1)). The two functions corresponding to the same value of k (band number) are given the same wavenumber and the same full width at half height. A simulated spectrum $F(\nu)$ for N observed bands is as following[14]:

$$F(\nu) = \sum_{k=1}^{k=N} I_k \left(p_k e^{-\left(4\ln 2\frac{(\nu-\nu_k)^2}{\sigma_k^2}\right)} + (1-p_k)\frac{\sigma_k^2}{\sigma_k^2+4(\nu-\nu_k)^2} \right) \quad (1)$$

In this model, each band is characterized by four parameters: the wavenumber (ν_k), the bandwidth (σ_k) or full width at half height, the symmetric band intensity (I_k) and weights of Gaussian (p_k) and Lorentzian ($1-p_k$).

Fitting procedure consists in optimization of the band parameters in the function $F(\nu)$ to have minimal value of the expression (2):

$$\sum_{i=1}^{N1} (S(\nu_i) - F(\nu_i))^2 \quad (2)$$

where $S(\nu_i)$ is an experimental spectrum curve and N1 is a number of points in this spectrum.

The computer program work is stopped when for all points of the treated spectral range the root-mean-square deviation of simulated spectral curve from the experimental one was less than 1/200 of the mean intensity value. Thus there is obtained the high quality simulated spectrum with the determined band parameter sets.

Before using this fitting procedure an establishing a number of elemental bands was produced. Fixing wavenumbers for established number of elemental bands was made after studying Fourier self-deconvoluted[15–16] spectra of studied lignins and spectra of second derivative. It was made by detail analysis and comparison of origin softwood lignin spectra with Fourier self-deconvoluted and second derivative spectra. As a result the set of elemental bands in spectra of softwood lignins of different isolation in spectral range 740–1840 cm^{-1} was established.

A peculiarity of this program was using second derivative of experimental spectral curve as for fixing band wavenumbers (ν_k), peak intensities (I_k), bandwidths (σ_k) and G/L contribution coefficients (p_k) so and at optimization process.

Results

In the earlier investigations of softwood lignin spectra the row of absorption bands was found: 1720, 1670, 1590, 1500, 1463, 1423, 1360, 1330, 1270, 1215, 1140, 1125, 1087, 1031, 970, 855 and 815 cm^{-1}.[1,17,19,20,29] Using mathematical treatment of lignin spectra together with these bands we revealed some new peaks.[18]

Results of spectroscopic investigations of different lignin samples allowed to establish a set of stable characteristic bands. This set consists of 26 elemental symmetric bands in the 740–1620 cm^{-1} spectral range and 2–4 bands in the 1620–1800 cm^{-1} one. The parameters of elementary bands of the set were optimized by specially created computer procedure for the best decomposition of experimental lignin FTIR spectra on their constituent bands. The computer program was stopped when simulated spectral curve was close to the experimental one in all considered spectral range. Thus for each studied lignin we have got the high quality simulated spectrum with the definite set of elementary bands. For example, the FTIR absorption spectrum of Bjorkman's lignin and its band model are presented in Figure 3.

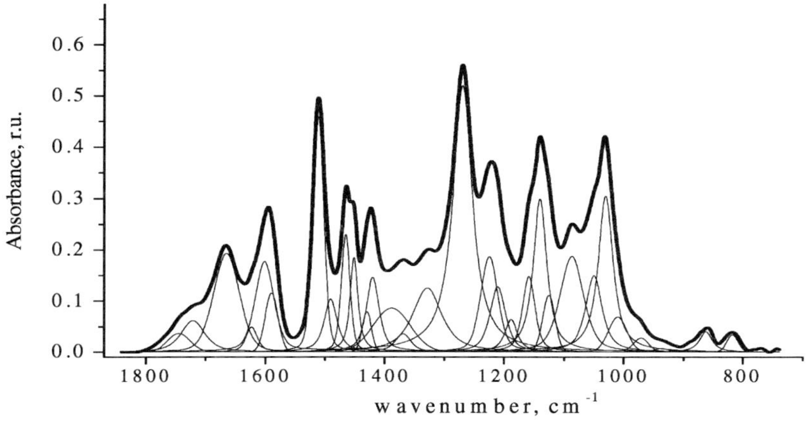

Figure 3.
The original spectral curve of Bjorkman's lignin (thick line) and its model components (thin line).

From physical point of view the elementary bands can be attributed to determined type of molecular vibrations of the lignin molecule units and existing interpretation of some bands can be found in literature.[1,17,19,20]

Analysis of the band parameters of the model after its applying to describe investigated lignins spectra showed that the band maximum positions in the 740–1620 cm^{-1} range did not change (the shift limits are $\pm$2 cm^{-1}), but values of bandwidths and peak intensities were varied in limits 15% and 32%.

This mathematical model is sensitive and it allows to assess small changes in the structure of softwood lignins which we observe by the molecular spectrum. To demonstrate high sensitivity of this approach an applying this spectral model to describe influence of reducing on lignin structure is presented.

So, for example, at studying process of reduction of Pepper's lignin sample small changes in its spectrum after reducing with $NaBH_4$ can be observed in Figure 4. The model band parameters showed that no significant changes occurred during this treatment. Reducing resulted to transformation of C=O groups to COH ones and it gave decreasing absorption bands at 1664 and 1720 cm^{-1}. Intensity of the model band at 1720 cm^{-1} dropped more than two times. Sarkanen *et al.*[21] assigned the first peak to νC=O of conjugated groups and that at 1720 cm^{-1} was assigned to nonconjugated moieties. It is worth to note that there was observed more than two times increasing intensity of the elemental model band near 1010 cm^{-1}. From this observation we propose that this band can be assigned partly to aliphatic hydroxyl groups vibrations.

Detail analysis of molecular spectra of three mild isolated lignins (lignins of Bjorkman, Pepper and Freudenberg) using their model band parameters allowed to reveal following spectral differences. For example, isolation procedure for Freudenberg's lignin is so that elementary bands of Freudenberg's lignin are broader and it results to decreasing intensities of the strong bands near 1510, 1465, 1270, 1140, 1030 cm^{-1}. These bands are the most intense ones in the molecular spectra of the G-type (guaiacyl kind) lignins. Peculiarity of Freudenberg's lignin spectrum is that the band at 1010 cm^{-1} is more than two times higher in comparison to two other mild isolated softwood lignins. Lignin of Bjorkman spectrally is more close to Pepper's lignin but some discrepancies are observed. It is possible that the enhanced band near 1660 cm^{-1} showing higher content of C=O groups conjugated with aromatic rings leads to some redistribution

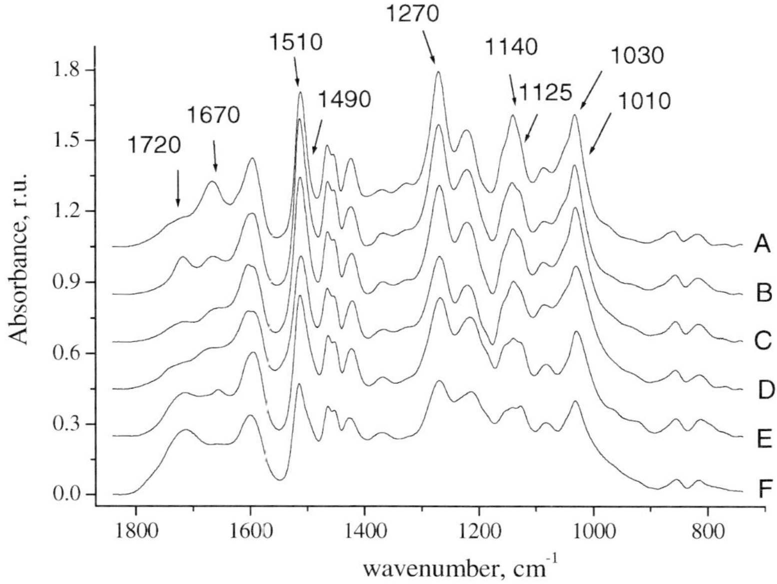

Figure 4.
FTIR spectra of extracted softwood lignins in the diapason 740–1840 cm^{-1}. From top to bottom: Bjorkman's lignin (A); Pepper's lignin (B), reduced Pepper's lignin (C), lignin of Freudenberg (D), dissolved Freudenberg's lignin (E) under the kraft cooking treatment, technical lignin dissolved after 150 minute kraft cooking (F).

in strong bands intensities - decreasing intensity of the bands near 1510, 1450, 1210 cm^{-1} and increasing intensities of the bands at 1590 and 1140 cm^{-1}.

So, there were revealed spectral differences in molecular spectra of three mild isolated lignins showing influence isolation method on lignin structure.

Studying influence of alkali cook conditions allowed to obtain new information about spectral manifestation of destruction processes of lignin.

Influence of degradation process on polymer structure was studied by comparing molecular spectra of mild isolated lignins and technical ones. Special laboratory treatment of Freudenberg's lignin close to traditional kraft cooking was done. Figure 4 shows the FTIR spectra of Freudenberg's lignin (curve D) and dissolved Freudenberg's lignin under the kraft cooking treatment (curve E).

Although many changes are seen not enough clearly in the spectra, the parameters of spectral model bands showed essential changing structure during degradation. These data showed increasing intensities of the bands at 1664 and 1713 cm^{-1}. These bands arise due to destruction process of lignin macromolecules and were attributed to vibration of C=O bonds in carbonyl and carboxyl groups.[1,21]

This alkali treatment gives more than two times increasing intensity of the band near 1490 cm^{-1}, some increasing bands at 1450, 1210, 1188 cm^{-1} and significant dropping the bands at 1140, 1050 and 1030 cm^{-1}.

It is well known that the β-O-4 type of structure (structure (1) in Figure 5) constitutes the major type of structural element in lignins. About 40–60% of all interunit linkages in lignin is via this ether bond.

As it was established[13,22–24] that base reaction at alkali treatment consists in cleavage of aryl-ether bonds what brings about an increase of phenolic hydroxyl content. These reactions lead to destruction of lignin macromolecules. A proposed mechanism for the alkali treatment is shown in Figure 5. After this reaction of degradation of lignin

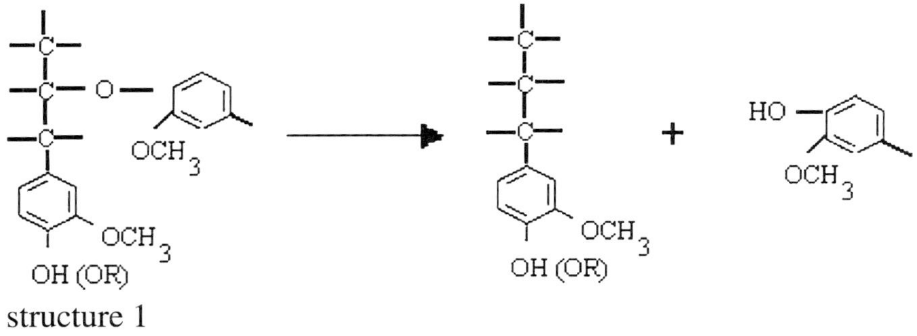

Figure 5.
A proposed mechanism for the alkali pulping. The major types of structural element in lignins – structure 1.

macromolecules the possible process is process of forming carbonyl and carboxyl groups. For example, appearing coniferaldehydes after cleavage of alkyl - aryl ether bonds was observed.[1]

Chemical analysis of the Freudenberg's lignin[13,24] showed that it has 52 aryl-ether bonds and 18 phenolic hydroxyl groups per 100 phenyl propane units (per C9-units). The dissolved during kraft treatment Freudenberg's lignin contains 22 aryl-ether bonds and 52 phenolic hydroxyl groups per C9-units. So we see that treated Freudenberg's lignin has approximately in 2.8 times more content of phenolic hydroxyl groups and approximately in 2.4 times less aryl-ether bonds.

From these chemical data changes we can possibly conclude about assigning the band near 1490 cm^{-1} to phenolic ring vibration and the bands at 1450, 1210, 1188 cm^{-1} to phenolic ring or phenolic hydroxyl groups vibrations. Decreasing aryl-ether bonds can result in dropping the bands at 1140, 1050 and 1030 cm^{-1}.

Analysis of spectra of dissolved lignins at various stages of cooking showed the same changing in theirs spectra as in spectrum of dissolved Freudenberg's lignin under the kraft cooking treatment.

It is worth to notice that elemental bands of technical dissolved lignins are broader and comparatively lower absorption at 1510, 1465, 1270, 1140, 1030 cm^{-1} was observed. In comparison with band parameters of mild isolated lignins there was revealed that degradation of lignin structure resulted to increasing bands connected with carbonyl and carboxyl groups and bands near 1490, 1450, 1210, 1188 cm^{-1} and lowering intensity of 1140, 1050 and 1030 cm^{-1} bands. There was revealed that similar directed changes of band parameters in the spectra of dissolved technical lignins depended on temperature and cook time.

It can be suggested that the spectral variations expose the changes in relative content of phenol-hydroxyl and aryl-ether bonds in lignins extracted at different stages of cooking. These results are well confirmed by investigation of structure of dissolved lignin by chemical methods which showed that cleavage of aryl-ether bonds brings about an increase of phenolic hydroxyl content.[13,22–24]

Chemical analysis of native lignin sample (lignin in wood) showed that it has 66 aryl-ether bonds and 13 phenolic hydroxyl groups per 100 phenyl-propane units but the dissolved technical lignin[13,24] contained 12 aryl-ether bonds and 66 phenolic hydroxyl groups per 100 C9-units.

It is interesting that the relation of intensities of two bands originating from the same vibration mode for two different aromatic ring types at 1490 and 1510 cm^{-1} changes more than two times for the presented samples. The value of this relation is connected with the transformation of one type of non-phenol aromatic ring to phenol type during the investigated treatments and can be used to fast evaluation of phenol hydroxyl groups. Relation of intensities of two bands at 1140 and 1125 cm^{-1} changes more than two times. The band near 1140cm^{-1} perhaps can be attributed to vibration of aryl-ether bonds but the second band at 1125cm^{-1} may be assigned to vibration bonds in forming condensed structure.

Conclusions

By computer treatment of the spectral data a mathematical model of softwood lignins was established. Its using for many of softwood lignin samples investigation showed a good reproducability and accuracy.

The model set of bands allowed to reveal discrepancies in spectral parameters of three mild isolated softwood lignins (lignins of Bjorkman, Pepper and Freudenberg) showing influence of isolation procedure on their structures.

The band parameters cleared changes during degradation treatment. It was shown that increasing and decreasing intensities of some bands go directly with increasing treatment time and temperature.

Significant discrepancies in the content and state of carbonyl and carboxyl groups were found in spectral region 1660–1720 cm^{-1}. This spectral range allowed evaluating degree of degradation of each studied lignin sample.

This method allows to systematizate lignins after theirs structure.

[1] K. V. Sarkanen, C. H. Ludwig, *Lignins: Occurrence, Formation, Structure and Reactions*, Wiley-Interscience, New-York **1971**.
[2] K. Freudenberg, A. C. Neish, *Constitution and Biosynthesis of Lignin*, Springer Verlag, Berlin, Heidelberg, New York **1968**.
[3] D. Fengel, G. Wegener, *Wood chemistry, ultrastructure and reactions*, Walter de Gruyter, Berlin, Germany **1989**.
[4] A. Sakakibara, *Wood Science and Technology* **1980**, *14*, 89.
[5] W. G. Gllasser, H. R. Gllasser, *Holzforschung* **1974**, *28*, 5.
[6] E. Adler, *Wood Sci. Technol.* **1977**, *11*, 169.
[7] R. H. Atalla, U. P. Agarwal, *Science* **1985**, *227*, 636.
[8] M. Akerholm, L. Salmén, *Polymer* **2001**, *42*, 963.
[9] D. A. Sukhov, O. Derkacheva, E. I. Evstigneyev, S. A. Kazanskiy, D. M. Kheyfetz, Proceed. of the 8th ISWPC, **1995**, *3*, 79.
[10] . Bjorkman, *Svensk Papperstidn.* **1956**, *59*, 477.
[11] J. M. Pepper, M. Siddiqueullan, *Can. J. Chem.* **1961**, *39*, 1454.
[12] G. F. Zakis, Functional analysis of lignin and its derivatives (in Russian), Zynante, Riga. **1987**.
[13] E. I. Evstigneev, H. D. Maiyorova, A. U. Platonov, *Tappi* **1992**, *75*, 177.
[14] Spectroscopic Software OPUS 1991, BRUKER ANALYTISCHE MESSTECHNIK GMBH.
[15] J. K. Kauppinen, D. F. Moffat, H. M. Mantsch, D. G. Cameron, *Appl.Spectr.* **1981**, *35*, 271.
[16] J. K. Kauppinen, D. F. Moffat, H. M. Mantsch, D. G. Cameron, *Anal.Chem.* **1981**, *53*, 1453.
[17] O. Faix, *Holzforshung* **1986**, *40*, 273.
[18] D. A. Sukhov, E. I. Evstigneyev, O. Yu. Derkacheva, S. A. Kazanskiy, D. M. Kheyfetz, Third European Workshop on Lignocellulosics and Pulp. Stockholm. Extended Abstracts, 1994, 11.
[19] O. Faix, *Holzforschung* **1991**, *45*, 21.
[20] O. Faix, B. Anderson, G. F. Zakis, *Holzforschung* **1998**, *52*, 268.
[21] K. V. Sarkanen, H. M. Chang, G. G. Allan, *Tappi* **1967**, *50*, 583.
[22] G. Gellerstedt, E.-L. Lindfors, *Holzforschung* **1984**, *38*, 151.
[23] J. Gierer, *Holzforschung* **1982**, *36*, 43.
[24] E. I. Evstigneev, H. D. Maiyorova, A. U. Platonov, Proc. Of 6th Int. Symp. On Wood and Pulping Chem. - Melbourne, Australia **1991**, V2, 131.
[25] K. V. Sarkanen, H. M. Chang, G. G. Allan, *Tappi* **1967**, *50*, 572.
[26] G. Gellerstedt, E.-L. Lindfors, C. Lapierre, B. Monties, *Svens Papperstidn.* **1984**, *87*, 61.
[27] P. Mansson, *Holzforschung* **1983**, *37*(3), 143.
[28] P. Mansson, *Tetrahedron Lett.* **1982**, *23*(17), 1845.
[29] T. P. Schultz, M. C. Templeton, G. D. McGinnis, *Anal. Chem.* **1985**, *57*, 2867.

Solid-State NMR Studies of Polysaccharide Systems

Jiří Spěváček, Jiří Brus*

Summary: Two polysaccharide systems were studied by solid-state NMR methods: (*i*) *Chitin/glucan complexes.* The ^{13}C NMR spectra have shown that in samples isolated from the mushroom *Pleurotus* sp., the glucan content was always higher in stems than in pilei. While carbonyl lineshape in complex isolated from *Aspergillus niger* mycelium shows similar hydrogen bonding as in neat chitin, a significantly higher amounts of hydrogen bonding between carbonyl groups of chitin and hydroxy groups of glucan was found in complexes isolated from *Pleurotus* sp. (*ii*) *Biodegradable starch/polycaprolactone (PCL) blends.* From the relaxation times $T_1(H)$ and $T_{1\rho}(H)$ it follows that blends starch/PCL, starch/ester oligomers and starch formate/ester oligomers are phase-separated even on the scale 20–110 nm. On the contrary, starch formate/PCL blend is phase-separated on the scale 1–9 nm but homogeneously mixed on the scale 20–90 nm. Therefore formylation of starch significantly improves its miscibility with PCL.

Keywords: biodegradable polymers; chitin/glucan complex; polysaccharides; solid-state NMR; starch/polycaprolactone blends

Introduction

Polysaccharides represent a significant group of biopolymers and solid-state NMR can provide important information in their structural characterization, similarly as with synthetic polymers. Two different systems containing polysaccharides which fall in the category of glucans (polymers of glucose) were investigated by solid-state NMR methods in the present study. The first one are chitin/glucan complexes isolated from the fruit bodies of edible mushroom *Pleurotus* sp. It is well known that fungal cell walls contain, besides chitin, hemicelluloses and mannans, also β-glucans. In last 15 years increased attention has been devoted to β-glucans isolated from the cell walls of fungi, which can act as nonspecific modulators of the immune system and have found applications as immunoadjuvants, antitumor and radioprotective agents.[1] Such structural features as β-(1$\rightarrow$3)linkages in the backbone of the glucan, and additional β-(1$\rightarrow$6) branching points are needed for antitumor activity. Higher-molecular-weight glucans have been reported to be more effective against tumors than those of low molecular weight.[2,3] Significant reduction of cholesterol and glucose in blood due to β-glucans has been also described.[4] In fungi glucans form water-insoluble complexes with chitin. Until now very little information exists on the structure of these complexes.[5] Therefore, in the present study we tried to contribute to the structural characterization of insoluble chitin/glucan complexes isolated from the mushroom *Pleurotus* sp. and *Aspergillus niger* mycelium by solid-state ^{13}C NMR spectroscopy.

The second system discussed in the present article are biodegradable starch/polycaprolactone (PCL) blends. Research on biodegradable polymers has received increasing attention in recent years and PCL is one of the most often investigated systems. However, since this polymer is still expensive and does not match all the

Institute of Macromolecular Chemistry AS CR, v. v. i., 162 06 Prague 6, Czech Republic
Fax: +420 296809410;
E-mail: spevacek@imc.cas.cz

technical requirements for possible applications, its blends with starch, which is a cheap abundant resource, are also investigated (cf. references in ref. [6]). The miscibility in these blends is one of the most relevant parameters related to blend properties. In the present study, the results of the measurements of ^{1}H spin-lattice relaxation times T_1(H) and ^{1}H spin-lattice relaxation times in the rotating frame $T_{1\rho}$(H), as obtained using ^{13}C CP/MAS NMR spectroscopy on starch/PCL and starch formate (starch modified by formylation)/PCL blends,[6] are discussed from the point of view of the blend morphology (domain size). Similar approach was recently used to characterize the morphology in starch/crosslinked poly(acrylic acid) blends[7] or in investigations of the blends of PCL with other synthetic polymers.[8–10]

Experimental Part

Samples

Chitin/glucan complexes were isolated from the mushroom *Pleurotus* sp. as insoluble fraction after extraction according to the modified procedure of Freimund et al.,[11] which involves extraction with 80% ethanol, boiling water and subsequent extraction with a 1M NaOH.[12] Dried mycelium of *Aspergillus niger* was deproteinised and then deacetylated under alkaline conditions. Samples of neat α-chitin (Fluka), chitosan (Fluka) and β-glucans isolated from the yeast were also studied for comparison.

Native wheat starch was provided by Roquette (France). Starch formate with degree of substitution ~1.2 was prepared from the dry native wheat starch and 99% of aqueous formic acid (Sigma Aldrich) by the procedure described elsewhere.[13,14] Polycaprolactone (PCL) Capa 6800 (molecular weight $M_w = 80\,000$) was provided by Solvay. Ester oligomers (1,6-hexane-diol adipate and phthalate; molecular weight $M_w = 2\,700$) were provided by Durez. The starch (or starch formate)/PCL (or ester oligomers) blends were extruded with a twin screw extruder (Brabender, DSK 42/6) controlled by a Lab-Station with a screw rotation speed of 30 rpm and temperatures of respectively 85, 90 and 95 °C from hopper to slot die (4'50 mm^2). The neat starch, starch formate as well as starch (starch formate)/PCL (ester oligomers) blends were dried under vaccuum for 8 h at 50 °C before measurements. After this treatment, a water content close to 1% was measured by modified Karl Fischer method (ISO 5381:1983) both in the neat starch (starch formate) and blend samples studied.

NMR Measurements

Solid-state ^{13}C CP/MAS (cross polarization/magic angle spinning) NMR spectra were measured at ambient temperature on a Bruker Avance 500 spectrometer at 125.8 MHz with spinning frequency 8–11 kHz and contact time 2 ms, i.e., under conditions that allow quantitative analysis.[15,16] All investigated systems are rather rigid and the only component where some interference between the higher mobility and CP might hypothetically exist is amorphous phase of PCL or ester oligomers. However, the fact that for the neat PCL the ^{13}C CP/MAS NMR spectrum is very similar to the fully relaxed ^{13}C MAS spectrum measured without CP with 45° pulses and 1200 s relaxation delay[17] shows that this is not the case. Chemical shifts in the ^{13}C NMR spectra were referred to the carbonyl line of glycine (with a signal at 176.0 ppm from TMS) by sample replacement. The ^{1}H rotating-frame spin-lattice relaxation times $T_{1\rho}$(H) and spin-lattice relaxation times T_1(H) were measured at ambient temperature via ^{13}C detection from ^{13}C CP/MAS NMR spectra. The experimental scheme with a variable spin-lock time in the range 0.1–10 ms after the proton signal excitation followed by constant contact time was used in $T_{1\rho}$(H) measurements; the proton spin-locking field in frequency units was 80 kHz. T_1(H) values were measured using the combination of cross-polarization and saturation recovery pulse sequence.

Results and Discussion

Chitin/Glucan Complexes

Figure 1 shows ^{13}C CP/MAS NMR spectra of the neat α-chitin (Fluka) and neat β-glucan isolated from the yeast. The assignment of signals to various carbon types follows that in the literature.[18,19] The signal at 33 ppm is due to impurities, residual proteins and/or lipids.[19,20] In consequence of the fact that chitin is also polymer of glucose derivative, N-acetylglucosamine, the range of the resonances of C1-C6 carbons (55–104 ppm) is similar in both systems, though not identical. However, for chitin it is characteristic the presence of acetyl groups on C2 carbons which are manifested by the separate signals of methyl and carbonyl carbons. These signals then can be used to determine the composition of chitin/glucan complexes.

^{13}C CP/MAS NMR spectra of chitin/glucan complexes, isolated from the stem and pileus of the mushroom *Pleurotus* sp., are shown in Figure 2. The comparison of these spectra with those obtained for neat α-chitin, chitosan and β-glucans, as well as with the spectra of neat chitin and β-glucans recently reported in the literature,[18–20] confirmed that investigated samples are really chitin/glucan complexes and enabled us to determine their composition. Higher relative intensities of methyl and carbonyl signals found for the sample isolated from

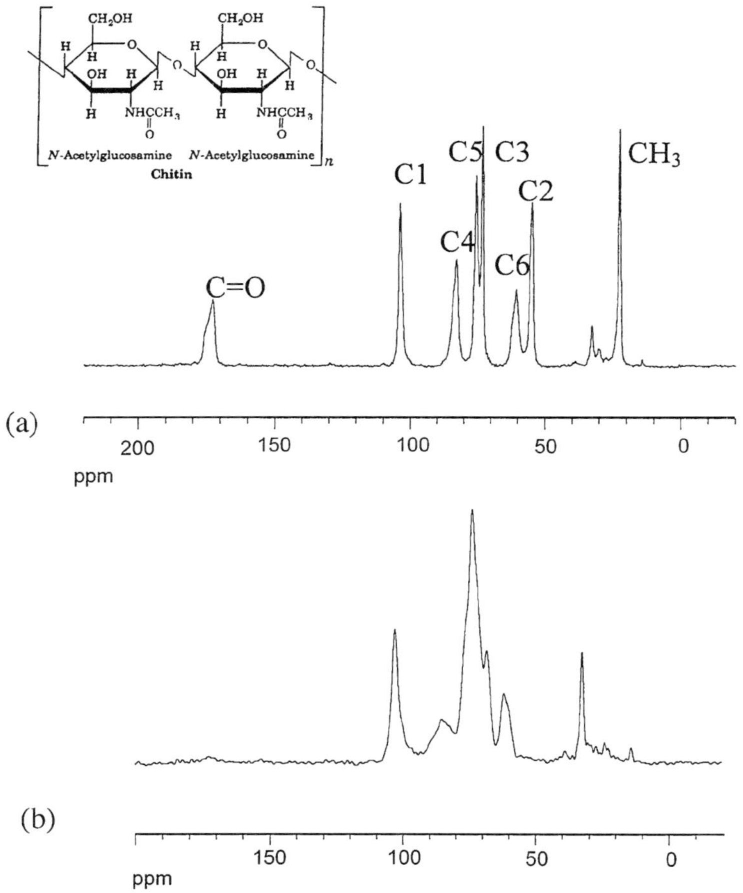

Figure 1.
^{13}C CP/MAS NMR spectra of neat α-chitin (Fluka) (a) and β-glucan isolated from yeast (b).

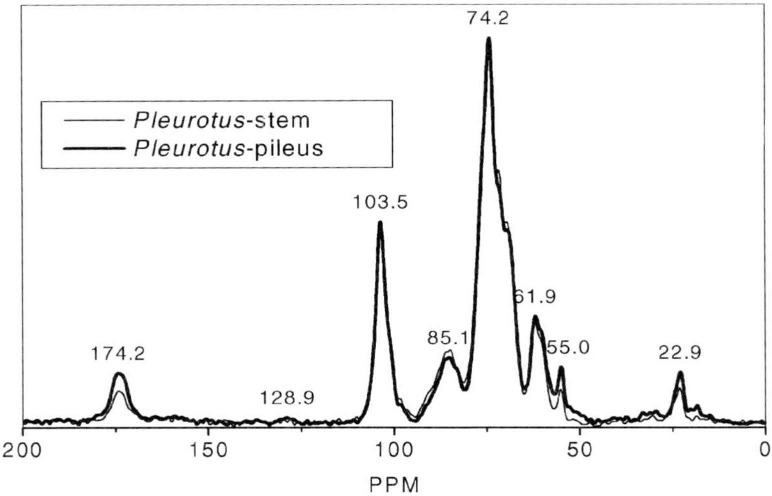

Figure 2.
^{13}C CP/MAS NMR spectra of chitin/glucan complexes isolated from the stem (thin line) and from the pileus (thick line) of the mushroom *Pleurotus* sp.

the pileus indicate a higher amount of the chitin component, i.e., a higher amount of β-glucans exists in the stem of *Pleurotus* sp. mushrooms. In the ^{13}C CP/MAS NMR spectrum of neat chitin measured with contact time 2 ms (cf. Figure 1a) the integrated intensities of all carbon types, including methyl carbons, are shown correctly. We assumed that this holds also for chitin/glucan complex and its composition was calculated using a formula

$$\text{mol\% chitin units } (N\text{–acetylglucosamine}) = [I_{CH3}/(I_{\Sigma}/6)] \times 100 \qquad (1)$$

where I_{CH3} is the integrated intensity of the signal of methyl carbons at 22.9 ppm and I_{Σ} is the integrated intensity of the range 50–110 ppm where C1-C6 carbons of both components resonate. For the samples isolated from the stem and pileus, as shown in Figure 2, we obtained 12 mol% and 25 mol% of the chitin units, respectively. The finding that the content of β-glucans is higher in stems than in pilei was confirmed by ^{13}C NMR for a series of samples isolated from the mushrooms *Pleurotus* sp. and *Agaricus blazei* and it is in accord with results of enzymatic analysis (Megazyme enzymatic kit).[21]

Interesting information on hydrogen bonding and structures in investigated samples follows from the expanded carbonyl region of ^{13}C NMR spectra as depicted in Figure 3. Already for the neat α-chitin the asymmetric lineshape indicates the presence of two components with chemical

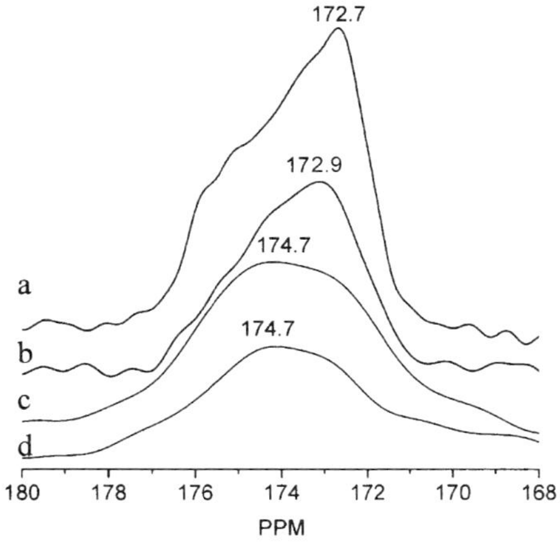

Figure 3.
Expanded carbonyl region in ^{13}C CP/MAS NMR spectra of neat α-chitin (a), chitin/glucan complex isolated from the *Aspergillus niger* mycelium (b), chitin/glucan complex isolated from the pileus (c) and stem (d) of the mushroom *Pleurotus* sp.

shifts that differ by ~2 ppm. Kameda et al. assigned these two components to carbonyl groups hydrogen bonded exclusively to the NH groups (more intensive component at 172.7 ppm) and to carbonyl groups hydrogen bonded also to hydroxy groups (less intensive component with larger chemical shift).[18] Assuming Lorentzian lineshapes, the ratio 0.38 : 0.62 was obtained by us after deconvolution for integrated intensities of both components, in accord with ref.[18] The carbonyl lineshape detected for the chitin/glucan complex isolated from the *Aspergillus niger* mycelium is similar to that for the neat chitin showing that hydrogen bonding is not significantly influenced by the presence of the glucan component. This result can indicate for this complex the grafted structure as suggested recently by Tarabukina et al.,[5] where both components are probably linked by 1,3-β-D-glucopyranoside covalent bonds, without forming mutual hydrogen bonds significantly. On the contrary for the samples of chitin/glucan complexes isolated both from the stem and pileus of the mushroom *Pleurotus* sp. the intensity of the carbonyl component with 2 ppm larger chemical shift is by 34% higher in comparison with the neat chitin; the ratio 0.51 : 0.49 was obtained for integrated intensities of both carbonyl components after deconvolution. This result suggests that in this case the carbonyl groups of the chitin component take part in the stabilization of the chitin/glucan complex by forming hydrogen bonds with hydroxy groups of the glucan component in significant amount. A larger density of hydrogen bonds between chitin and glucan also suggests a rather parallel arrangement of both components in the respective chitin/glucan complex. The obtained results indicate relatively large structural diversity of chitin/glucan complexes in fungal cells of various origin.

Biodegradable Starch/PCL Blends

In Figure 4 ^{13}C CP/MAS NMR spectra of the native starch, starch formate, starch/PCL (40/60) blend, starch formate/PCL (40/60) blend and neat PCL are shown. The signals of the starch and PCL were assigned in accord with the literature.[8–10,22,23]

The signal of C4 carbons of starch is reported to be characteristic for the amorphous glassy phase because it was not detected in hydrated, highly crystalline starch samples.[23] For neat PCL and PCL component in blends, signals of methylene carbons I and II are split into two lines corresponding to crystalline and amorphous components.[17] From integrated intensities of two signal components of PCL I and II carbons it follows that degree of crystallinity of PCL component in starch/PCL and starch formate/PCL blends is virtually the same as in neat PCL and amounts to ~63–66%. For neat PCL this value agrees well with the degree of crystallinity determined by DSC,[9] so supporting that ^{13}C CP/MAS NMR spectra show essentially correct signal intensities (cf. Experimental Part, NMR Measurements). From Figure 4 it follows that in blend samples the signals of starch and PCL components are well separated and therefore ^{1}H spin-lattice relaxation times T_1(H) and spin-lattice relaxation times in the rotating frame $T_{1\rho}$(H) can be measured via ^{13}C detection from a series of ^{13}C CP/MAS NMR spectra.

The approach used to characterize the molecular morphology (domain size) of investigated blend samples is based on the ^{1}H spin-diffusion. The maximum diffusive path length L by spin-diffusion in three dimensions for a time T_i is approximately given as[24]

$$L = (6DT_i)^{1/2} \quad (2)$$

where D is the spin-diffusion coefficient and T_i is either $T_{1\rho}$(H) or T_1(H) according to the relaxation experiment. For the spin-diffusion coefficient a value of 8×10^{-16} m^{2}s^{-1} has been found in poly(methyl methacrylate)/polystyrene block copolymers; this value is expected to be typical for rigid organic systems.[25] For the spin-diffusion coefficient in mobile amorphous PCL (in rubbery state at room temperature) a lower value can be expected.[24,26] A value $D = 0.5 \times 10^{-16}$ m^{2}s^{-1} is probably a reasonable estimate.[27]

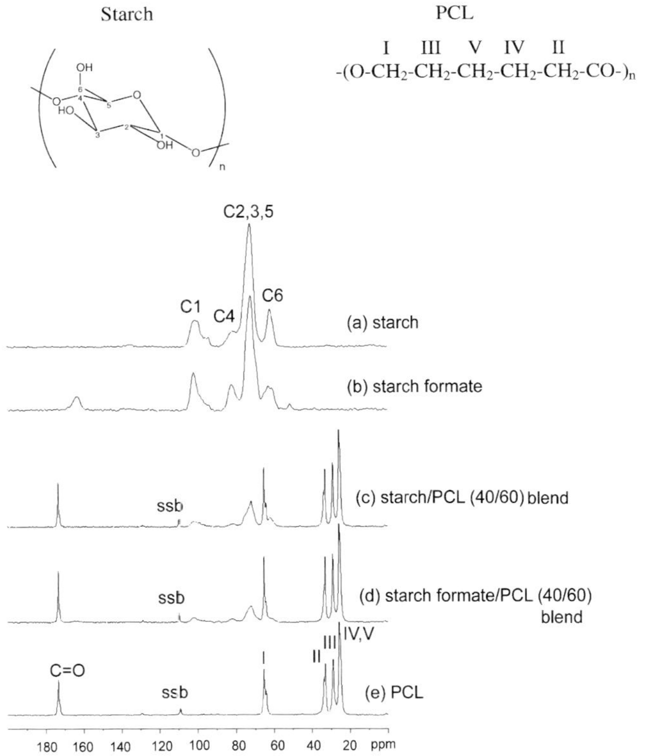

Figure 4.
^{13}C CP/MAS NMR spectra of the native starch (a), starch formate (b), starch/PCL (40/60) blend (c), starch formate/PCL (40/60) blend (d) and neat PCL (e); ssb means spinning sideband.[6]

In Table 1 the values of spin-lattice relaxation times in the rotating frame $T_{1\rho}(H)$ are shown. We always could fit the relaxation curves by one component. $T_{1\rho}(H)$ values determined from an analysis of various carbon lines of the starch or starch formate were equal within experimental error due to the spin-diffusion mechanism[6] and therefore only the mean values as obtained for the signals of C1, C2,3,5, C4 and C6 carbons are shown in Table 1. From Table 1 it follows that in blends the $T_{1\rho}(H)$ values of the starch formate component (and to a certain extent also of the starch component) are shorter in comparison with values in neat samples. This indicates that molecular mobility of the starch formate (and perhaps also of the starch) is somewhat higher in blends in comparison with neat samples. For PCL carbons only the mean $T_{1\rho}(H)$ values as obtained for III and IV,V carbons are shown because for I and II carbons, where separated signals were detected for the crystalline and amorphous

Table 1.
^{1}H spin-lattice relaxation times in the rotating frame $T_{1\rho}(H)$ of starch (or starch formate) and PCL (or ester oligomers) in investigated blends and neat components

Sample	$T_{1\rho}(H)^{a)}$ (ms)	
	Starch	PCL
starch	6.9	–
starch formate	4.7	–
PCL	–	24.5
ester oligomers	–	46.5
starch/PCL (60/40)	5.3	29.2
starch/PCL (40/60)	4.9	24.4
starch formate/PCL (60/40)	2.5	29.0
starch formate/PCL (40/60)	2.3	28.3
starch/ester oligomers (40/60)	4.8	34.0
starch formate/ester oligomers (40/60)	2.6	37.1

a) Estimated error ±10%.

phase, the different $T_{1\rho}(H)$ values (~35 ms and 12 ms) were found for the crystalline and amorphous phase, respectively, both for neat PCL and PCL component in blends.[6] This means that the sizes of the crystalline and amorphous domains of PCL are larger than the maximum diffusive path length L and therefore the $T_{1\rho}(H)$ values are not equilibrated over the whole system. For the $T_{1\rho}(H)$ experiment a scaling factor 1/3 should appear in parenthesis in Eq. (2), due to spin-lock field and faster spinning[28], and therefore $L \cong 8$ nm and 1 nm for the crystalline and amorphous PCL phase, respectively. Similar situation follows from Table 1 for domain size in investigated blend samples. From Table 1 it follows that $T_{1\rho}(H)$ values are markedly different when determined from analysis of signals of starch (or starch formate) carbons in comparison with those determined from PCL (or ester oligomers) carbon signals. These results show that in all investigated blends both components are separated into domains of which the length scale exceeds the maximum diffusive path length $L \approx 1$–9 nm.

To investigate the molecular miscibility on the larger scale, the spin-lattice relaxation times $T_1(H)$ were measured, again using a selective carbon detection. Also in this case the relaxation curves were well fitted by one component. The values $T_1(H)$ as determined from relaxation curves of various starch or starch formate carbons do not significantly differ; the same holds for $T_1(H)$ determined from signals of various PCL (or ester oligomers) carbons[6] showing that the spin-diffusion process is effective within domains formed by single components of the blends. Therefore only the mean $T_1(H)$ values as obtained for starch (starch formate) C1-C6 carbons and PCL (ester oligomers) I–V carbons are shown in Table 2. From Table 2 it follows that for the blends starch/PCL, starch/ester oligomers and starch formate/ester oligomers $T_1(H)$ of both components differ from one another. These results suggest that in these blends both components are phase separated even on the larger scale of 20–110 nm (cf. Eq. (2)). A different behaviour was found from

Table 2.
^1H spin-lattice relaxation times $T_1(H)$ of starch (or starch formate) and PCL (or ester oligomers) in investigated blends and neat components

Sample	$T_1(H)^{a)}$ (s)	
	Starch	PCL
starch	1.4	–
starch formate	2.6	–
PCL	–	1.1
ester oligomers	–	1.5
starch/PCL (40/60)	1.6	1.1
starch formate/PCL (40/60)	1.6	1.6
starch/ester oligomers (40/60)	1.8	1.5
starch formate/ester oligomers (40/60)	2.5	1.6

a) Estimated error ±10%.

$T_1(H)$ values for starch formate/PCL blend. From Table 2 it follows that in this blend the same $T_1(H)$ values were determined for starch formate and PCL components. At the same time, the $T_1(H)$ value found from signals of starch formate carbons is shorter in comparison with the neat starch formate and the value found from signals of PCL carbons is longer in comparison with the neat PCL, showing that the spin-diffusion process is sufficient to equilibrate $T_1(H)$ values. Therefore it is possible to conclude that in starch formate/PCL blend both components are homogeneously mixed on the scale 20–90 nm and that formylation of starch significantly improves its miscibility with PCL.

Conclusion

Two polysaccharide systems were studied by solid-state NMR methods which we applied to investigate their composition and interactions on the one hand (chitin/glucan complexes) and phase structure on the other hand (biodegradable starch/PCL blends):

Chitin/glucan complexes. The ^{13}C CP/MAS NMR spectra enabled us to characterize the composition and structure of the complexes. In samples isolated from the mushroom *Pleurotus* sp., the glucan content was always higher in stems than in pilei. Analysis of the carbonyl region of the chitin component in complex isolated from the *Aspergillus niger* mycelium corresponds to the grafted structure where both components might be linked virtually only by covalent bonds.[5] By contrast, analysis of the same region in samples isolated from the mushroom *Pleurotus* sp. shows higher amounts of hydrogen bonding between carbonyl groups of chitin and hydroxy groups of glucan, indicating a rather parallel arrangement of both components in the complex.

Biodegradable starch/PCL blends. ^{1}H spin-lattice relaxation times T_1(H) and spin-lattice relaxation times in the rotating frame $T_{1\rho}$(H) were measured via ^{13}C detection from ^{13}C CP/MAS NMR spectra to characterize the blend morphology (domain size). From the $T_{1\rho}$(H) and T_1(H) values it follows that blends starch/PCL, starch/ester oligomers and starch formate/ester oligomers are phase-separated even on the scale 20–110 nm. On the contrary, starch formate/PCL blend is phase-separated on the scale 1–9 nm but homogeneously mixed on the scale 20–90 nm. Therefore formylation of starch significantly improves its miscibility with PCL.

Acknowledgements: This work was supported by the Grant Agency of the Czech Republic (grant 525/05/0273). The authors thank A. Synytsya and J. Čopíková for samples of chitin/glucan complexes and T. Divers and Y. Grohens for samples of starch/PCL blends.

[1] J. Šandula, G. Kogan, M. Kačuráková, E. Machová, *Carbohydr. Polym.* **1999**, *38*, 247.
[2] T. Mizuno, *Food Rev. Int.* **1995**, *11*, 7.
[3] T. Mizuno, *Int. J. Med. Mushrooms* **1999**, *1*, 9.
[4] P. Manzi, L. Pizzoferrato, *Food Chem.* **2000**, *68*, 315.
[5] E. B. Tarabukina, N. A. Kalinina, A. V. Adamov, V. A. Petrova, L. A. Nudga, S. I. Klenin, *Vysokomol. Soedin., Ser. A* **2005**, *47*, 778.
[6] J. Spěváček, J. Brus, T. Divers, Y. Grohens, *Eur. Polym. J.* **2007**, *43*, 1866.
[7] D. Ameye, E. Pringels, P. Foreman, J. P. Remon, P. Adriaensens, L. Storme, J. Gelan, *Polymer* **2005**, *46*, 2338.
[8] Z. Zhong, Q. Guo, Y. Mi, *Polymer* **1999**, *40*, 27.
[9] C. De Kesel, C. Lefèvre, J. B. Nagy, C. David, *Polymer* **1999**, *40*, 1969.
[10] J. Wang, M. K. Cheung, Y. Mi, *Polymer* **2002**, *43*, 1357.
[11] S. Freimund, M. Sauter, O. Käppeli, H. Dutler, *Carbohydr. Polym.* **2003**, *54*, 157.
[12] K. Míčková, P. Blafková, J. Černá, J. Čopíková, A. Synytsya, A. Synytsya, J. Spěváček, I. Jablonský, V. Erban, *Chem. Listy* **2005**, *99*, 666.
[13] T. Divers, I. Pillin, J. F. Feller, G. Levesque, Y. Grohens, *Starch/Stärke* **2004**, *56*, 389.
[14] I. Pillin, T. Divers, J. F. Feller, Y. Grohens, *Macromol. Symp.* **2005**, *222*, 233.
[15] R. Voelkel, *Angew. Chem., Int. Ed. Engl.* **1988**, *27*, 1468.
[16] J. Spěváček, J. Straka, B. Schneider, *J. Appl. Polym. Sci., Appl. Polym. Symp.* **1991**, *48*, 371.
[17] H. Kaji, F. Horii, *Macromolecules* **1997**, *30*, 5791.
[18] T. Kameda, M. Miyazawa, H. Ono, M. Yoshida, *Macromol. Biosci.* **2005**, *5*, 103.
[19] L. Johansson, P. Tuomainen, M. Ylinen, P. Ekholm, L. Virkki, *Carbohydr. Polym.* **2005**, *58*, 267.
[20] M. Fernández Cervera, J. Heinämäki, M. Räsänen, S. L. Maunu, M. Karjalainen, O. M. Nieto Acosta, A. Iraizoz Colarte, J. Yliruusi, *Carbohydr. Polym.* **2004**, *58*, 401.
[21] K. Míčková, A. Synytsya, J. Čopíková, J. Černá, M. Maryška, M. A. Coimbra, J. Spěváček, *Chem. Listy* **2006**, *100*, 846.
[22] R. P. Veregin, C. A. Fyfe, R. H. Marchessault, M. G. Taylor, *Macromolecules* **1986**, *19*, 1030.
[23] T. Y. Bogracheva, Y. L. Wang, C. L. Hedley, *Biopolymers* **2001**, *58*, 247.
[24] A. Asano, K. Takegoshi, in: "*Solid State NMR of Polymers*", I. Ando, T. Asakura, Eds., Elsevier **1998**, p. 361.
[25] J. Clauss, K. Schmidt-Rohr, H. W. Spiess, *Acta Polym* **1993**, *44*, 1.
[26] J. Straka, P. Schmidt, J. Dybal, B. Schneider, J. Spěváček, *Polymer* **1995**, *36*, 1147.
[27] S. Spiegel, K. Schmidt-Rohr, C. Boeffel, H. W. Spiess, *Polymer* **1993**, *34*, 4586.
[28] Q. Chen, K. Schmidt-Rohr, *Solid State Nucl. Magn. Reson.* **2006**, *29*, 142.

In situ ATR-FTIR and SFM Studies on the Influence of Adsorption time on Deposition and Nanostructure of Poly(ethyleneimine)/Poly(acrylic acid) Multilayers

M. Müller, S. Paulik*

Summary: The deposition and the nanostructure of polyelectrolyte multilayers (PEM) of branched poly(ethyleneimine)/poly(acrylic acid) (PEI/PAC) was studied in dependence of the adsorption time (t_{ADS}) of the individual steps. PEM were reproducibly deposited applying up to $z = 20$ adsorption steps at the fixed pH combination of 10/4 and polyelectrolyte concentration $c_{PEL} = 0.005$ M in a flow cell using an automated valve system. in situ ATR-FTIR spectroscopy and SFM were used for quantitative determination of deposited amount and thickness, respectively. A linear relation between PEL band integrals and thickness of thin PEM films was found. Varying t_{ADS} from 0.5 to 5 min in each of the adsorption steps resulted in a steep rise of the deposited PEM amount. For $t_{ADS} > 5$ min the deposition did only marginally increase. Evidence for the release of outermost located PEI upon PAC immersion (even step) and of outermost PAC upon PEI immersion (odd step) was obtained. SFM images on consecutively deposited PEM-6 showed a slight increase in structure size and roughness for increasing t_{ADS}. These studies help to prepare polyelectrolyte based films with controlled thickness for the interaction with biofluids in the biomedical and food field.

Keywords: ATR-FTIR; deposition; poly(acrylic acid); poly(ethyleneimine); polyelectrolyte multilayers; SFM

Introduction

Polyelectrolyte (PEL) multilayers (PEM) can be used as a simple surface modification technique for curved and planar substrates based on aqueous systems. PEM were introduced by Decher[1] in the early nineties and have become an interesting topic of colloid and surface science, which is comprehensively reviewed therein.[2–4] PEM have wide application fields in biomedicine,[5–8] diagnostics and sensorics,[9–11] and separation technology.[12–14] In principle, the PEM deposition process is based on the consecutive adsorption of polycations (PC) and polyanions (PA), typically on charged substrates beginning with the oppositely charged PEL. Although the preparation of PEM can be easily performed, fundamental issues like overcompensation and growth mechanism, location of the counterions, internal PEL order and composition, surface morphology and longterm stability of PEM are still not completely resolved. Concerning PEM growth mechanism the traditional picture is based on the linear relation between the UV absorbance of one PEM component like poly(styrene sulfonate) and the bilayer number, from which a well defined regular PEL uptake of a constant thickness increment was derived.[1] From the beginning of 2000 this picture was partly reviewed by the observation of exponential growth in certain systems, especially when charged polypeptides like poly(L-lysine) or poly (L-glutamic acid) were involved,[15,16] which was explained by diffusion of PEL "in" and "out" of the PEM. A three zone model of

Leibniz Institute of Polymer Research Dresden e.V. (IPF Dresden), Department of Surface Modification, Hohe Straße 6, 01069 Dresden, Germany
E-mail: mamuller@ipfdd.de

PEM[17] was propagated claiming a non-stoichometric zones for the surface (I) and the outermost region (III) and a 1:1 stoichometry in the core zone II between I and III. Recently, the PEM deposition process was found to consist of both uptake of the incoming immersed PEL and rupture of the previously adsorbed PEL by the respective oppositely charged PEL, as it was convincingly shown by Cohen-Stuart.[18] While they used reflectometry to show that upon immersing the new PEL solution an initial overall thickness decrease occurs, systematic studies are lacking on molecular composition changes.

In this paper we focus on the deposition and molecular composition of a PEM system consisting of commercially available branched poly(ethyleneimine) (PEI) and poly(acrylic acid) (PAC) using quantitative *in situ* attenuated total reflexion Fourier transform infrared (ATR-FTIR) spectroscopy and scanning force microscopy (SFM). While accumulated knowledge from the literature is available on the influence of parameters like chain charge,[19] ionic strength[20–22] and pH,[23,24] on PEM deposition the influence of adsorption time has not been systematically studied yet. Herein the influence of the adsorption time in every individual adsorption step is focussed at. Therefore, as main parameter the adsorption time of each of the respective deposition steps was varied, while pH and PEL concentration were kept constant.

Experimental Part

Materials

Commercial branched poly(ethyleneimine) (PEI, 750.000 g/mol, branching degree: 65%[25]) from Sigma-Aldrich (Steinheim, Germany) and linear poly(acrylic acid) (PAC, 50.000 g/mol) from Polysciences Inc. (Warrington, U.S.A.) were used. Polyelectrolyte (PEL) solutions were prepared by dissolving in Millipore water at concentrations of $c_{PEL} = 0.005$ M. The pH values of the unbuffered polyelectrolyte solutions were 10.0 ± 0.1 for PEI and 4.0 ± 0.2 for PAC solution, respectively. Trapezoidal silicon internal reflection elements (IRE, $50 \times 20 \times 2$ mm^3) were purchased from Komlas GmbH (Berlin, Germany), which were at first placed in a mixture of H_2SO_4/H_2O_2 (50/50 v/v%), then in water, alcohol and chloroform and finally were cleaned by UV plasma under reduced pressure (plasma cleaner/sterilizer PDC-32 G, Harrick, Ossining, NY, U.S.A.) to remove organic impurities and create reproducible surface properties.

PEM Depositon

PEM were prepared by injecting consecutively in that sequence PEI solution at pH = 10, pure water, PAC solution at pH = 4 at a concentration of $c_{PEL} = 0.005$ M into the S compartment (see below) of the *in-situ*-ATR cell as it is shown in the Figure 1.

An automated valve system[26] (M.M., W. Jenschke, IPF Dresden e.V., Wünschmann GmbH, Dresden) triggering the IR spectroscopic measurements was used, by which flow (mL/min) and adsorption time (min) could be varied according to the following protocol: (i) Injection of PEI solution at a flow of 40 mL/min for 10 sec, then further adsorption at a flow of 1 mL/min for 5 min, (ii) injection of Millipore water (flow: 40 mL/min) for 10 sec, then further rinsing (flow: 1 mL/min) for 150 sec

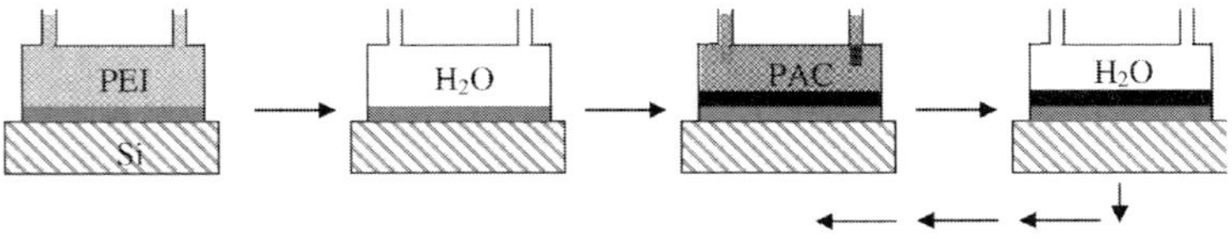

Figure 1.
Polyelectrolyte multilayer (PEM) deposition in the *in-situ* ATR cell (grey: polycation solution (PC), black: polyanion solution (PA), white: Millipore water (H_2O).

and IR measurement, (iii) injection of PAC solution (flow: 40 mL/min) for 10 sec, then further adsorption (flow: 1 mL/min) for 5 min, (iv) injection of Millipore water (flow: 40 mL/min) for 10 sec, then further rinsing (flow: 1 mL/min) for 150 sec and IR measurement, (v) = (i) (iterations).

Scanning Force Microscopy

SFM measurements were performed on the Ultramiscroscope consisting of an optical microscope and SFM attachment (Nanostation II, SIS GmbH, Herzogenrath, Germany). Silicon probe tips from Nanosensors (Darmstadt, Germany) were used having radii of around 10 nm. The measurements were performed in the "non-contact mode" directly on the dry PEM samples on Si-IRE (internal reflection elements) used in the ATR-FTIR measurements under room conditions. The SFM images were recorded in topography, error and phase mode. As soon as artefacts (e.g. triangles as convolution of the tip with the object) appeared, the tip was immediately replaced. The scanning parameters were optimized by minimizing the signal in the error mode. Surface profiles were generated from SFM raw data by the SISCANPro software. The RMS roughness values were determined from the respective SFM images in the topography mode. Film thickness was measured based on topography images (32×32 μm) of zones treated by careful scalpel cuts considering the height difference between undamaged film and bare silicon.

in situ ATR-FTIR Spectroscopy

The principle of in situ attenuated total reflection Fourier transform infrared (ATR-FTIR) spectroscopy is given in the Figure 2. ATR-FTIR measurements were performed on the in situ ATR-FTIR apparatus (Optispec, Zürich, Switzerland), which was installed on a FTIR spectrometer (IFS 55, Bruker Optik GmbH, Leipzig, Germany). This attachment consists of a special mirror setup and a transparent *in situ* sorption cell (M.M., IPF Dresden e.V.) using a plasma cleaned silicon internal reflection element (Si-IRE). ATR-FTIR absorbance spectra were recorded by the single-beam-sample-reference (SBSR) technique[27] based on probing separately the upper sample (S) and lower reference (R) half of a trapezoidal Si-IRE ($50 \times 20 \times 2$ mm^3, N = 11 active reflections on the shorter side), which was sealed by O-rings and front and back side of the in-situ cell, by a single IR beam. Wavenumber dependent intensities recorded for S $I_S(\nu)$ were devided by those for R $I_R(\nu)$ and the absorbance was computed by $A_{SBSR} = -\log\ (I_S(\nu)/I_R(\nu))$. Typically, the in situ ATR-FTIR measurements on the consecutive deposition of PEI/PAC were triggered on-line by the automated

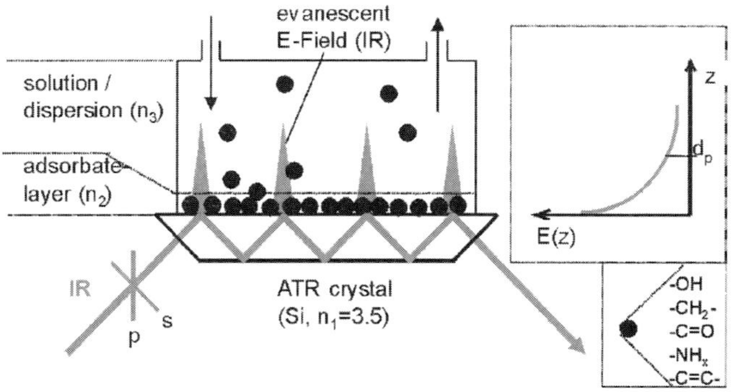

Figure 2.
Principle of in-situ ATR-FTIR spectroscopy featuring exponentially decaying evanescent waves at the optically dense/rare medium interface to probe surface attached organic material. (From[28] with kind permission).

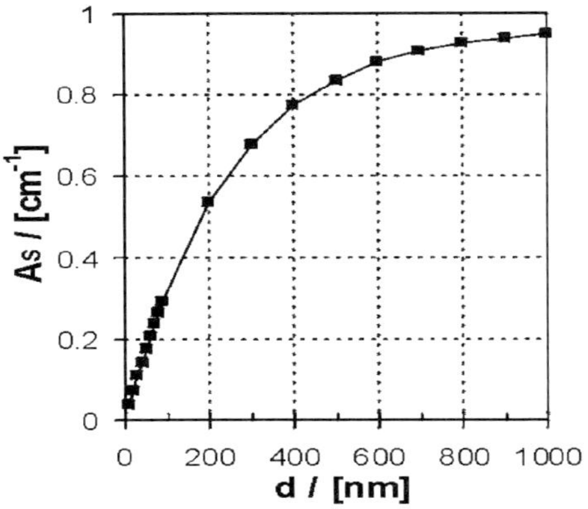

Figure 3.
Course of the ATR measured absorbance A_S in dependence of the variable layer thickness d according to Equation (1) and (2) using arbitrary constant values for N, ε and c. (From[28] with kind permission).

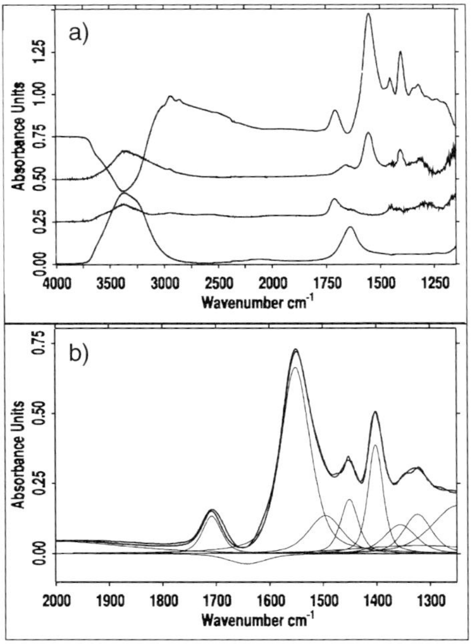

Figure 4.
a) ATR-FTIR spectra of liquid water (scaled by 0.5), 1 M PAC at pH = 2, 1 M PAC at pH = 10 and PEM-10 in contact to Si-IRE (bottom to top). b) Lineshape analysis results on a typical PEM spectrum (PEM-10) using nine positive and one negative Lorentzian/Gaussian (50/50%) component at 1643 cm^{-1} (δ(OH)) (constraint fit).

valve system. ATR-FTIR spectra (50 scans were coadded) were always recorded after rinsing the respective PEI or PAC solution out of the sample compartment by Millipore water (see protocol above). The spectral resolution was 2 cm^{-1}. No significant spectroscopic changes were observed comparing ATR-FTIR spectra recorded in the presence of PEI or PAC solution with those recorded after rinsing and in the presence of Millipore water.

Processing of the ATR-FTIR Data

In the ATR-FTIR spectra on (especially thick) PEM films in liquid water at the silicon IRE surface a negative ν(OH) stretching band at around 3380 cm^{-1} as well as a negative δ(OH) band at 1643 cm^{-1} appeared shown in the Figure 4a, which is due to the lower concentration or removal of liquid water at the modified IRE surface, as it will be pointed out later. (Other spectra in Figure 4a are explained below).

Since especially the negative δ(OH) band overlaps the diagnostic ν(C=O) and ν(COO^-) bands of poly(acrylic acid) (see below), this band has to be taken into account for an exact data analysis. This problem was addressed by lineshape analysis (LSA) in the range from 2000–1250 cm^{-1}, representing the lineshape by the superposition of ten Lorentzian/Gaussian (50%/50%) bands. At first LSA was performed under free relaxation of the parameters (wavenumber) position, halfwidth and intensity for nine components and all PEM-z spectra (z = 2 to 20). As tenth component a negative δ(OH) band at 1643 cm^{-1} with a halfwidth of 85 cm^{-1}, whose parameters were found by fitting a bulk water spectrum (Figure 4a, bottom spectrum), was kept fixed and only the (negative) intensity was relaxed. This procedure evidenced nine further positive components at 1250, 1323, 1354, 1402, 1451, 1490, 1552, 1707 and 2000 cm^{-1}, which could be also identified by visual inspection. Note that the 1250 and 2000 cm^{-1} components were used as "dummy" components to avoid boundary effects on the intensities of the critical 1707, 1643 and 1552 cm^{-1} components. Then the respective positions and halfwidths of all peaks were averaged and in a second run all these spectra were fitted fixing the averaged values of position and halfwidth and relaxing solely the intensities (constraint fit). A representative constraint fit result is shown for PEM-10 in the Figure 4b, featuring the original and simulated lineshape and the ten overlapping components.

Quantitative Analysis of the ATR-FTIR Data

Exact quantitative ATR-FTIR analysis is based on a modified Lambert-Beer law given in Equation (1):

$$A_S = N\,\varepsilon\,c\,d_{E,S} \qquad (1)$$

including the integrated absorbance of a given IR band measured in s-polarization A_S [cm^{-1}], number of active reflections N, absorption coefficient ε [cm/Mol], concentration c [Mol/cm^3], and the effective thickness $d_{E,S}$ [cm^{-1}] due to Harrick,[29] which is given by:

$$d_{E,S} = n_2\,d_p\,E_Y^2/(2\cos\theta\,n_1) * [1 - \exp(-2d/d_p)] \qquad (2)$$

$d_{E,S}$ is a function of the refractive indices of the involved media n_1 (Si, 3.5), n_2 (PEM, 1.5) and n_3 (water, 1.33), incident angle of IR beam θ (45°), thickness d and the >penetration depth d_P ($\approx$260 nm at 3000 cm^{-1}, 460 nm at 1700 cm^{-1} and 510 nm at 1550 cm^{-1}) and the relative electrical field component in y-direction $E_Y = 2\cos\theta/(1-(n_3/n_1)^2)^{1/2}$. The surface concentration Γ [Mol/cm^2] can be obtained from by inserting c from (1) into following Equation (3):

$$\Gamma = c\,d \qquad (3)$$

A typical course of the measured absorbance A_S of an organic layer system in dependence of its thickness d is given in the Figure 3. Two points are characteristic: At first for thin films between $d = 0$ up to 300 nm a quasi linear dependence of A_S on d can be approximated. Secondly, for thick films A_S runs into a saturation, meaning that the ATR-FTIR method gets increasingly insensitive to additional layer deposition or the nature of outer film regions, when d exceeds penetration depth $dp \approx 500$ nm. For the semiquantitative determination of Γ the integrated areas of typical polyelectrolyte (PEI, PAC) IR bands A, which were measured by unpolarized IR beam, were considered. For the determination of PEI surface concentration the integral of the ν(CH) band at 2855 cm^{-1} (limits: 2890–2820 cm^{-1}) was used denoted as A_{CH} ($=A_{PEI}$). For the PAC surface concentration the integral of the ν(C=O) component ($A_{C=O}$) at 1707 cm^{-1} and of the $\nu(COO^-)$ component (A_{COO^-}) at 1552 cm^{-1} were used according to the relation $A_{PAC} = F \cdot A_{C=O} + A_{COO^-}$. In that work $F = 2.24$ was used, which is due to the ratio of the absorption coefficients of $\nu(COO^-)$ and ν(C=O) band. This ratio can be obtained from the spectra of 1 M PAC at $pH = 2$ (only ν(C=O)) and at $pH = 10$ (only $\nu(COO^-)$), respectively, which are given in the Figure 4a (above). These integrals A_{PEI} and A_{PAC} are approximately proportional to the surface concentration (denoted further as deposited amount) of PEI and PAC, respectively, if the overall thickness of the PEM film $d < 300$ nm (see Figure 3). Otherwise for $d > 300$ nm the PEL deposited amount is increasingly underestimated for increasing PEM thicknesses. For relative comparisons normalized values of A^N_{PEI} and A^N_{PAC} were used, whereby the actual A_{PEI} or A_{PAC} value was devided by the corresponding value obtained for an adsorption step $z \gg 10$, where the respective band integral was in saturation (no further signal increase in dependence of z).

Results and Discussion

In the following results on the characterization of consecutively adsorbed PEM consisting of branched PEI and PAC at silicon IRE at a fixed pH setting and constant c_{PEL}, which were obtained by

ATR-FTIR spectroscopy and SFM, will be presented. The main parameter studied was the adsorption time t_{ADS}. Branched PEI was chosen, since it was expected to result in higher deposited amounts compared to linear PEI. ATR-FTIR spectroscopy was used to determine deposited amount and chemical composition in dependence of t_{ADS}, while SFM performed on the samples previously measured by ATR-FTIR gave informations on the formed nanostructures in dependence of t_{ADS}.

The paper is structured as follows: At first typical *in-situ*-ATR-FTIR spectra on consecutive PEI/PAC deposition in dependence of the adsorption step z are presented and assignments of typical IR bands related selectively to PEI and PAC, respectively, are given. Secondly, deposition profiles, i.e. courses of typical band integrals in dependence of t_{ADS} and interpretations of their non trivial courses are given. Thirdly, SFM images on equivalent samples previously characterized by ATR-FTIR are presented illustrating the influence of t_{ADS} on PEM nanostructure. Finally, the results are discussed and conclusively summarized.

ATR-FTIR Spectra

Typical ATR-FTIR spectra on the consecutive deposition of PEI and PAC on Si-IRE are given in the upper part of Figure 5 for a single PEI layer ($z = 1$, PEM-1, bottom) up to twelve consecutively adsorbed PEI and PAC layers ($z = 12$, PEM-12).

As fixed parameters settings $c_{PEL} =$ 0.005 M and pH $\approx$ 10 (PEI) and $\approx$4 (PAC) were chosen, since under these conditions the deposited amount was found to be highest [30]. In these original uncorrected PEM spectra the increasing overall intensity and the changes of both the ν(C=O) band (carbonyl stretch) at around 1707 cm^{-1} and the $\nu(COO^-)$ band (asymmetric carboxylate stretch) at 1552 cm^{-1}, respectively, in dependence of z are most significant, which both can be exclusively attributed to the PAC component. Moreover, a weak ν(CH) band between 2890–2820 cm^{-1} is important, which can be exclusively assigned to the CH and CH_2 moieties of PEI (Figure 5, right part). This was recently confirmed by ATR-FTIR spectra on bulk 1 M PEI compared to 1 M PAC solution therein.[30] Furthermore, an increasing negative ν(OH) band at around 3380 cm^{-1} shows up, which reflects the more distant location of water in contact to the PEM film (S compartment) compared to the bare Si-IRE (R compartment) with respect to the Si surface. Hence the ν(OH) band senses film formation as the removal of water from the surface in

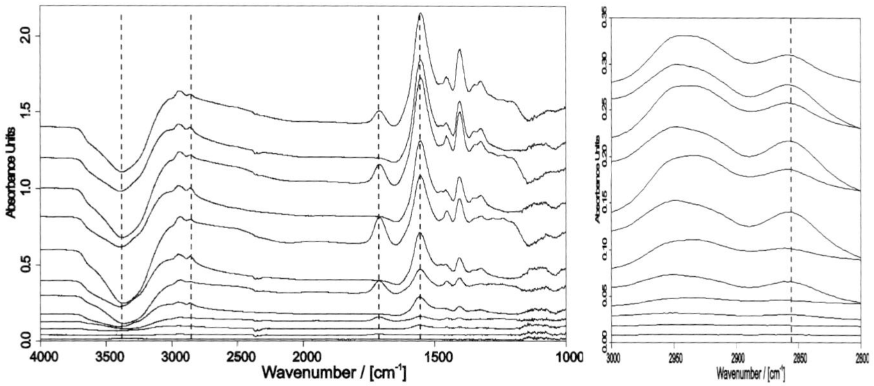

Figure 5.
Left: Original *in-situ*-ATR-FTIR spectra (uncorrected data, see Experimental) on the consecutive deposition of PEM from solutions of PEI at pH $\approx$ 10 and PAC at pH $\approx$ 4 for $c_{PEL} = 0.005$ M and $t_{ADS} = 5$ min onto Si-IRE. PEM-z are shown from z = 1 to 12 from bottom to top. Right: respective extended ν(CH) spectral regions (PEI). Typical IR bands used for further analysis are indicated by vertical broken lines (see text).

Table 1.
Assignment of IR bands and lineshape components appearing in ATR-FTIR spectra of PEM-PEI/PAC.

Wavenumber/[cm^{-1}]	Assignment	Component
3380	ν(OH)	H_2O
2955	ν(CH)	PEI
1707	ν(C=O)	PAC
1643	δ(OH)	H_2O
1552	$\nu_a(COO^-)$	PAC

dependence of the adsorption step z. A summary of selected IR bands or the fitted band components (see above) used in this study and their assignment is given in the Table 1.

Deposition Profiles

The integrated areas of the selected IR bands were used to quantify PEM-PEI/PAC deposition. Most relevantly, the ν(CH) band at 2855 cm^{-1} (A_{CH}) can be used as a direct diagnostic measure for the deposited amount of PEI, while for the deposited amount of PAC the fitted band components ν(C=O) ($A_{C=O}$) and $\nu_a(COO^-)$ (A_{COO^-}) have to be considered. In the Figure 6 typical courses of the integrals of $\nu(COO^-)$, ν(C=O), ν(CH) and ν(OH) band versus the adsorption step z are given, which are discussed in the following:

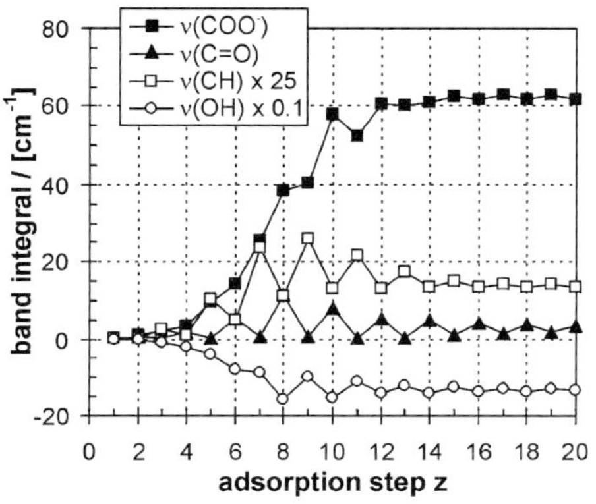

Figure 6.
Courses of the integrals of the ν(CH) band at 2855 cm^{-1} (white cubes) assigned to PEI, the fitted ν(C=O) and $\nu(COO^-)$ band components at 1707 and 1552 cm^{-1} (black triangle, cube) assigned to PAC and the ν(OH) band (white circle) of water versus z measured during consecutive PEM-z deposition from PEI/PAC solutions for pH $\approx$ 10/4, c_{PEL} = 0.005 M and t_{ADS} = 5 min onto Si-IRE. The given band integrals are based on ATR-FTIR sprectra, which were corrected for the negative δ(OH) band at 1640 cm^{-1} (see Experimental).

$\nu(COO^-)$ Band Component

The course of the $\nu(COO^-)$ band component (A_{COO^-}) due to PAC shows after a certain quasi linear lag phase until around z = 4 a steep rise and an asymptotic behaviour for the adsorption steps z = 5–20. This is primarily due to the ATR-FTIR mode, which senses lateral PEM deposition as a linear function of A versus z, while vertical PEM deposition is sensed as an exponentially damped function given in Equation (1) and drawn in the Figure 3 (Experimental section). Hence, we conclude a short initial phase of mainly lateral growth up to z = 4, followed by a mainly vertical PEM deposition from z = 5–20.

ν(C=O) Band Component

For the ν(C=O) band components ($A_{C=O}$) due to PAC a non trivial course was found. In the modulated course of $A_{C=O}$, the values for even steps were always higher compared to the odd step before and after. This is generally due to a periodic uptake and loss of carboxylic acid (COOH) groups. Thereby the COOH group decrease in the odd steps can be caused either by deprotonation or also by the loss of PAC molecules. Since COOH and COO^- groups are complementary, the deprotonation of COOH should result in a gain of COO^- groups. However, since e.g. the decrease of $A_{C=O}$ from z = 8 to z = 9 is not paralleled by a corresponding rise of A_{COO^-}, we conclude, that predominantly protonated PAC is released in every PEI step. Hence from both modulated courses PEM loss tendencies were found, when either PEI or PAC is immersed on the actual PEM-z to create PEM-(z + 1). Furthermore $A_{C=O}$ shows a transient behaviour with a maximum at z = 8, which is due to the ATR-FTIR sensing characteristics: i.e. at

$z = 8$ there is a maximum of COOH groups sensed, while for subsequent steps $z = 10$, 12–20, even assuming a similar number of COOH groups in the outermost region, a smaller $A_{C=O}$ value was measured. From that we conclude on the one hand a considerable accumulation of PAC at the outermost PEM surface region. On the other hand for $z = 8$ the roughness of the PEM film could enable the PAC solution to contact PEM regions of small thickness or even the bare silicon surface, so that the evanescent field for PEM-8 senses PAC stronger compared to PEM-20, where accumulated PAC is no longer near the silicon surface or in the region of stronger evanescent field (see also ν(CH) band below).

ν(CH) Band

For the band integrals of ν(CH) due to PEI (further denoted as A_{PEI}) also a non trivial course was found. First of all strong modulation features can be identified, so that for PEI in every odd step A_{CH} was higher compared to the even step before and after. This modulated course of A_{CH} is due to a periodic uptake of PEI material for the odd PEI steps $z = 1$–19 and a respective loss of PEI material for the even PAC steps $z = 2$–20. From this we conclude that the initially bound PEI is partly released by the oppositely charged PAC into solution. Secondly very interestingly, in the Figure 6 the maximum A_{CH} value for $z = 9$ ($\approx 26\ cm^{-1}$) is higher compared to the saturation value ($\approx 14\ cm^{-1}$) at $z = 20$ by a factor of nearly 2. Hence, considering a thickness increment from 0 to 500 nm, which is about in the range of the penetration depth (see Experimental), a higher PEI amount can be concluded for thinner PEM-9 at 0.005 M, where mainly the outermost bound PEI amount is sensed, compared to thicker PEM-19, where mainly the internally bound PEI amount is sensed. Analogously to the findings for A_{PAC}, this means that in the outermost regions of e.g. PEM-19 a similar large PEI uptake may take place, but within the given thickness increment this is no longer detected by ATR-FTIR. At that point it has to be noted, that the difference of the refractive indices for water ($n_3 = 1.33$) dominating for PEM-9 and for the PEM film ($n_2 = 1.5$) dominating for PEM-20 could not cause this absorbance difference. According to Harrick[29] an increase in n_2 should rather result in a higher A_{CH} value by some factor of 1.09. These results can be interpreted based on two models, which may interfere somehow. On the one hand the outermost region of this PEM type may always consist of a huge amount of accumulated excess PEI, which is exceeding the internally bound amount by far. On the other hand the PEM may always be in a highly eroded state, so that until a certain z the PEI solution may still contact regions of bare silicon substrate sensed maximum by ATR-FTIR. The first interpretation would be in line with an accepted three zone model,[17,31] where the outermost zone is claimed to be highly nonstoichiometric, while in the internal bulk zone 1:1 stoichiometry prevails. In the respective subsequent steps this excess PEL amount is desorbed to a high degree, which can be seen by the huge modulation amplitude for PEI/PAC. However, the true picture accounts additionally for highly eroded PEM (second interpretation), where the three zone model is only locally valid. This point will be addressed in the morphology part below.

ν(OH) Band

Finally, in the Figure 6 also the modulated course of the negative ν(OH) band is given, which reflects the location of water with respect to the bare silicon surface or the sensing evanescent field. Significantly, in the odd PEI steps the integral of the ν(OH) band (A_{H2O}) is always higher compared to the even PAC steps before and after. This tendency can be due to two contributions: Either outermost PEI layers are more hydratable compared to outermost PAC layers, or in the PEI steps the thickness or deposited amount of the whole PEM is always smaller compared to the PAC step before and after. A detailed interpretation will be published therein.[30]

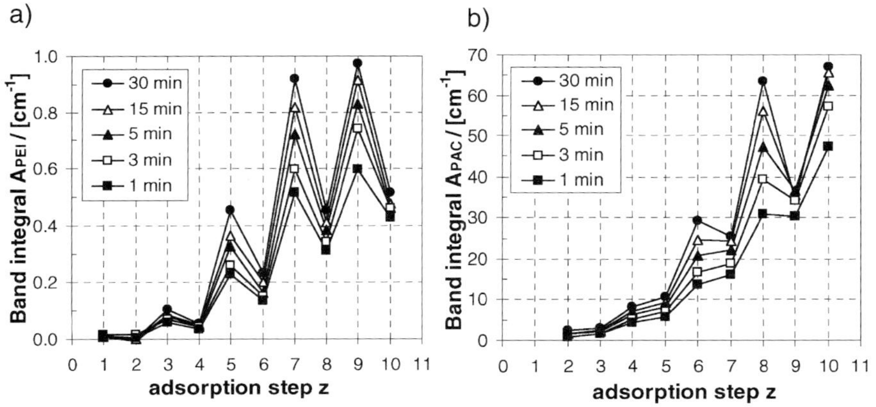

Figure 7.

a) A_{PEI} of PEM-z plotted versus adsorption step z = 1 to 10 for t_{ADS} = 0.5, 1, 3, 5, 15, 30 min (indicated). b) A_{PAC} of PEM-z plotted versus adsorption step z = 1 to 10 for t_{ADS} = 0.5, 1, 3, 5, 15, 30 min (indicated).

Influence of Adsorption Time t_{ADS}

Furthermore, PEM deposition was studied varying the adsorption time t_{ADS}. In the Figure 7a the courses of A_{PEI} and in the Figure 7b those of A_{PAC} are shown in dependence of the adsorption step for various t_{ADS} values related to the individual adsorption steps. A_{PAC} was calculated according to $A_{PAC} = F \cdot A_{C=O} + A_{COO^-}$ (see Experimental) to address the total PAC amount. The A_{PEI} and A_{PAC} courses reflect the actual uptake and release of PEI and PAC, respectively, at the support surface. In both sets of deposition profiles a significant increase of both the overall deposited amount and of the uptake/release amplitudes in dependence of t_{ADS} is seen. These uptake/release tendencies for PEI and PAC are in line with results from studies of Cohen-Stuart and coworkers [18], who reported similar tendencies for a different system. As it was also claimed therein, obviously, the PEM deposition process is kinetically controlled.

As a summary, Figure 8a shows the direct dependence of the band integrals A_{PEI} and A_{PAC} measured for PEM-6 on t_{ADS}. The courses take initially a steep ascent until $t_{ADS} \approx 5$ min, followed by a less steep one. Practically, this means

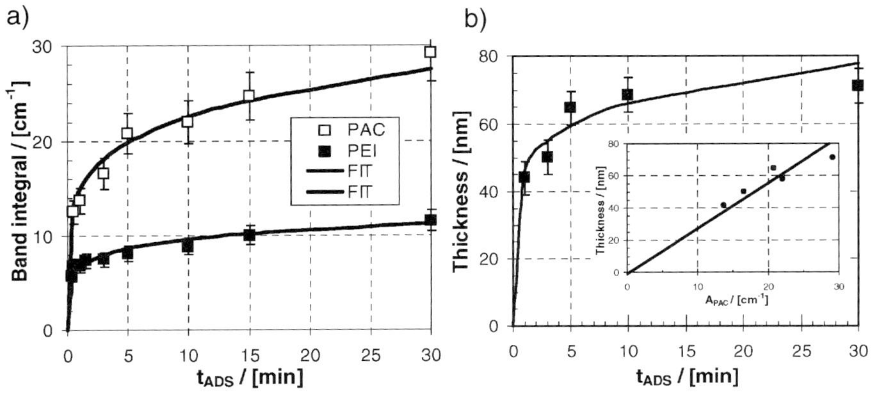

Figure 8.

a) A_{PEI} (×50) and A_{PAC} of PEM-6 plotted versus t_{ADS}. The full lines are fitting curves using Equation (4). b) Thickness values of PEM-6 (SFM line cut method) plotted versus t_{ADS} (full lines: Fit by Equation (4)). Inlet: thickness (SFM plotted versus A_{PAC} (ATR-FTIR).

that $t_{ADS} \approx 5$ min is sufficient for that PEM type under the chosen pH and c_{PEL} conditions to obtain an effective PEM deposition. Empirically, in the given range of $t_{ADS} = 0.33$–30 min the deposited amount scales with t_{ADS} by an empirical Smoluchowski like function (4) suggesting an aggregation related deposition process:

$$A = A_0 \, t_{ADS}{}^b \qquad (4)$$

Fitting the experimental data of both band integrals A_{PEI} and A_{PAC} empirically by Equation (4) results in approximately similar b values for PEI ($b = 0.15 \pm 0.02$) and PAC ($b = 0.18 \pm 0.01$). The fitted curves are shown as black lines in Figure 8a. From that generally an aggregation type of deposition process might be speculated on and similar incorporation rates of PEI and PAC into the PEM might be concluded.

The results of Figure 8a obtained by ATR-FTIR were checked with the SFM line cut method as another sensitive analytical method delivering thickness values of the PEM films. For that the same

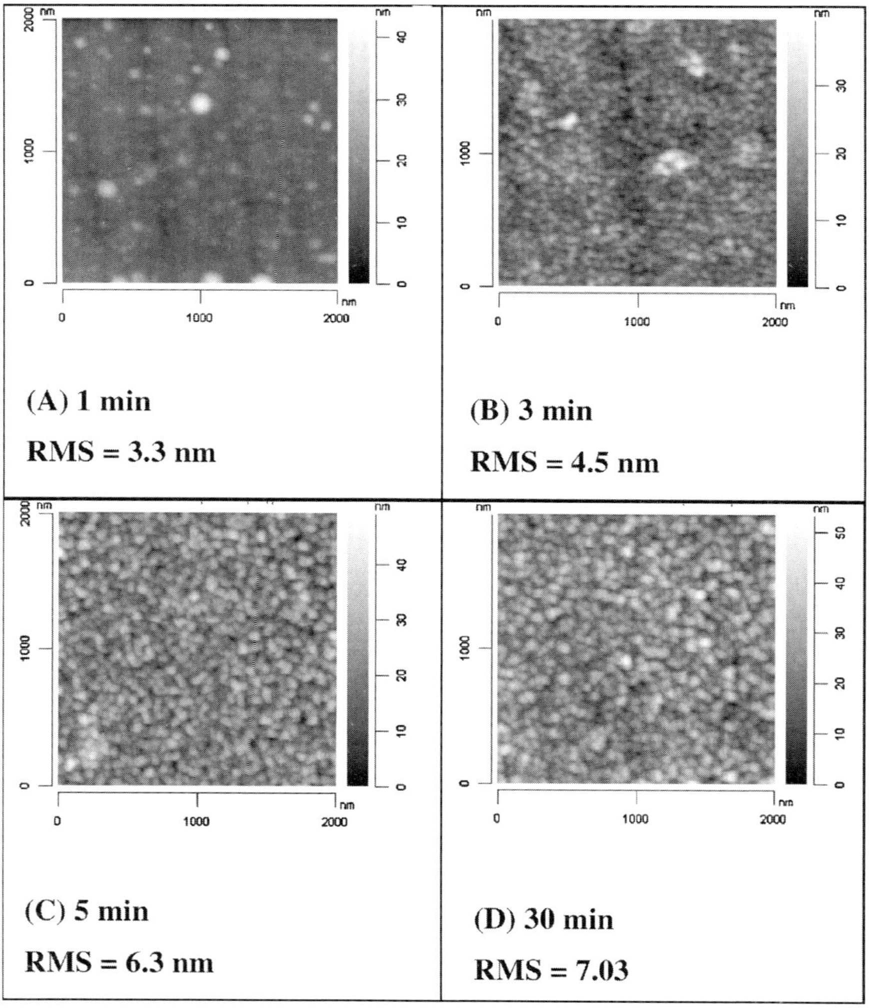

Figure 9.
Typical SFM images (topography) on PEM-6 samples deposited under constant pH setting of 10/4 (PEI/PAC) and $c_{PEL} = 0.005$ M for various adsorption times t_{ADS}.

PEM-6 samples as they were used in the ATR-FTIR analysis were considered. The results are shown in the Figure 8b. Significantly, a similar trend of PEM deposition in dependence of t_{ADS} was found: after an additional steep rise of the thickness until $t_{ADS} = 5$ min the further thickness increase with t_{ADS} is only low. Hence, practically the PEM deposition process for PEI/PAC at $c_{PEL} = 0.005$ M does not alter that much applying $t_{ADS} >$ 5 min. Hence, this value seems to be sufficient for an effective and quick process. Fitting the thickness data with Equation (4) replacing A and A_0 by d and d_0 led to $b = 0.15 \pm 0.03$, which confirms the ATR-FTIR data (the fitted curve is shown as full line). From this result it could be additionally concluded, that if thicknesses within the range 0–300 nm are studied, the band integrals obtained by ATR-FTIR analysis are approximately linear with the "true" thickness values obtained by SFM line cut method. This observation is supported by the inlet in the Figure 8b, where the thickness found by SFM line cut method is plotted versus the A_{PAC} values of the ATR analysis. From this plot an approximate linear relationship between the observables of both methods was confirmed, which seems to be the more exact the lower the thicknesses were.

PEM Nanostructure

The equivalent PEM samples investigated by ATR-FTIR spectroscopy were studied by scanning force microscopy (SFM) aiming at the surface morphology and roughness in depen-dence of t_{ADS}. For that purpose only PEM-6 samples deposited under constant pH (10/4) and c_{PEL} (0.005 M) were considered. Four PEM-6 samples were chosen deposited during $t_{ADS} = 1$, 3, 5 and 30 min, and in the Figure 9 A, B, C typical SFM images of these samples are shown.

There the structures and especially their sizes were slightly changing in dependence of t_{ADS}. For $t_{ADS} = 1$ min (A) individually appearing dots on a relative smooth base layer were found, which changed to more continuous layers of crowded dots for $t_{ADS} = 3$ min (B). For $t_{ADS} = 5$ and 30 min (A, D) the dots were more densely packed with a tendency to merging. Generally an enlargement of the surface structure in dependence of t_{ADS} could be derived. Included in the Figure 9 are roughness (RMS) values, which increase with increasing t_{ADS} and correlate qualitatively with the found deposited amounts given in Figure 8b. Presumably, the more PEM grow in the vertical direction with increasing adsorption time, the more roughness is created.

Conclusion

Quantitative ATR-FTIR spectroscopy and SFM were applied on deposition of PEM of PEI and PAC in dependence of adsorption time (t_{ADS}) at fixed pH and PEL concentration c_{PEL} providing for selective informations on PEI (ν(CH)) and PAC (ν(COO^-), ν(C=O)) uptake and release, respectively, as well as on surface morphology.

Plotting band integrals of these bands versus t_{ADS} showed, that the deposited amount of PEI and PAC increased steeply between $t_{ADS} = 0$ to 5 min followed by a small further increase for $t_{ADS} > 5$ min. This was confirmed by SFM profile measurements on line cuts through the PEM films providing for the PEM thickness (d_{SFM}). Within a thickness range from 0–100 nm the band integrals related to PEI and PAC (ATR-FTIR) were shown to be approximately linear with d_{SFM} (this would be not the case for thicknesses exceeding the ATR penetration depth (see Experimental)). An empirical Smoluchowski like analytical function represented the data quite sufficiently, suggesting an aggregation type of deposition process.

From oscillations features in the deposition profiles, the release of PEI upon PAC immersion (even steps) and of PAC upon PEI immersion (odd steps) could be claimed for all applied t_{ADS}. From that a competition between PEL surface uptake and PEL solution complexation formation can be concluded in line with studies of Cohen-Stuart and coworkers.[18]

SFM data on PEM-6 samples deposited at constant pH = 10/4 and $c_{PEL} = 0.005$ M in dependence of t_{ADS} showed isolated structures for $t_{ADS} = 1$ min, moderately fused ones for $t_{ADS} = 3$ min and further fused ones for $t_{ADS} \geq 5$ min. Additionally, a subsequent roughness increase from around 3 to 7 nm was found.

These studies help to prepare polyelectrolyte based films with controlled thickness for the interaction with biofluids in the biomedical and food field.

Acknowledgements: This work has been financially supported by Deutsche Forschungsgemeinschaft (DFG) within Sonderforschungsbereich (SFB) 287 (B5).

[1] G. Decher, J. D. Hong, J. Schmitt, *Thin Solid Films* **1992**, *210*, 831.
[2] P. Bertrand, A. Jonas, A. Laschewsky, R. Legras, *Macromol. Rapid Commun.* **2000**, *21*, 319–348.
[3] M. Schönhoff, *J. Phys.: Condensed Matter* **2003**, *15*, 1781.
[4] G. Decher, J. B. Schlenoff, Eds., *Multilayer Thin Films-Sequential Assembly of Nanocomposite Materials*, Wiley-VCH, Weinheim **2003**.
[5] H. Kim, M. W. Urban, *Langmuir* **1998**, *14*, 7235.
[6] M. Müller, T. Rieser, K. Lunkwitz, S. Berwald, J. Meier-Haack, D. Jehnichen, *Macromol. Rapid Commun.* **1998**, *19*(7), 333.
[7] D. L. Elbert, C. B. Herbert, J. A. Hubbell, *Langmuir* **1999**, *15*, 5355.
[8] E. Brynda, M. Houska, M. Jirouskova, J. E. Dyr, *J. Biomed. Mater. Res.* **2000**, *51*, 249.
[9] A. Wu, D. Yoo, J. K. Lee, M. F. Rubner, *J. Am. Chem. Soc.* **1999**, *121*, 4883.
[10] E. Brynda, M. Houska, A. Brandenburg, A. Wikerstal, J. Skvor, *Biosensors and Bioelectronics* **1999**, *14*, 363.
[11] F. Caruso, F. N. Furlong, K. Ariga, I. Ichinose, T. Kunitake, *Langmuir* **1998**, *14*, 4559.
[12] P. Stroeve, V. Vasquez, M. A. N. Coelho, J. F. Rabolt, *Thin Solid Films* **1996**, 284-285, 708.
[13] F. van Ackern, L. Krasemann, B. Tieke, *Thin Solid Films* **1998**, *327–329*, 762.
[14] W. Lenk, J. Meier-Haack, *Desalination* **2002**, *148*, 11–16.
[15] E. Hübsch, V. Ball, B. Senger, G. Decher, J. C. Voegel, P. Schaaf, *Langmuir* **2004**, *20*, 1980.
[16] P. Lavalle, V. Vivet, N. Jessel, G. Decher, J. C. Voegel, P. J. Mesini, P. Schaaf, *Macromolecules* **2004**, *37*, 1159.
[17] G. Ladam, P. Schaad, J. C. Voegel, P. Schaaf, G. Decher, F. Cuisinier, *Langmuir* **2000**, *16*, 1249.
[18] N. G. Hoogeveen, M. A. Cohen-Stuart, G. J. Fleer, M. R. Böhmer, *Langmuir* **1996**, *12*, 3675.
[19] R. Steitz, W. Jaeger, R. v. Klitzing, *Langmuir* **2001**, *17*(15), 4471–4474.
[20] G. Decher, J. Schmitt, *Progr. Colloid. Polym. Sci.* **1992**, *89*, 160.
[21] X. Arys, A. M. Jonas, B. Laguitton, R. Legras, E. Wischerhoff, *Prog. Org. Coat.* **1998**, *34*, 108.
[22] M. Lösche, J. Schmitt, G. Decher, W. G. Bouwman, K. Kjaer, *Macromolecules* **1998**, *31*, 8893.
[23] D. Yoo, S. S. Shiratori, M. F. Rubner, *Macromolecules* **1998**, *31*, 4309.
[24] S. S. Shiratori, M. F. Rubner, *Macromolecules* **2000**, *33*, 4213.
[25] D. Appelhans, IPF Dresden, private communication.
[26] Patent in preparation.
[27] U. P. Fringeli, in: *Encyclopedia of Spectroscopy and Spectrometry*, J. C. Lindon, G. E. Tranter, J. L. Holmes, Eds., Academic Press, New York 2000, pp. 58–75.
[28] M. Müller, B. Keßler, W. Ouyang, *Z. Phys. Chem.* **2007**, *221*, 127–138.
[29] N. J. Harrrick, *Internal Reflection spectroscopy*, Harrick Sci. Corp., Ossining, New York 1979.
[30] M. Müller, B. Keßler, *Langmuir* 2007, (submitted).
[31] M. Castelnovo, J. F. Joanny, *Langmuir* 2000, *16*(19), 7524.

Phase Behaviour of Poly(styrene-*co*-methacrylic acid)/ Poly(styrene-*co*-*N,N*-dimethylacrylamide)/ Poly(styrene-*co*-4-vinylpyridine) Ternary Blends by DSC and FTIR

Khaled ElMiloudi,[1,2] *Said Djadoun**[2]

Summary: we have investigated by DSC and FTIR the miscibility and phase behaviour of binary and ternary blends of different ratios of poly(styrene-*co*-methacrylic acid) containing 15 mol% of methacrylic acid (SMA15) with poly(styrene-*co*-N,N-dimethylacrylamide) containing 17 mol% of N,N,-dimethylacrylamide (SAD-17) and poly(styrene-*co*-4-vinylpyridine) containing 15 mol% of 4-vinylpyridine. SMA15 is miscible with both SAD17 and S4VP15 and interacts more strongly with S4VP15 than with SAD17 as evidenced by the positive deviations from linear average line observed with these blends and the appearance of new bands in the 1800–1550 cm^{-1} region. This behaviour is known as ΔK effect. The FTIR study confirms that though the specific intermolecular interactions that occurred with each pair of the SMA15/S4VP15 and SMA15/SAD17 binary components are of different strength, they still exist in the ternary blend. Even though the three binary polymer pairs are individually miscible, the ternary system of SMA15/S4VP15/SAD17 exhibits only partial miscibility with small loop of immiscibility due to a significant ΔK effect. These results obtained by DSC and FTIR are in a fair agreement with theoretical prediction applying the Painter-Coleman association model.

Keywords: DSC; FTIR spectroscopy; poly(styrene-*co*-4-vinylpyridine); poly(styrene-*co*-methacrylic acid); poly(styrene-*co*-*N,N*-dimethylacrylamide); ternary blend

Introduction

It is well known that most polymer pairs do not mix in the absence of specific interactions. The introduction of hydrogen bonding between the two constituents of the blend is now a known method used to enhance the miscibility of pairs of polymers initially immiscible and increasing number of miscible binary polymer blends are reported.[1–4] Though ternary miscible polymer blends offer many opportunities to elaborate new polymer materials with improved properties, based on recent studies,[5–9] only few ternary polymer blends have been reported to be truly miscible over the entire range of composition in comparison to binary blends.

It has been reported by Jo *et al.*[6] that ternary blends of poly(hydroxyl ether of bisphenol A) /PMMA/PEO, composed of three miscible binary blends showed a closed immiscibility loop phase diagram. Similar observations were also reported by Chang *et al.*[7] and by W.P.Hsu[8] in their studies of the ternary phenolic resin PEO/PCL and the PMMA/PVPh/poly(vinyl cinnamate) blends, respectively. An unusual completely miscible ternary blend of phenolic/phenoxy/PCL was recently reported.[9]

[1] Faculté des Sciences et Sciences de l'Ingénieur, Université Hassiba Benbouali, B.P.151, Chlef, Algeria
[2] Laboratoire des Matériaux Polymères, Faculté de Chimie, Université des Sciences et de la Technologie Houari Boumediene, B.P. 32, El Alia, Algiers 16111, Algeria
E-mail: matpolylab@yahoo.fr

Coleman and Painter[10] stated that because of the $\Delta\chi$ and ΔK effects, truly miscible ternary blends are obtained only in very limited cases where these two effects are finely balanced. The first part of this work is an attempt to dilute the interactions that occurred between poly(methacrylic acid) PMA with poly(4-vinylpyridine) P4VP and poly(methacrylic acid) with poly(N,N-dimethylacrylamide) PDMA that led to their complexation, by introduction of relatively high amounts of styrene moieties within these interacting polymers. The specific interactions that occurred between each binary miscible blend and the phase behaviour of their ternary blends were investigated by DSC and FTIR spectroscopy. Theoretical prediction of the phase behaviour of the SMA15/S4VP15/SAD17 ternary system was performed using the Painter-Coleman association model.

Synthesis and Characterization of Copolymers

Poly(styrene-*co*-methacrylic acid) containing 15 mol% of methacrylic acid (SMA15), poly(styrene-*co*-4-vinylpyridine) containing 15 mol% of 4-vinylpyridine (S4VP15) and poly(styrene-*co*-N,N-dimethylacrylamide) containing 17 mol% of N,N-dimethylacrylamide (SAD17) were prepared by free radical polymerization using azo-bis-isobutyronitrile as initiator at 60 °C. They were then purified by dissolution/precipitation and dried under vacuum at 60 °C for several days. They were characterized in a similar way as previously described[4] by UV spectroscopy and proton NMR.

Table 1.
Polymer characteristics.

Copolymers	Styrene content	M_w	M_w/M_n	Tg
	mol%	Kg.mol^{-1}		K
SMA15	85	65.55	2.74	410
S4VP15	85	06.62	2.98	377
SAD17	83	12.26	2.84	379

The average molecular weights of these copolymers were determined by GPC using a Perkin Elmer 1100 HPLC. The copolymer characteristics are listed in Table 1.

Thermal Analyses

The thermal stability of the as-cast copolymers was checked by TGA. These copolymers were stable up to 350 °C, as displayed from their TGA and *d*TGA curves in Figure 1, determined using a Mettler Toledo 851^e at a heating rate of 10 °C/min under nitrogen flow.

Films of binary SMA15/S4VP15, SMA15/SAD17, S4VP15/SAD17 and ternary SMA15/S4VP15/SAD17 blends were prepared by casting from THF solutions with evaporation of solvent.

The glass transition temperature T_g of these copolymers and of their binary and ternary blends of different ratios was measured using Mettler Toledo 821^e DSC, at a heating rate of 20 °C/min. Since these polymers contain hydrophilic groups and in order to minimize water absorption, they were accordingly kept under reduced pressure in a vacuum oven at 60 °C for several days prior to their use. All the thermograms recorded during the second heating scan were considered to obtain the T_g values.

FTIR Measurements

Thin films of SMA15, S4VP15 and SAD17 copolymers and of their binary SMA15/S4VP15, SMA15/SAD17, S4VP15/SAD17 and ternary SMA15/S4VP15/SAD17 blends of different ratios for FTIR measurements were prepared by casting from a 2–3% w/v solution in THF onto KBr disks. The solvent was first evaporated at room temperature. In a similar way as with DSC, the disks were then dried in vacuum oven at 60 °C for several days. Infrared spectra were recorded at room temperature on a SHIMADZU FTIR spectrometer with

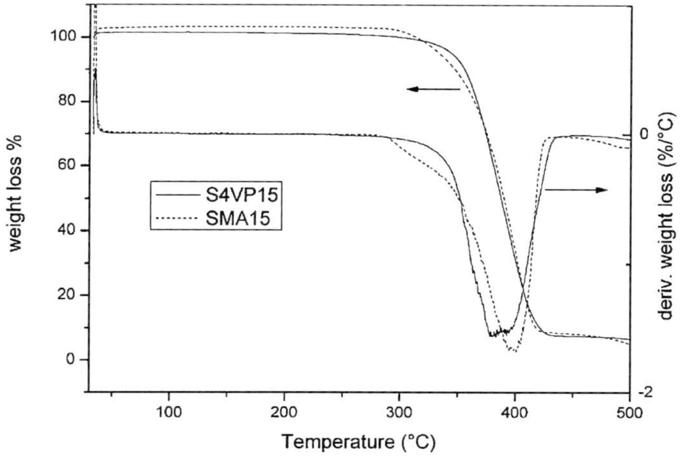

Figure 1.
TGA and *d*TGA curves of the SMA15 and S4VP15.

spectral resolution of 1 cm^{-1} and 64 scans were signal averaged.

Theoretical Background

At a temperature above the glass transition temperature, the competing dynamic equilibria between the specific groups in SMA15/S4VP15/SAD17 blend system can be depicted by the Scheme 1–3 below:

Free carboxylic acid groups (B_1) are assumed to self-associate in a cyclic dimeric form (B_2) and inter-associate both with free amide groups (A_1) to form a 1:1 complex (BA) and with free pyridine groups (C_1) to form 1:1 complex (BC).

According to the association model, described in detail in the monograph of Coleman *et al.*[11], the self-association equilibrium constant (Scheme 1) is given by:

$$K_2 = \frac{\phi_{B_2}}{\phi_{B_1}^2}\left[\frac{1}{2}\right] \quad (1)$$

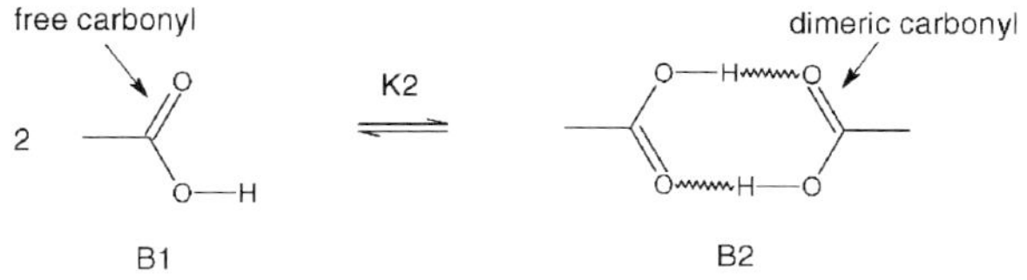

Scheme 1.

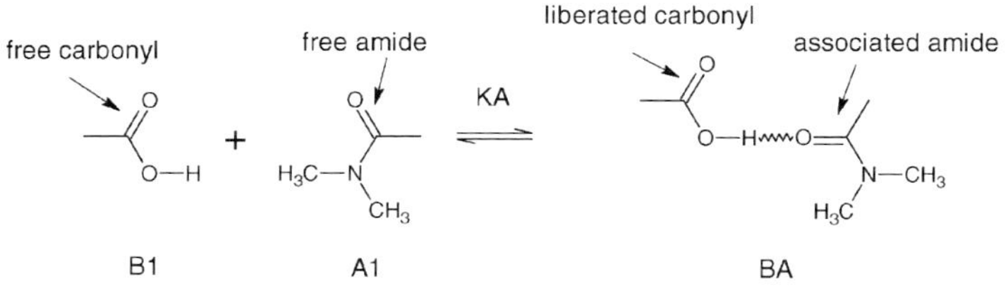

Scheme 2.

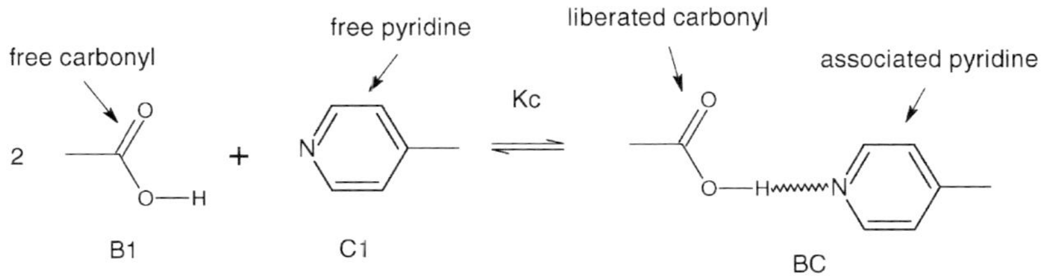

Scheme 3.

The inter-association equilibrium constant (Scheme 2) is given by:

$$K_A = \frac{\phi_{BA}}{\phi_{B_1}\phi_{A_1}}\left[\frac{r_A}{1+r_A}\right] \quad (2)$$

And the inter-association equilibrium constant (Scheme 3) is given by:

$$K_C = \frac{\phi_{BC}}{\phi_{B_1}\phi_{C_1}}\left[\frac{r_C}{1+r_C}\right] \quad (3)$$

The stoichiometry of the system is given by

$$\phi_{B_1} + \phi_{B_2} + \phi_{A_1} + \phi_{BA} + \phi_{C_1} + \phi_{BC} = 1 \quad (4)$$

$$\Phi_B = \phi_{B_1}\left[1 + \frac{K_A\phi_{A_1}}{r_A} + \frac{K_C\phi_{C_1}}{r_C}\right] + 2K_2\phi_{B_1}^2 \quad (5)$$

$$\Phi_A = \phi_{A_1}(1 + K_A\phi_{B_1}) \quad (6)$$

$$\Phi_C = \phi_{C_1}(1 + K_C\phi_{B_1}) \quad (7)$$

Where ϕ_i, Φ_i and $r_{A\text{ or }C}$ are the volume fraction of the polymer segment containing the i species, volume fraction of interacting polymer in the blend and the ratio of the molar volumes $V_{A\text{ or }C}/V_B$, respectively. The index B and A relates to self-associated polymer (B) which can inter- associate both with polymer (A) and polymer (C), respectively.

As the specific groups are diluted in styrene co-monomer, equilibrium constants can be transferred to a standard value by the relation:

$$K^{copolymer}V_B = K^{std}100 \quad (8)$$

100 cm^3 is taken as a standard volume.

The Painter-Coleman association model describes the free energy of mixing of ternary polymer blends when strong specific interactions are present by the following expression:

$$\frac{\Delta G_M}{RT} = \frac{\Phi_A}{M_A}\ln\Phi_A + \frac{\Phi_C}{M_C}\ln\Phi_C + \frac{\Phi_B}{M_B}\ln\Phi_B + \Phi_A\Phi_B\chi_{AB} + \Phi_C\Phi_B\chi_{CB} + \Phi_A\Phi_C\chi_{AC} + \frac{\Delta G_H}{RT} \quad (9)$$

The Flory-Huggins parameter χ is given by

$$\chi = (\delta_A - \delta_B)^2\frac{V_{ref}}{RT} \quad (10)$$

Where δi is the solubility parameter that can be determined by group molar contribution.

The term $\frac{\Delta G_H}{RT}$ is the favourable contribution to the change in the free energy of mixing due to the presence of intermolecular hydrogen bonding interactions and is given by

$$\frac{\Delta G_H}{RT} = \Phi_B\ln\left(\frac{\phi_{B_1}}{\phi_{B_1}^0}\right) + \frac{\Phi_A}{r_A}\ln\phi_{A_1} + \frac{\Phi_C}{r_C}\ln\phi_{C_1} + \Phi_B\left\{\left[K_2\left(\phi_{B_1} - \phi_{B_1}^0\right)\right] + \left[(1 - K_2\Phi_{B_1})\left(\frac{\xi+\zeta}{1+\xi+\zeta}\right)\right]\right\} - \left[\left(1 - K_2\phi_{B_1}^0\right)\Phi_B\ln\Phi_B + \frac{\Phi_A}{r_A}\ln\Phi_A + \frac{\Phi_C}{r_C}\ln\Phi_C\right] \quad (11)$$

Where $\zeta = K_A\Phi_{A1}/r_A \quad \xi = K_A\Phi_{A1}/r_A$

The criteria for thermodynamic stability for ternary polymer mixtures are given by:

$$\frac{\Delta G_M}{RT} < 0, \frac{\partial^2(\Delta G_M)}{\partial^2\Phi_A} > 0 \text{ and } \left[\frac{\partial^2(\Delta G_M)}{\partial^2\Phi_A}\right]\left[\frac{\partial^2(\Delta G_M)}{\partial^2\Phi_B}\right] - \left[\frac{\partial^2(\Delta G_M)}{\partial\Phi_A\partial\Phi_B}\right] > 0 \qquad (12)$$

Results and Discussion

Thermal Analysis

We have in a first step, carried out a DSC analysis for the three different SMA15/SAD17, SMA15/S4VP15 and S4VP15/SAD17 binary blends and confirmed their miscibility from the single glass transition temperature. The miscibility of SMA15/SAD17 and SMA15/S4VP15 is mainly due to the specific interactions that occurred between the carboxylic acid groups and the pyridine and carbonyl amide groups, respectively. As it can be seen in Figure 2 that shows the variation of T_g of the blend with SMA15 composition for these binary systems, positive deviations from linear average line were observed with all these blends.

A simplified form of the well-known Kwei equation, for a binary blend, was used to estimate the specific interactions that occurred within these blends from the constant q obtained from:

$$T_g = w_1 T_{g1} + w_2 T_{g2} + q\, w_1\, w_2 \qquad (13)$$

Where w_i and T_{gi} are the weight fraction and glass transition temperature of the i constituent.

The Kwei constant q measures the strength of the specific interactions.

The obtained q values shown in Figure 2 confirm that stronger interactions occurred within the SMA15/S4VP15 blends between the carboxylic and the pyridine groups. The Kwei constant obtained with the SMA15/SAD17 is in good agreement with that previously reported for blends composed of poly(styrene-co-acrylic acid) containing 18 mol% of acrylic acid and the same SAD17.[12]

The T_g of S4VP15 and SAD17 are very close. A single T_g varying with the composition and slightly higher than the linear average line with these blends, indicates the presence of very weak interactions compared to the blends above.

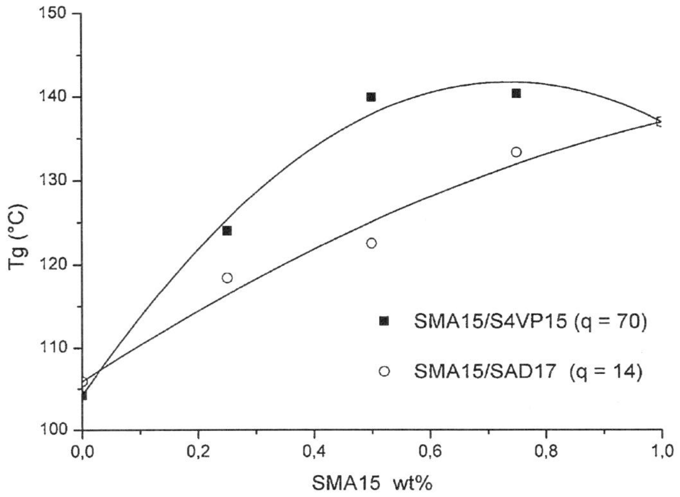

Figure 2.
T_g-composition of the binary blends.

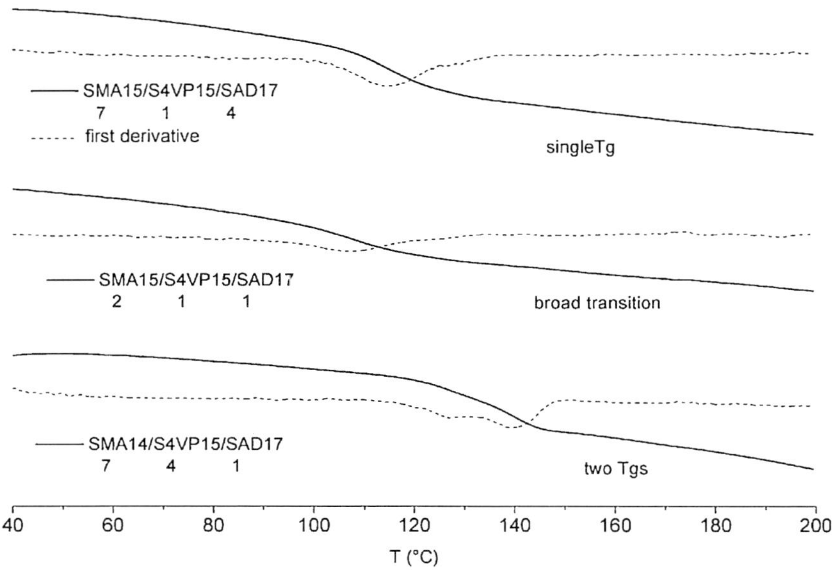

Figure 3.
Selected thermograms of SMA15/S4VP15/SAD17 blends.

Figure 3 shows selected second scan DSC thermograms of SMA15/S4VP15/SAD17 ternary blends with various compositions. Different miscibility behaviour is observed with these ternary polymer blends.

Depending on the blend composition, a single T_g or two T_gs are observed. A single Tg increasing with SMA15 content is observed with ternary blends containing a constant ratio of S4VP15/SAD17 equal or higher than 1:1. The ternary 4/7/1 and 7/4/1 blends are among those that exhibited two T_gs indicating that phase separation occurred. Their two phases are composed of miscible binary blends. Due to the stronger interactions that occurred within the SMA15/S4VP15 blends, the phase of high T_g contains mostly SMA15 and S4VP15.

Assuming that no hydrogen bonding interactions occurred between S4VP15 and SAD17 the modified Kwei equation below[13] was used to describe the ternary blend:

$$T_g = w_1 T_{g1} + w_2 T_{g2} + w_3 T_{g3} + q_{12}\, w_1\, w_2 + q_{13}\, w_1\, w_3 \qquad (14)$$

The values of the experimental T_g obtained by DSC and those calculated from the equation above and the well-known Fox relation are summarized in Table 2. As illustrated in Figure 4, the predicted values using the modified Kwei equation are in a fair agreement with those obtained experimentally.

Figure 5 displays a triangular phase diagram of the ternary polymer blend based on the DSC analyses.

FTIR Analysis

FTIR spectroscopy was first used to confirm the presence of the specific interactions that occurred within the blends between the acidic copolymer SMA15 and basic S4VP15 or SAD17 copolymers in the region where significant changes were observed. Figure 6 shows as an example the FTIR spectra of SMA15/SAD17 and SMA15/S4VP 1:1 ratio blends recorded at room temperature in the 1800–1550 cm^{-1} region. For brevity, the spectra of the other blends are not shown here. The FTIR spectrum of the SMA15/S4VP15 blend shows in Figure 6 in the 1650–1550 cm^{-1},

Table 2.
Experimental and predicted T_gs

Blend ratio			T_g exp	ΔT_g	T_g lin	T_g Fox	T_g Kwei
SMA15	S4VP15	SAD17	°C	°C	°C	°C	°C
0	1	0	104	9	104	104	104
0	0	1	106	9	106	106	106
0	1	3	106	16	105	105	108
0	1	1	110	11	105	105	109
0	3	1	109	13	105	105	107
1	1	10	**102**	16	108	108	111
1	4	7	101	12	108	107	113
1	7	4	104	16	107	107	114
1	10	1	**116**	25	107	106	113
1	1	4	105	17	111	110	116
1	4	1	110	18	110	109	120
2	5	5	105	18	110	109	119
1	0	3	118	14	114	112	116
1	3	0	124	24	112	111	125
1	1	2	96	20	113	112	121
1	2	1	**114**	26	113	111	124
1	1	1	**111**	20	116	114	126
4	1	7	110	14	116	114	121
4	7	1	137 . . . 120	25	115	113	–
5	2	5	**123**	14	119	117	127
5	5	2	130 . . . 116	22	118	116	–
1	0	1	123	17	121	119	125
1	1	0	140	11	121	118	138
2	1	1	**108**	21	121	119	132
7	1	4	114	19	124	122	130
7	4	1	140 . . . 127	19	123	121	–
4	1	1	141	22	126	124	136
3	0	1	133	18	129	128	132
3	1	0	140	13	129	127	142
10	1	1	137	17	132	130	137
1	0	0	137	7	137	137	137

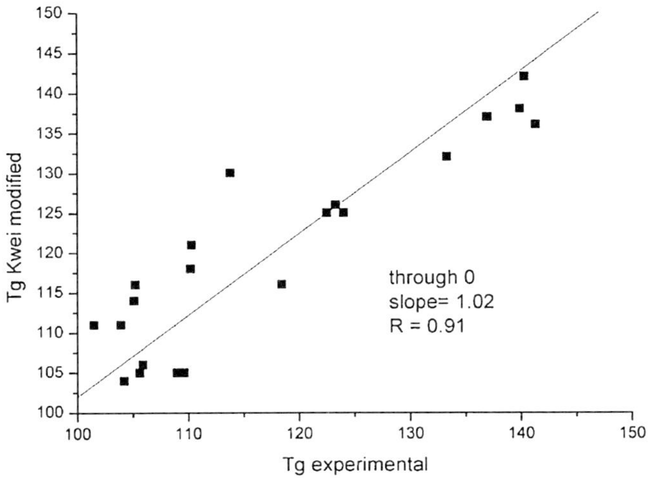

Figure 4.
Tg exp-Tg predicted by modified Kwei equation.

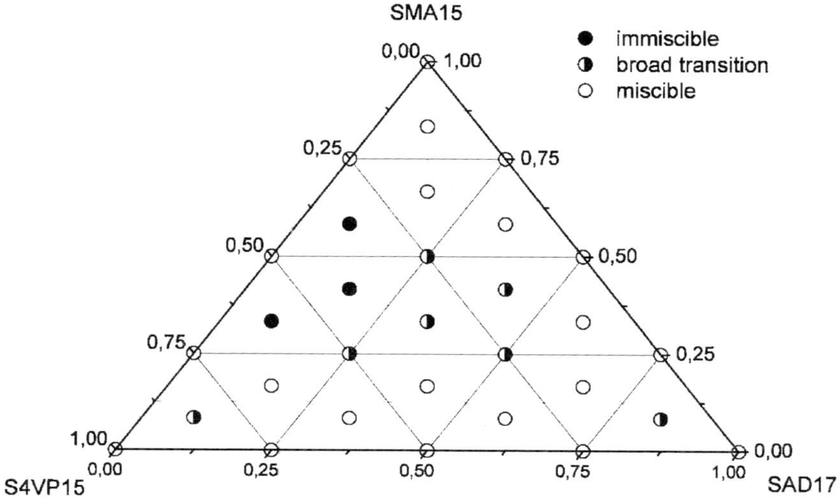

Figure 5.
Experimental phase diagram.

three contributions at 1597 cm^{-1}, 1607 cm^{-1} and 1638 cm^{-1}, attributed to free pyridine ring, hydrogen bonded pyridine "associated" and pyridinium cation, respectively. Similar band were reported in the literature.[14] SMA15 develops specific interactions with both S4VP15 and SAD17 as evidenced by the appearance of new bands as illustrated in Figure 6. Qualitatively, stronger interactions occurred between the carboxylic groups of SMA15 and 4-vinylpyridine of S4VP15 as illustrated in the figure above that compares the 1:1 ratios of these two binary blends from the high fraction of dimers observed with the SMA15/SAD17 blends and the presence of partial ionization that occurred at 1638 cm^{-1} with the other system.

We have carried out a quantitative analysis and calculated the equilibrium self-association constant for the acidic copolymer K_2 and its inter association constant with 4-vinylpyridine K_C[4] and with *N,N*-dimethylacrylamide K_A.[12]

The DSC results are in good agreement with those obtained in previous studies and confirm from the K_C value larger than K_A that stronger interactions occurred within the SMA15/S4VP15.

The values of these constants, together with the different parameters such as molar volume, solubility parameters that were used to calculate the free energy of mixing and phase diagrams for these ternary systems using the PCAM are given in Table 3.

Figure 7 shows the phase behaviour of this ternary system using the Painter-Coleman association model that predicts an immiscibility closed loop in a fair agreement with the variation trend of the experimental T_g values determined by DSC shown in the phase diagram of Figure 4 even in the presence of partial ionization of pyridine.

Figure 8 displays, as an example, the FTIR spectra of some of these ternary blends reflecting the different phase behaviours.

The fractions of the different species involved in these ternary blends through the various specific interactions were calculated using adequate curve fitting procedures. Due to the significant overlapping of the numerous specific interactions that occurred within the ternary blends, adequate band separation is rather very difficult to process particularly in the 1680–1550 cm^{-1} region as illustrated in Figure 9.

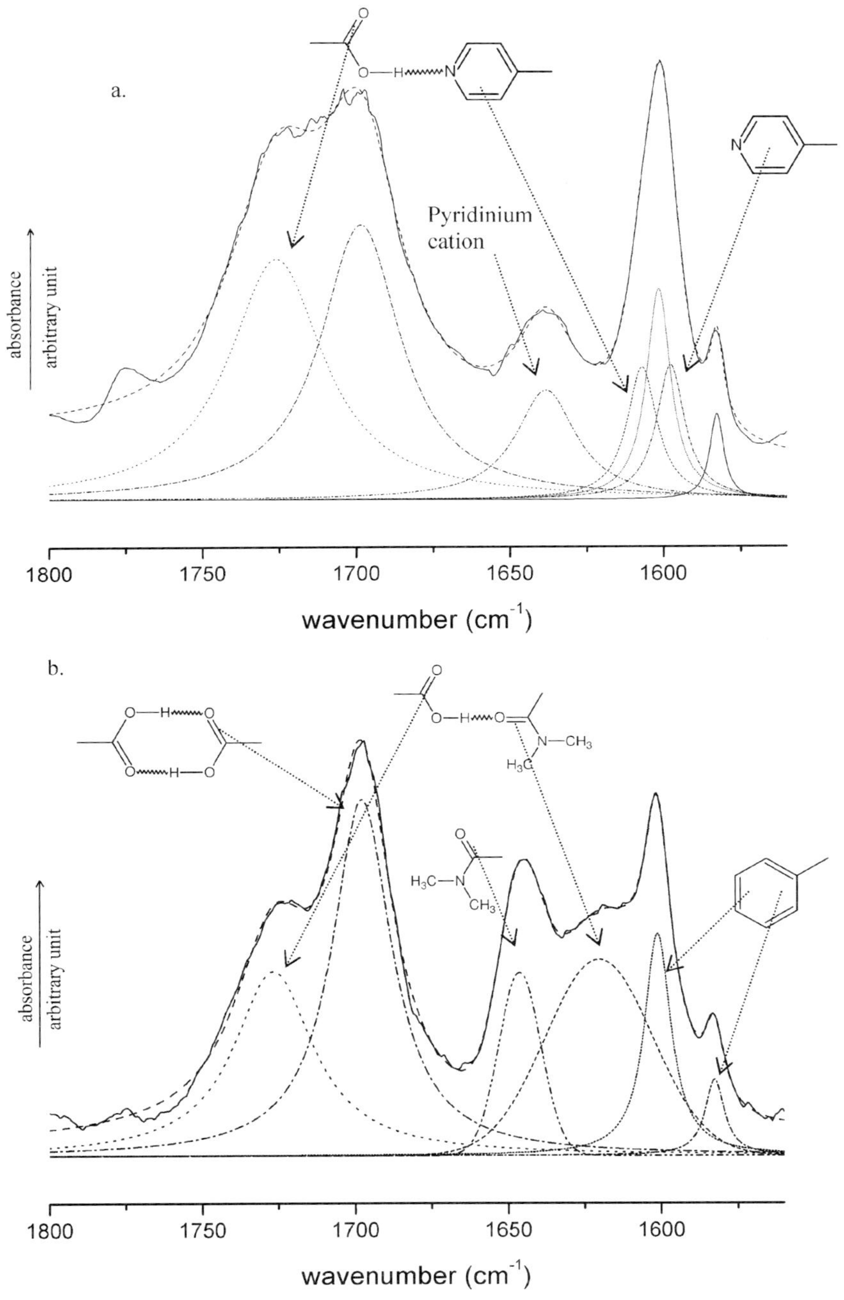

Figure 6.
FTIR spectra of SMA15/S4VP15 blends (a) and of SMA15/SAD17 blends (b).

A semi-quantitative analysis was carried out only in the carbonyl 1800–1650 cm^{-1} region. For a constant 1:1 ratio of the binary S4VP15/SAD17, the fraction of free carbonyl groups of the SMA15, increased with the SMA15 content. Figure 9 confirms the

Table 3.
Thermodynamic parameters

Copolymers	Molar volume	Solubility parameters	Equilibrium constants
	$cm^3 \cdot mol^{-1}$	$(cal \cdot cm^{-3})^{0.5}$	
SMA15	597	9.635	$K_2 = 163$
S4VP15	617	9.695	$K_C = 1300$
SAD17	547	9.534	$K_A = 200$

presence of the different species due to the specific interactions that occurred within the ternary blends. The intensity of such interactions was found to vary with the blend composition and the corresponding phase separation of the 4/7/1 and the miscibility of the 4/1/7 blends in a good agreement with both the experimental DSC study and the theoretical prediction using the Painter-Coleman association model.

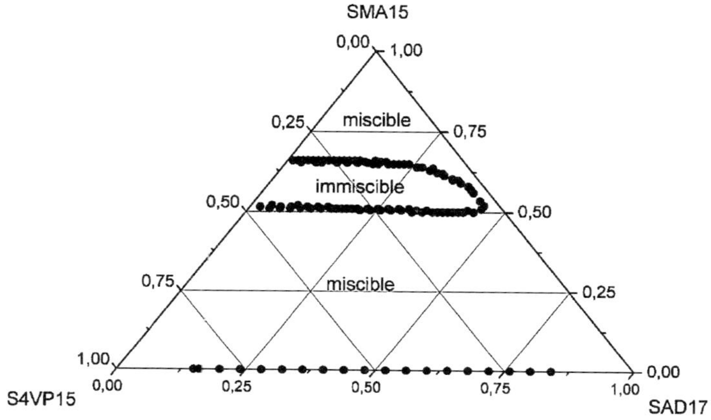

Figure 7.
Ternary phase diagram calculated using the Painter-Coleman association model.

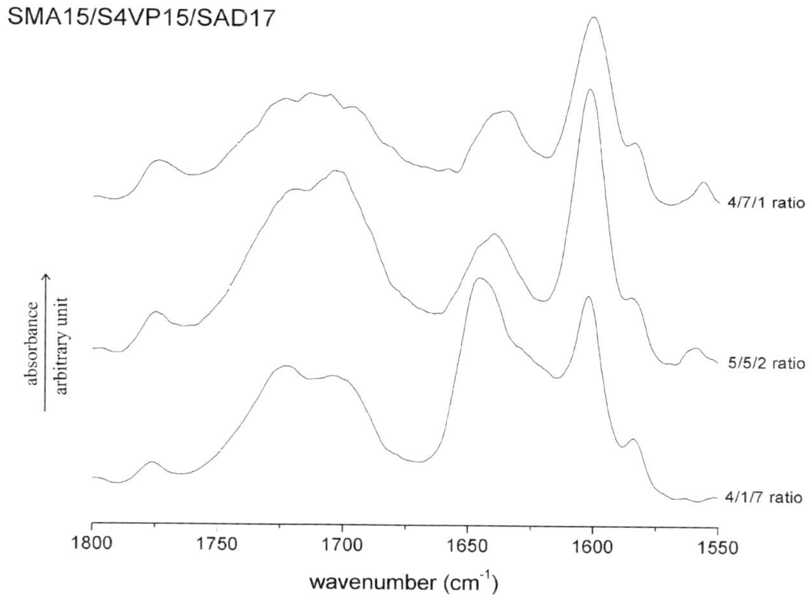

Figure 8.
FTIR spectra of selected SMA15/S4VP15/SAD17 ternary blends in the 1800–1550 cm^{-1} region.

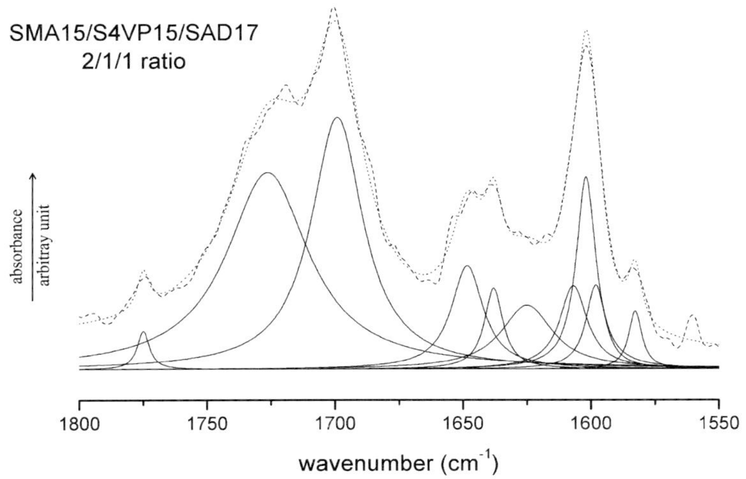

Figure 9.
Illustration of expected curve fitting of SMA15/S4VP15/SAD17 2/1/1 ternary blend.

Conclusion

The phase behaviour of the ternary system, poly(styrene-*co*-methacrylic acid) containing 15 mol% of methacrylic acid (SMA15), poly(styrene-*co*-N,N-dimethylacrylamide) containing 17 mol% of N,N,-dimethylacrylamide (SAD-17) and poly(styrene-*co*-4-vinylpyridine) containing 15 mol% of 4-vinylpyridine, was studied by both DSC and FTIR spectroscopy. The results showed that this blend exhibits only partial miscibility with a closed loop of immiscibility even though the binary polymer pairs composed of any two of these three constituents were miscible. This is due to the significant difference in the strength of the interactions that SMA15 developed with S4VP15 and SAD17, respectively. The use of the Painter-Coleman association model to this ternary system, in which both partial ionization and hydrogen bonding interactions occurred, seems to predict fairly well the observed phase behaviour analyzed by DSC.

Acknowledgements: The authors thank Professors Serge Geribaldi and Nicolas Sbirrazzuoli from the CMOM, university of Nice, France for use of their thermal analysis facilities.

[1] E. Eastwood, S. Viswanathan, C. P. O'Brien, D. Kumar, M. D. Dadmun, *Polymer* **2005**, *46*, 3957.
[2] Y. He, B. Zhu, Y. Inoue, *Prog.Polym.Sci.* **2004**, *29*, 1021.
[3] M. Rodriguez-Castro, L. C. Cesteros, I. Katime, S. M. Nuno-Donlucas, *J.Polym.Sci. part B; Polymer Physics* **2006**, *44*, 2404.
[4] K. ElMiloudi, M. Benygzer, S. Djadoun, N. Sbirrazzuoli, S. Geribaldi, *Macromol. Symp.* **2005**, *230*, 39.
[5] Z. Benabdelghani, A. Etxeberia, S. Djadoun, J. J. Iruin, C. Uriarte, *J. Chromatography A* **2006**, *1127*, 237.
[6] B. K. Hong, J. W. Kin, W. H. Jo, S. C. Lee, *Polymer* **1997**, *38*, 4373.
[7] S. W. Kuo, C. L. Lin, F. C. Chang, *Macromolecules* **2002**, *35*, 278.
[8] W. P. Hsu, *Thermochimica Acta* **2007**, *454*, 50.
[9] S. W. Kuo, S. C. Chan, H. D. Wu, F. C. Chang, *Macromolecules* **2005**, *38*, 4729.
[10] H. Zhang, D. E. Bhagwagar, J. F. Graf, P. C. Painter, M. M. Coleman, *Polymer* **1994**, *35*, 5379.
[11] M. M. Coleman, J. F. Graf, P. C. Painter, "*Specific interactions and the miscibility of polymer blends*", Technomic Publishing Co., Lancaster, PA 1991.
[12] A. S. Hadj Hamou, S. Djadoun, *J. Appl. Sci.* **2007**, *103*, 1011.
[13] W. P. Hsu, *J. Appl. Sci.* **2006**, *101*, 643.
[14] V. Villar, L. Irusta, M. J. Fernandez-Berridi, J. J. Iruin, M. Iriarte, L. Gargallo, D. Radie, *Thermochimica acta* **2003**, *402*, 2765.

 DOI: 10.1002/masy.200850511

In situ Fluorescence Study of Swelling, Sorption and Desorption Processes in and out of PAAm Gels

Gülşen Akın Evingür,[*1] *Kadir Karslı,*[1] *Önder Pekcan*[2]

Summary: Small molecule sorption and desorption in and out of polyacrylamide (PAAm) gels were studied at various temperatures. Pyranine (P_y) dissolved in water used as a probe. Fluorescence emission intensity, I_p from P_y was monitored for studying sorption and desorption processes. Scattered light intensities, I_{sc} from PAAm gel was also monitored to observed structural variations during sorption and desorption process. Li-Tanaka model was applied to produce the swelling time constants, τ_c and sorption coefficients, D_c for the swelling processes. On the other hand, sorption and desorption processes were studied and coefficients were produced by using Fickian model. Related activation energies were also calculated from the corresponding physical processes.

Keywords: desorption; fluorescence; hydrogels; sorption; temperature

Introduction

The sorption, swelling and desorption kinetics of physical and chemical gels are very important in various fields of industry: in the pharmaceutical industry in designing desorption devices for drugs, in the agricultural industry for producing storable foods, and in medical industry in developing artificial organs.

The imperfections in the hydrogel structures have been known to influence the solvent permeability, the sorption of small and large molecules and more indirectly the swelling properties of gels. In the swollen state these imperfections manifest themselves in a nonuniformity of polymer concentration. Considerable work has been done on the characterization of the gel inhomogeneities. It was shown that high permeability of polyacrlamide (PAAm) gels is related to the inhomogeneous cross-link distribution.[1] The effect of inhomogeneities of the polymer network on the swollen state of acrylamide gels and on the diffusion of water molecules within the gels were examined.[2] When an ionized acrylamide gel is allowed to swell in water, an extremely interesting pattern appears on the surface of the gel and the volume expansion increases by adding some amount of sodium acrylate.[3] If acrylamide gels are swollen in acetone-water mixture, gel aging time plays an important role during collapse of the network.[4] The kinetics of swelling of acrylamide gels was studied by light scattering and the cooperative diffusion coefficient of the network was measured.[5,6] Small angle x-ray and dynamic light scattering were used to study the swelling properties and mechanical behavior of acrylamide gels.[7]

The swelling process of chemically cross-linked gels can be understood by considering the osmotic pressure versus the restraining force.[8–12] Two polymeric hydrogels containing (PVP) - crosslinked polyacrylamide and poly (vinyl alcohol) (PVA)-crosslinked polyacrylamide were loaded with the sulfamethoxazole drug and their swelling and drug-release dynamics were investigated at a fixed pH and at room temperature (27 °C).[13] Swelling, deswelling and drug release properties were studied on polyDEAAm-based and

[1] Istanbul Technical University, Department of Physics, Maslak, 34398 Istanbul, Turkey
Fax: +90-212- 285 63 86;
E-mail: evingur@itu.edu.tr
[2] Isık University, Department of Physics, Kumbaba-Şile, 34980 Istanbul, Turkey

polyNIPAAm hydrogels.[14] The use of hydrophobically modified hydrogels for drug release was examined.[15] Swelling and controlled- release behaviors of hydrophobically modified poly (methacrylic acid) hydrogels were investigated.[16]

Photon transmision technique was used to study gelation and swelling process of PAAm where transmitted light intensity decreased and increased during gel formation and swelling processes respectively.[17–19] Similar technique was employed to study N-isopropylacrylamide gels.[20,21] On the other hand, time resolved and steady state (SSF) fluorescence techniques were employed for studying swelling and diffusion processes in PMMA gels.[22,23] Steady state (SFF) fluorescence techniques were employed for studying slow release in various solvents.[24] By using the same method, small molecule diffusion into hydrogels at various temperatures was obtained in our laboratory.[25]

In this work, sorption, swelling and desorption kinetics of PAAm at various temperatures was studied by using steady-state fluorescence (SSF) technique. Pyranine (P_y) dissolved in water was used as fluorescence probe to study sorption and desorption process. Pyranine emission intensity, I_p was monitored to investigate these processes at various temperatures. Scattered light intensities, I_{sc} from PAAm gel was elaborated for studying swelling processes. Li-Tanaka model was used to determine the swelling time constants, τ_c and cooperative sorption coefficients, D_{sc} for the swelling gels. Fickian model was employed to produce sorption and desorption coefficients. The corresponding activation energies were also obtained.

Theoretical Considerations

Kinetics of Swelling

Swelling experiments of disc shaped gels have shown that the relative changes of diameter and thickness are the same, indicating that the gel-swelling processes are not pure diffusional processes. In fact the equality of the relative changes of diameter thickness stems from the none zero shear modulus, μ which results; the change of the total shear energy in response to any small change in shape that maintains constant volume element within the gel should be zero. The high friction coefficient, f between the network and the solvent overdamps the motion of the network, resulting in diffusion like relaxation. The equation of the motion of a network element during the swelling can be given by[8]

$$\frac{\partial \vec{u}}{\partial t} = D_C \vec{\nabla}^2 \vec{u} \tag{1}$$

where $\vec{u}$ is the displacement vector measured from the final equilibrium location after the gel is fully swollen ($u = 0$ at $t = \infty$). $D_c = (K + 4\ \mu/3)/f$ is the collective diffusion coefficient. Here t denotes the time and K is the bulk modulus. Equation (1) has been used with some success to study the swelling of gels.[22] However, these studies did not properly treat the shear deformation that occurs within a gel during swelling, and, hence, can not explain, for example, the isotropic swelling of a cylindrical gel. This shortcoming was due to the shear modulus of the network keeping the system in shape by minimizing the non-isotropic deformation. For a disc shaped gel, any change in diameter is coupled to a change in thickness. The total energy of a gel can be separated into bulk energy and shear energy. The bulk energy is related to the volume change, which is controlled by sorption. The shear energy, F_{sh} on the other hand, can be minimized instally by readjusting the shape of the gel[8]

$$\delta F_{sh} = 0 \tag{2}$$

Simultaneous solution of Equation (1) and (2) produces the following equation for the swelling of a gel disc in terms of vapor and solvent uptakes W and W_∞ at time t and at equilibrium respectively as follows

$$\frac{W_\infty - W}{W_\infty} = \sum_{n=1}^{\infty} B_n \exp(-t/\tau_n) \tag{3}$$

In the limit of large t, or if τ_c is much larger than the rest of τ_n, all higher terms ($n \geq 2$) in Equation (3) can be omitted and the swelling kinetics is given by the following relation

$$\left(1 - \frac{W}{W_\infty}\right) = B_1 \exp(-t_s/\tau_c) \tag{4}$$

It should be noted from Equation (3) that $\Sigma B_n = 1$, therefore B_1 should be less than 1. B_1 is related to the ratio of the shear modulus, μ and longitudinal osmotic modulus, $M = (K + 4\,\mu/3)$. Hence, once the value of B_1 is obtained, one can determine the value of $R = \mu/M$. Here we have to note that Equation (4) can also be obtained by using the theoretical results,[5] in the case of $R \rightarrow 3/4$ ($\mu/K \rightarrow \infty$), time constant $\tau_c \approx (3/4 - R)^{-1}$ goes to infinity and all B_n's go to zero except B_1, which goes to unity. The dependence of B_1 on R for a disc can be found in the literature.[22] τ_c is related to the collective diffusion coefficient D_c at the surface of a gel disc by

$$D_c = \frac{3a^2}{\tau_c \alpha_1^2} \tag{5}$$

where α_1 is a function R only and is given the literature,[5] and a stands for the half thickness of the gel in the final equilibrium state. Hence, D_c can be calculated, if one obtains τ_c values from the swelling experiments.

Fickian Model

According to Fick's law, the equation for diffusion in one dimension, when the diffusion coefficient D is constant, is expressed[26,27] as

$$\frac{\partial c}{\partial t} = \frac{\partial}{\partial x}\left(D \frac{\partial c}{\partial x}\right) = D \frac{\partial^2 c}{\partial x^2} \tag{6}$$

where c is the concentration of diffusing species at time, t. For a plane sheet geometry and keeping constant the initial concentration of water, the solution of the Fick equation is given by the equation

$$\frac{M}{M_\infty} = 1 - \sum_{n=0}^{\infty} \frac{8}{(2n+1)^2 \pi^2} \exp\left(\frac{-D(2n+1)^2 \pi^2 t}{d^2}\right) \tag{7}$$

where d is the thickness of the specimen and M and M_∞ are the masses of water sorbed or desorbed at times t and ∞, respectively.

Experimental Part

Each gel was prepared by using 0.71×10^{-3}kg AAm (Acrylamide) and 0.008×10^{-3}kg APS (Ammonium per sulfate) as an initiator by dissolving them in 5×10^{-3}l of water in which 0.5 μl of TEMED(tetramethylethylenediamine) was added as an accelerator. 0.01×10^{-3}kg BIS (N′-methylenebis (acrylamide)) was used as a crosslinking agent. This stock solution was deoxygenated by bubbling nitrogen through it for 10 min just before the polymerization process. It was transferred into round glass tubes of 0.99×10^{-2}m internal diameter. Pyranine concentration was kept at 10^{-4}M for all samples. Free radical crosslinking copolymerization (FCC) was performed (for each sample) at room temperature.

Perkin Elmer LS 50 spectrophotometer equipped with temperature controller was used for fluorescence studies. All measurements were made at the 90° position and slit widths were kept at 5 nm. 340 nm was chosen for excitation light and emission was detected at 512 nm during sorption, swelling and desorption experiments. Cell positions are presented in (Figure 1a and b), during sorption and desorption experiments respectively. Sorption and swelling experiments were designed by placing pyranine solution in water, on top of the gel at room temperature. Incident light, I_0 was directed into the gel to monitor sorption of pyranine molecules into PAAm gel. Scattered light intensities, I_{sc} from PAAm gel was also monitored to observed structural variations during sorption which corresponds to swelling process. After the sorption process is completed i.e. gel was

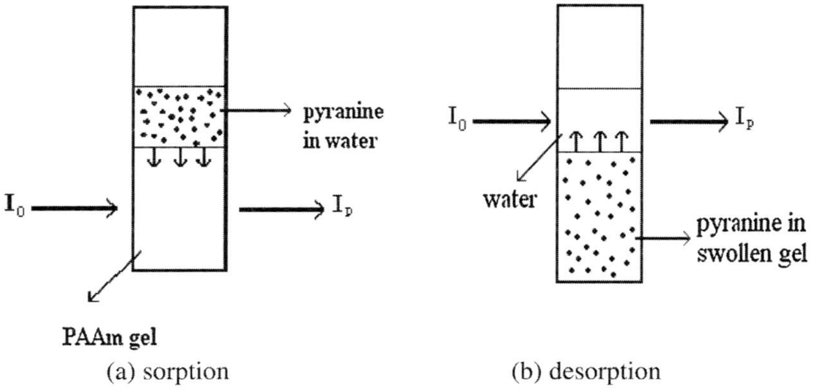

Figure 1.
Position of the gel during a-sorption, b- desorption.

saturated by pyranine molecules, pure water was then placed on top of the pyranine saturated gel to monitor desorption process. This time incident intensity is directed to water to detect the releasing pyranine molecules out of the gel.

Results and Discussion

Scattering intensity, I_{sc} versus swelling time, t is plotted in (Figure 2) for the gels swell at various temperatures. The numbers in (Figure 2) correspond to temperatures in °C. It is seen in (Figure 2) that, I_{sc} decrease by increasing swelling time, indicating that lattice heterogeneities disappear during water uptake of the gel under consideration. It should be noted that water uptake is much faster in gels at high temperatures. These results can be quantified by assuming that the water uptake, W is proportional to the transmitted light intensity, $I_{tr} = 1 - I_{sc}$. During swelling, as more water molecules enter into the gel, structural heterogeneities are disappeared as a result I_{sc} decreases as W increases.

(Figure 3) presents the results of the sorption experiments where P_y intensity, I_p increases as sorption time increased for the all gel samples. Here numbers represent the temperatures. Since I_p is directly proportional to the number of P_y molecules enter into the gel, the behavior of the intensity

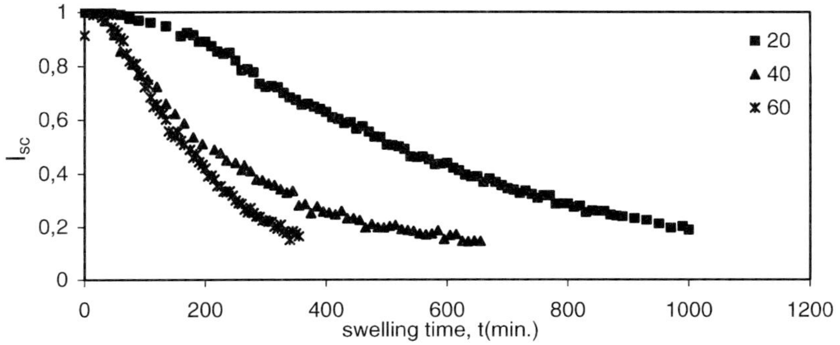

Figure 2.
Scattered light intensity, I_{sc} curves against swelling time, t, for the gels kept at 20 °C, 40 °C, and 60 °C temperatures.

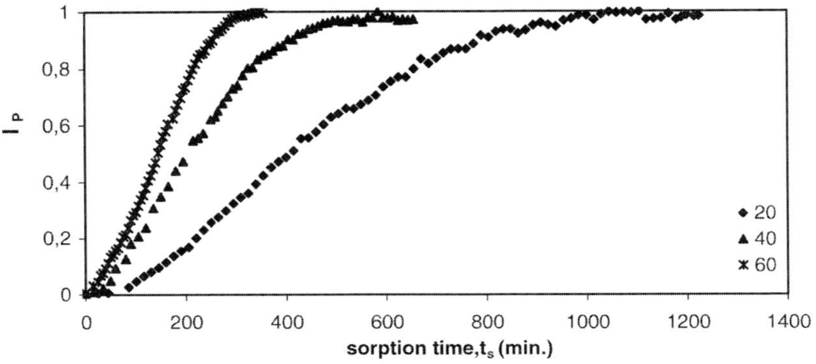

Figure 3.
The behavior of P_y intensity, I_p versus time of P_y sorption in of PAAm gels, kept at 20, 40, and 60 °C temperatures.

curves in (Figure 3) suggest that P_y molecules are absorbed much faster at higher temperatures.

Under this picture the sorption process can be treated using the Fickian diffusion model. The plot of I_p versus desorption time t_{sl} for desorption processes are plotted in (Figure 4). Desorption curves in (Figure 4) present continuous increase, till they saturate.

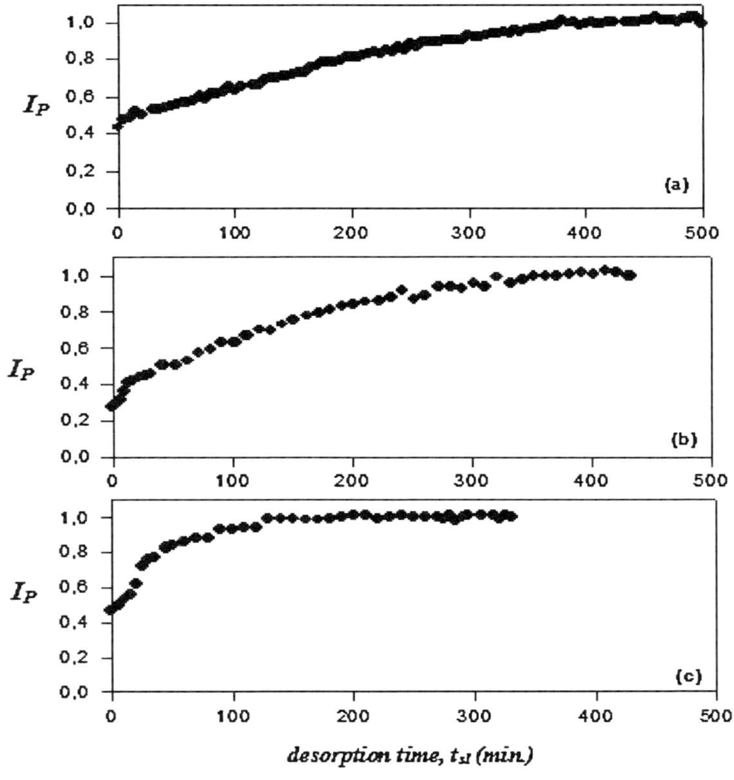

Figure 4.
Plot of P_y intensity, I_p during desorption at different temperatures a-20 °C, b-40 °C, and c-60 °C. $\mathbf{t_{sl}}$ represents the desorption time.

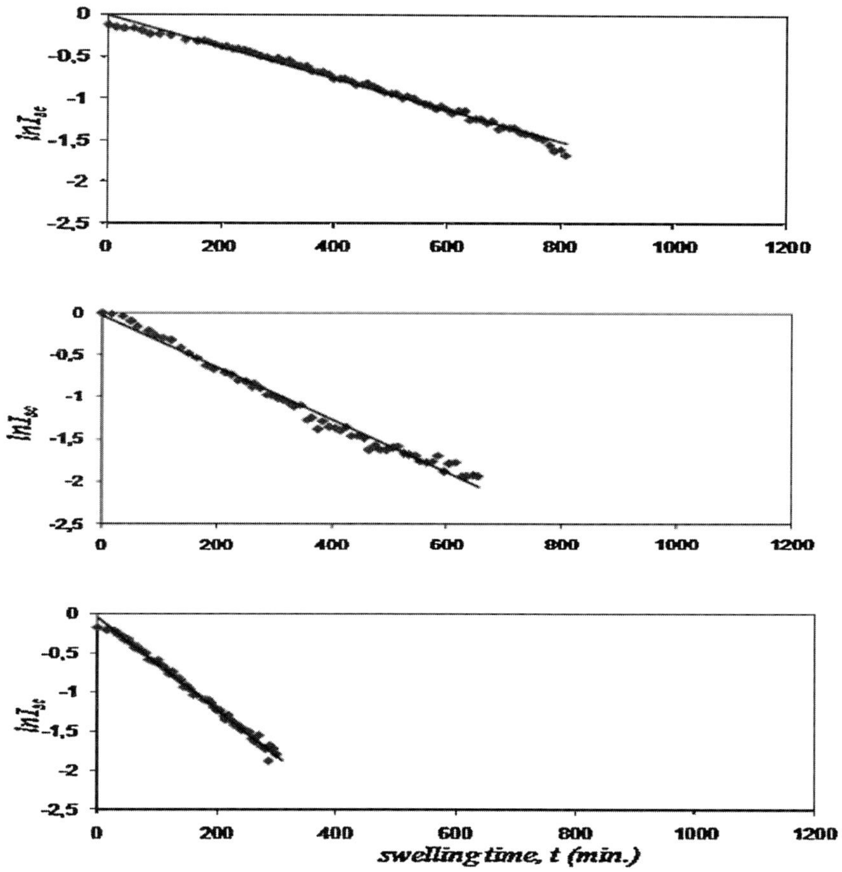

Figure 5.
Linear regression of the curves given in (Figure 2). The intercept and the slope of the curves produce B_1 and τ_c parameters.

In order to model the swelling processes, Equation (4) can be written in terms of I_{sc} in the logarithmic form, i.e.

$$\ln(I_{sc}) = \ln B_1 - t/\tau_c \tag{8}$$

Linear regression of curves in (Figure 5 a, b and c), using Equation (8) provides us with B_1 and τ_c values as shown in (Figure 5). Taking into account the dependence of B_1 on R one obtains R values and from $\alpha_1 - R$ dependence, α_1, values were produced.[8] Then using Equation (5) cooperative diffusion coefficients, D_{sc} were determined for PAAm gels. Experimentally obtained parameters, τ_c and B_1 together with a and D_{sc} values are summarized in (Table 1) for various temperatures. Here one should have noticed that measured D_{sc} values present larger numbers at high temperature. This result is expected, since swelling

Table 1.
Experimentally produced swelling time constants, τ_c preexperimental factors, B_1, and collective sorption coefficients, D_{sc}. a is the half thickness of PAAm gels at the final equilibrium state.

T (°C)	τ_c (s)	B_1	$D_{sc} \times 10^{-8}$ (m^2s^{-1})	$a \times 10^{-2}$ (m)
20	526.31	0.98	0.864	0.51
30	434.78	0.98	0.905	0.55
40	322.50	0.97	0.918	0.56
50	285.71	0.96	1.045	0.62
60	169.49	0.94	1.378	0.625

process is much faster at higher temperatures.

In the sorption experiment, PAAm gels are assumed to be as thin slabs, then the logarithmic form of Equation (7) for n = 0 can be given as follows

$$\ln\left(1-\frac{I_p}{I_{p\infty}}\right)=\ln\left(\frac{8}{\pi^2}\right)-\frac{D\pi^2}{d^2}t \qquad (9)$$

The fit of the Equation (9) to the data in (Figure 3) are given in (Figure 6), from where D_s and D_l values are produced and are listed (Table 2). As expected, it is seen that D values for the gels at high temperature are much larger than they are at low temperature. Here it is seen in (Figure 6), that sorption of P_y molecules has two distinct regions, namely short and long times, produces two different sorption coefficients as D_s and D_l respectively.

Similarly desorption data in (Figure 4) are also treated in (Figure 7) by using Equation (9) and produced desorption coefficients, D_{sl} which are listed in (Table 3). Again, here D_{sl} values present

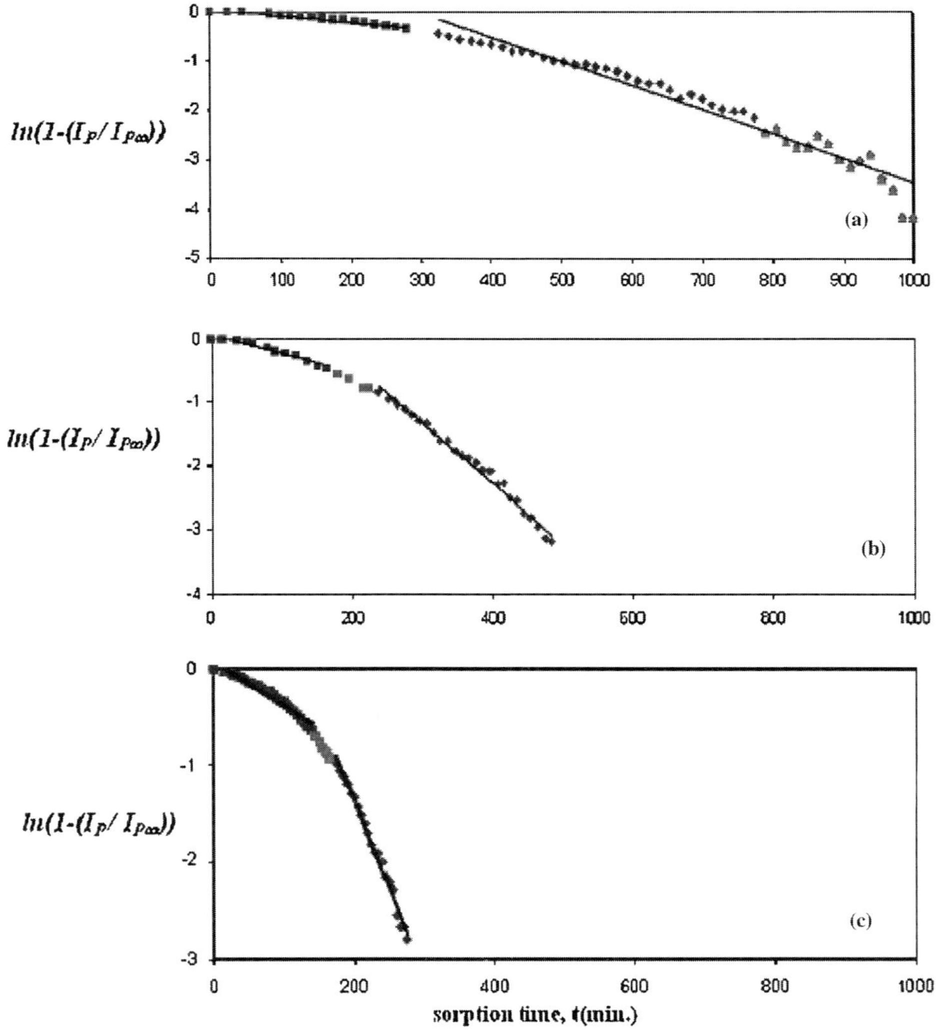

Figure 6.
Linear regression of the data in (Figure 3). The slopes of the straight lines produce D values.

Table 2.
Experimentally produced sorption coefficients, D_s and D_l. d is the thickness of PAAm gels at 20, 30, 40, 50 and 60 °C.

T (°C)	$D_{sc} \times 10^{-8}$ (m^2s^{-1})	$D_l \times 10^{-8}$ (m^2s^{-1})	$d \times 10^{-2}$ (m)
20	1.37	4.01	1.02
30	1.87	4.87	1.11
40	3.94	10.45	1.12
50	4.05	14.65	1.24
60	7.28	28.84	1.25

Table 3.
Experimentally produced desorption coefficients, D_{sl}. l is the initial thickness of the swollen PAAm gels at 20, 30, 40, 50 and 60 °C.

T (°C)	$D_{sl} \times 10^{-10}$ (m^2s^{-1})	$d \times 10^{-2}$ (m)
20	2.01	1.02
30	2.39	1.11
40	6.57	1.12
50	7.90	1.24
60	9.31	1.25

increase as the temperature is increased as expected. If one compares (Table 2) with (Table 3) it can be seen that desorption process is much slower than the two regions.

Arrhenius equation now can be used in all cases

$$D = D_0 e^{-\Delta E/kT} \tag{10}$$

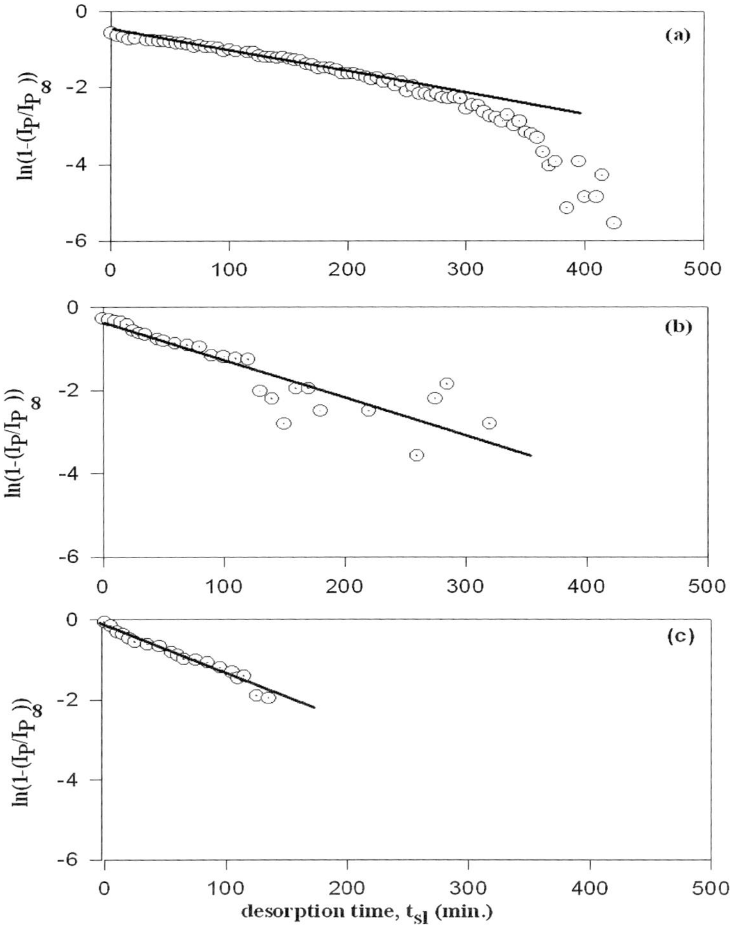

Figure 7.
Fits of the desorption data in (Figure 4). to Equation (9). The slope of the straight lines produce desorption coefficients, D_{sl} at a-20 °C, b-40 °C, c-60 °C.

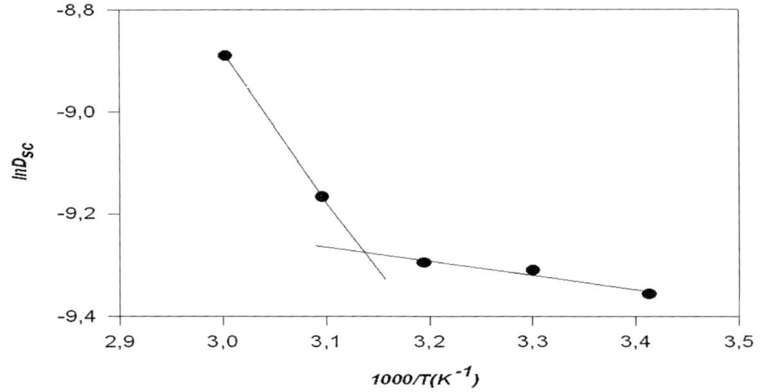

Figure 8.
Arrhenius treatment of D_{sc} values. The slopes of the straight line produce the two different activation energies, namely ΔE_1 and ΔE_2 for gel swelling.

to calculate activation energies, ΔE. (Figure 8) shows the Arrhenius treatment of D_{sc} from which the activation energies, ΔE_1 and ΔE_2 at low and high temperature regions for gel swelling are calculated, and found to be 24,69 $kJmol^{-1}$ and 2.29 $kJmol^{-1}$ respectively.

These values of activation energies are much smaller than the produced activation energy for PMMA in chloroform in the same temperature region.[22] From here one may conclude that the energy need for hydrogel swelling is very small compared to the organic solvent base systems. This

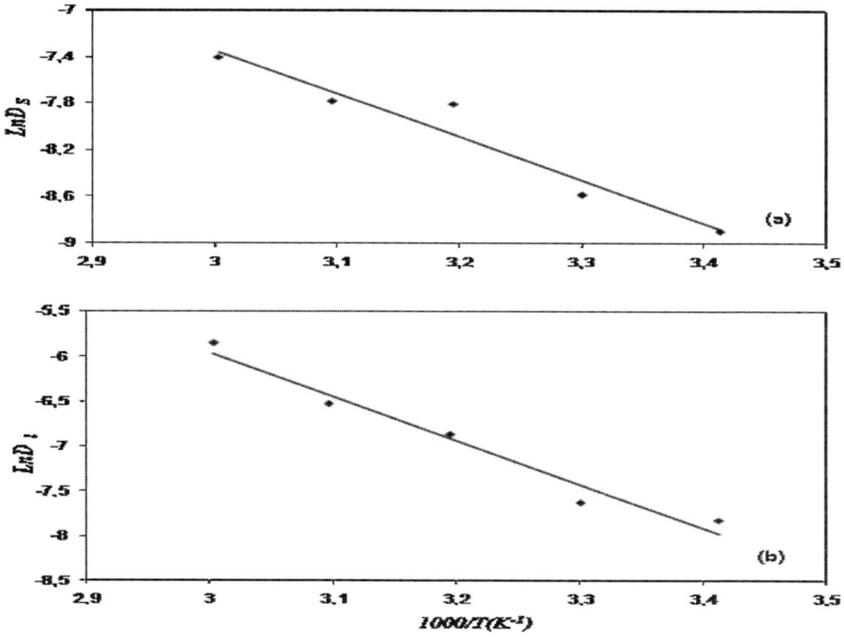

Figure 9.
Arrhenius treatment of a- short b-long time sorption coefficients which produced ΔE_s and ΔE_l activation energies, respectively.

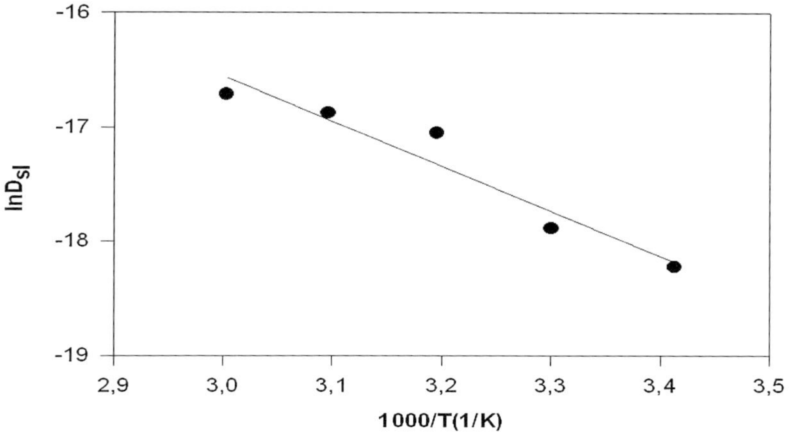

Figure 10.
Arrhenius treatment of desorption coefficients which produced ΔE_{sl} activation energy.

behavior of PAAm gel may be explained with its low T_g values. Two different values of activation energies may indicate two different regimes of swelling process. At low temperature region swelling needs much lower energy than at high temperature region, as expected.

Arrhenius treatment of the sorption coefficients produced two different sorption activation energies as shown in (Figure 9 a and b) from where ΔE_s and ΔE_l values were obtained and found to be 30, 9 kJmol^{-1} and 40, 6 kJmol^{-1} respectively.

Here one may suggest that P_y sorption into PAAm gel is a two stage mechanism. At short times, sorption is slow and needs less energy; however at longer times sorption speeds up with higher energy need. In both cases sorption activation energies are much higher than swelling activation energies indicating that these two processes have different origins.

Arrhenius treatment of the desorption data in (Table 3) is presented in (Figure 10) from where corresponding activation energy; ΔE_{sl} is produced and found to be 34.8 kJmol^{-1}. Here it is seen that desorption and sorption activation energies are comparable in the given regions indicating that both processes need almost same energies to be performed.

Conclusion

The produced results have shown that the steady state fluorescence method can be used for real time monitoring of the swelling, sorption and desorption processes. In this powerful method in situ fluorescence experiments are easy to performed and inexpensive for providing us with the important dynamic parameters such as swelling, sorption and desorption coefficients and activation energies.

[1] M. Weiss, T. Van Vliet, A. Silberberg, *J. Polym. Sci. Polym. Phys. Ed.* **1979**, *17*, 2229.
[2] T. P. Hsu, D. S. Ma, C. Cohen, *Polymer* **1983**, *24*, 1273.
[3] T. Tanaka, S. T. Sun, Y. Hirokawa, S. Katayama, J. Kucera, Y. Hirose, T. Amiya, *Nature* **1987**, *325*, 796.
[4] T. Tanaka, *Phys. Rev. Lett.* **1978**, *40*, 820.
[5] T. Tanaka, D. Filmore, *J. Chem. Phys.* **1979**, *20*, 1214.
[6] J. Bastide, R. Duoplessix, C. Picot, S. Candau, *Macromolecules* **1984**, *17*, 83.
[7] M. Ilavsky, *Macromolecules* **1982**, *15*, 782.
[8] Y. Li, T. Tanaka, *J. Chem. Phys.* **1990**, *92*(2), 1365.
[9] M. Zrinyi, J. Rosta, F. Horkay, *Macromolecules* **1993**, *26*, 3097.
[10] S. Candau, J. Bastide, M. Delsanti, *Adv. Polym. Sci.* **1982**, *7*, 44.
[11] E. Geissler, A. M. Hecht, *Macromolecules* **1980**, *13*, 1276.
[12] M. Zrinyi, F. Horkay, *J. Polym. Sci. Polym. Ed.* **1982**, *20*, 815.

[13] A. K. Bajpai, M. Rajpoot, *J. App. Poly. Sci.* **2001**, *81*, 1238.
[14] M. Panayiotou, R. Freitag, *Polymer* **2005**, *46*, 615.
[15] M. P. Mullarney, T. A. P. Seery, R. A. Weiss, *Polymer* **2006**, *47*, 3845.
[16] Y. Liu, W. Liu, W. Chen, L. Sun, G. Zhang, *Polymer* *48*, 2665.
[17] S. Kara, Ö. Pekcan, *Polymer* **2000**, *41*, 3093.
[18] Ö. Pekcan, S. Kara, *Polym. Plast. Technol. Eng.* **2002**, *41*, 573.
[19] Ö. Pekcan, S. Kara, *Polymer* **2000**, *41*, 8735.
[20] S. Kara, O. Okay, Ö. Pekcan, *J. App. Polym. Sci.* **2002**, *86*, 3589.
[21] Ö. Pekcan, S. Kara, *Phase Transitions* **2003**, *76*, 601.
[22] M. Erdoğan, Ö. Pekcan, *Polymer* **2001**, *42*, 4973.
[23] Ö. Pekcan, Y. Yılmaz, O. Okay, *J. App. Poly. Sci.* **1996**, *61*, 2279.
[24] Ö. Pekcan, M. Erdogan, *European Polymer Journal* **2002**, *38*, 1105.
[25] G. A. Evingür, K. Karslı, Ö. Pekcan, *Int. J. Pharm.* **2006**, *326*(1–2), 7.
[26] J. Crank, "*The Mathematics of Diffusion*", Clarendon Press, Oxford **1970**.
[27] J. Crank, G. S. Park, "*Diffusion in Polymer*", Acad. Press, London **1968**.

Vibrational Dynamics of Polyaniline Pernigraniline Base Form: A Conducting Polymer

Abhishek Kumar Mishra,[*1] *Poonam Tandon,*[2] *V. D. Gupta*[2]

Summary: Polyaniline is unique among conducting polymers in that its electrical properties can be reversibly controlled both by the oxidation state of the main chain and by protonation. In the present communication we report normal modes and their dispersion in polyaniline pernigraniline base (PANI PNB) form using Urey - Bradely force field which in addition to valence force field accounts for the non-bonded interactions in the *gem* and *cis* configurations and tension terms. Some of the distinguished features of the dispersion curves such as cross-over and Von Hove type of singularities are discussed and possible explanations are given in terms of symmetry considerations. Several new assignments, which are missing in earlier work, are reported.

Keywords: cross-over; density-of-states; dispersion curves; pernigraniline base form; polyaniline

Introduction

Since the discovery of metal like electrical conductivity in polyacetylene, there have been various studies on the physical and chemical properties of conjugated electro-active polymers. Among them, members of Polyaniline(PANI) have received considerable attention due to the controllable electrical conductivity[1], environmental stability[2], and interesting redox properties associated with the chain hetroatoms.[3]

Polyaniline, a green intractable powder, has a history beginning more than 120 years ago, when in 1862 Letheby[4] reported on the electrochemical oxidation product of aniline and in 1968 Surville et al.[5] on the high electrical conductivity and its pure electronic character in PANI, which depends on the "acidity, redox level & hydration"[5] of the polymer. PANI was rediscovered few years ago as a very important member of conducting polymers having wide potential technological applications.[6–11] Polyaniline is also used as an electrode material for semiconductor batteries,[9,12–16] as a semi-conducting material,[17] as a sensor material,[18] as an electrochromic-display material[19–21] and as a catalyst for photoelectrochemical processes.[21]

Polyaniline relates to a large class of polymers insofar as several forms of these compounds can be obtained. These different forms are described by two parameters: the average oxidation state[22,23] and the degree of protonation.[24] These conducting polymers differ substantially from polyacetylene and other aromatic and hetroaromatic polymers in the fact that their electronic structure is based on the overlap of alternating nitrogen and C6 rings. In its nonconductive undoped (or base) form, Polyaniline have the general formula $[(\text{–B–NH–B–NH–})_y\,(\text{–B–N=Q=N–})_{1\text{-}y}]_x$, (B and Q stands for benzenoid and quinoid structure respectively) where y can be varied continuously from one to zero (from completely reduced form to the completely oxidized one). The conversion to a conductive form can be accompolished by either protonic or electronic doping.[25]

[1] Department of Physics, Amity School of Engineering, Amity Institute of Applied Sciences, Amity University, Noida 201 301, India
E-mail: mishra_lu@hotmail.com
[2] Department of Physics, University of Lucknow, Lucknow-226007, India

The terms leucoemeraldine, emeraldine, and pernigraniline refer to the different oxidation states of the polymer where y = 1, ~0.5, and 0, respectively. In this work we have focussed our all attention on the fully oxidized (y = 0) non protonated form (Pernigraniline Base: PNB).

Since the electronic properties of the compounds are directly associated to the chemical bondings and the electron distribution along the polymer backbone, a clear knowledge of their vibrational properties can be of major importance. Vibrational spectroscopy plays an important role in the elucidation of polymeric structure. Normal mode analysis besides identification of various modes provides an insight into Infrared absorption (IR), Raman spectra and Inelastic Neutron Scattering (INS). An overall understanding of vibrational dynamics in a polymer involves calculation of the dispersion curves. These curves provide knowledge of degree of uninterrupted sequence lengths in an ordered conformation.

This is the first communication in which the complete planar and non-planar normal coordinate analysis of PANI Pernigraniline (PNB) form along with the dispersive behaviour of the normal modes is being reported. We present here a normal mode analysis of PANI PNB form with phonon dispersion in the first Brillouin Zone using Urey Bradley force field (UBFF).[26,27] Earlier works on normal mode analysis[28–30] of Lefrant and co-workers using valence force field suffers from the use of approximate force field resulting in the inaccurate and incomplete assignments specially in the lower region of spectra and lack of dispersion curves. For the full understanding of vibrational spectra, it is necessary to know the dispersive nature of normal modes. Such studies also provide information on the degree of coupling and dependence of the frequency of a given mode on the sequence length of the ordered conformation. These curves are also useful in calculating the density of vibrational states which in turn can be used for obtaining thermodynamic properties such as specific heat, entropy, enthalpy and free energy. Our calculations are based on UBFF, which in addition to valence force field accounts for the non-bonded interactions in the *gem* and *cis* configuration and the tension terms. In this force field the potential energy expression does not have quadratic cross terms. The force constants are supplemented by the repulsive forces between non-bonded atoms, which simulate the Van der Waals force[31] between them. It gives a better description of intra and inter unit interactions, and arbitrariness in choosing the force constants is reduced, thereby enabling us to arrive at a better unique force field.

This paper provides a better understanding of the PANI PNB spectra by assigning bands based on para-disubstituted benzene derivatives. It is often difficult to describe a vibration completely using symbols or descriptions; therefore describing the modes based on substituted benzene modes should give a complete understanding of the vibrational modes of PANI PNB form. Many authors have examined benzene derivatives [32–39] and made assignments that have been used in this work to assign the modes of PANI PNB form. One of the most comprehensive discussions is that of Varsanyi.[32] For purposes of assigning fundamental modes, PANI is considered to be "di-light", since N has an atomic weight below 14.

Theory

Calculation of Normal Mode Frequencies

Normal mode calculation for a polymeric chain was carried out using Wilson's GF matrix method[40,41] as modified by Higgs[42] for an infinite polymeric chain. The vibrational secular equation to be solved is?

$$|G(\delta)F(\delta) - \lambda(\delta)I| = 0 \quad 0 \leq \delta \leq \pi \qquad (1)$$

where δ is the phase difference between the modes of adjacent chemical units, $G(\delta)$ is the inverse kinetic energy matrix and $F(\delta)$ is the force field matrix for a certain phase value. The wavenumber $\bar{\nu}_i(\delta)$ in cm^{-1} are related to eigen values by $\lambda_i(\delta) = 4\pi^2c^2\,[\bar{\nu}_i(\delta)]^2$.

A plot of $\overline{\nu}_i(\delta)$ versus δ gives the dispersion curve for the ith mode. The use of the type of force field is generally a matter of one's chemical experience and intuition.[43] In the present work, we have used Urey-Bradley force field as it is more comprehensive then valence force field. The Urey-Bradley takes in to account both bonded and non-bonded interactions as well as internal tensions. Potential energy for this force field can be written as

$$V=\sum_{m,j,k} K'_{j,k} r_{j,k}^{(m)}\left(\Delta r_{j,k}^{(m)}\right)+K_{j,k}\left(\Delta r_{j,k}^{(m)}\right)^2/2+\sum_{m,i,j,k} H'_{i,j,k} r_{i,j}^{(m)} r_{j,k}^{(m)}\left(\Delta\alpha_{i,j,k}^{(m)}\right)+H_{i,j,k} r_{j,k}^{(m)}\left(\Delta\alpha_{i,j,k}^{(m)}\right)^2/2$$
$$+\sum_{m,i,j,k} F'_{i,k} q_{i,k}^{(m)}\left(\Delta q_{i,k}^{(m)}\right)+F_{i,k}\left(\Delta q_{i,k}^{(m)}\right)^2/2+\sum_j K_j^{\tau}(\Delta\tau_j)^2+\sum_j K_j^{\omega}(\Delta\omega_j)^2 \tag{2}$$

where the symbols have their usual meaning. The primed quantities are introduced as internal tensions. Non-bonded interactions involve attraction and repulsion of atoms due to the overlap of their electron shells. These effects are usually expressed by the 6-exp or 6-12 type potentials. The tension terms are assumed to be all zero.

Recently, spectroscopically effective molecular mechanics model have been used for inter and intra molecular interactions consisting of charges, atomic dipoles and Vander Waals (non bonded) interactions.[31]

The force constants, including those for the interaction of first and third non-bonded atoms, which give the "best fit", are given in the Table 1 and have been obtained by least squares fitting. In order to obtain the "best fit" with the observed wavenumbers the following procedure is adopted.

Starting with the approximate F matrix F_o and observed frequencies λ_{obs} (related through a constant), one can solve the secular matrix equation:

$$GF_0L_0 = L_0\lambda_0 \tag{3}$$

Let $\Delta\lambda_i = \lambda_{i_{obs}} - \lambda_{i_0}$ in the above equation. It can be shown that in the direct order of approximation

$$\Delta\lambda = J\Delta F \tag{4}$$

where J is computed from L_0. We wish to compute the corrections to F_0 so that the errors $\overline{\Delta\lambda}$ are minimized. We used the theory of least squares and calculate

$$J'P\overline{\Delta\lambda} = (J'PJ)\overline{\Delta F} \tag{5}$$

where P is the weighting matrix and J′ is the transposition of J. The solution of this equation is obtained by inverting J′PJ to give

$$\overline{\Delta F} = (J'PJ)^{-1}J'P\overline{\Delta\lambda} \tag{6}$$

If the number of frequencies is greater than the number of F matrix elements, the matrix J′PJ should be non-singular and we obtain the corrections ΔF, which will minimize the sum of the weighted squares of the residuals. This minimum sum provides the "best fit". If the corrections ΔF are fairly large, the linear relation between force constant and frequency term in the matrix Equation (3) breaks down. In such a situation, further refinement using higher order terms in the Taylor's series expansion of $\Delta\lambda_i$ is needed. King et al.[44] developed this procedure.

Table 1.
Internal Coordinates and force Constants for PANI PNB form (md/Å).

Internal coordinates	Force Constants	Internal coordinates	Force Constants
ν[C–C]*	5.05	φ[C–C–C]$	0.38(0.40)
ν[C–H]	4.74	φ[C–C–H]$	0.415(0.46)
ν[C=C]*	5.40	φ[C–C=C]$	0.42(0.35)
ν[C–N]	4.41	φ[H–C=C]$	0.20(0.35)
ν[N=C]	5.40	ω[C–H]	0.27
ν[C–C]$	4.14	ω[C–N]	0.368
ν[C=C]$	5.35	ω[N=C]	0.28
φ[H–C–C]*	0.30(0.30)	τ[C=C]*	0.033
φ[H–C=C]*	0.42(0.39)	τ[C–C]*	0.037
φ[C–C=C]*	0.69(0.50)	τ[C–N]	0.030
φ[C–C–N]	0.43(0.42)	τ[N=C]	0.030
φ[C=C–N]	0.43(0.42)	τ[C–C]$	0.032
φ[C–N=C]	0.425(0.40)	τ[C=C]$	0.042
φ[N=C–C]	0.68(0.58)		

Note: ν, φ, ω and τ denote stretch, angle bend, wag and torsion respectively. Non-bonded force constants are given in parentheses. * and $ corresponds to benzoid and quinoid rings respectively.

Results and Discussion

A chemical repeat unit of PANI PNB consists of one benzoid ring and one quinoid ring (Figure 1). The number of atoms in the repeat unit are 22 which give rise to 3N-4 = 62 normal modes which would be optically active and 4 acoustical modes which correspond to three translations and one rotation about the backbone. Initially Force constants from some similar molecules are transferred and later modified to give the "best fit" to the observed infrared and Raman frequencies.[29–30] The vibrational wave numbers were calculated for values of δ varying from 0 to π in steps of $.05\pi$. The assignments have been made on the basis of potential energy distribution (PED), absorption band shape, band intensity and absorption/scattering due to molecular groups placed in similar environments. Final sets of force constants along with the internal coordinate are given in Table 1. The force field (Table 1) thus obtained is also well supported with by the experimental spectra and one can get a very good fit frequencies without changing the off-diagonal elements. The observed Raman and IR[29,30] and calculated frequencies along with their potential energy distribution (PED) at the zone centre and zone boundary are given in Table 2. Since all the modes above 900 cm^{-1} are non-dispersive or have negligible dispersion, the dispersion curves are plotted only for the modes below 900 cm^{-1}. The lowest two branches of the dispersion curves correspond to the four acoustic modes. Out of these, three are due to the translation (one parallel and two perpendicular to the chain axis) and one due to rotation around the chain. Our spectral interpretation is also based on the fundamental p-disubstituted benzene modes and the assignments of the modes are discussed according to their nature of vibrations.

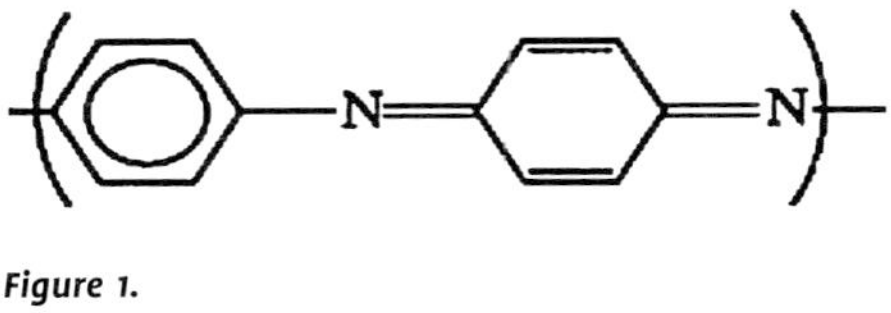

Figure 1.
One chemical repeat unit of Polyaniline PNB form.

C–H Stretching Vibrations

There C–H stretching vibrations of benzene derivatives generally appear above 3000 cm^{-1}. Vibrations 2, 7b, 20a and 20b are C–H stretching vibrations in p-disubstituted benzene derivatives and correspond to phase combinations: + + + +, + − + −, + − − + and + + + + respectively. By our l-vector calculations of PANI PNB form we found that calculated normal modes at 3087, 3086, 3085, 3084 corresponds to 20b, 20a, 2, 7b vibrations of the quinoid ring and 3072, 3069, 3068 and 3065 cm^{-1} corresponds to 20b, 20a, 2, 7b C–H stretching vibrations of the benzoid ring.

C–C Stretching Vibrations

Wilson no. 8a, 8b, 19a, 19b and 14 are associated with CC stretching vibrations. Two well resolved Raman bands at 1612 and 1579 cm^{-1} are assigned to the calculated normal modes at 1623 and 1580 cm^{-1} respectively. These two bands are found to be associated with CC stretching vibrations and can be seen as 8a vibrations of the benzoid and quinoid rings. The characteristic infrared band at about 1570 cm^{-1} is assigned to the quinoid unit (8b) while the 8b vibration of the benzoid ring can be seen as Raman peak at 1555 cm^{-1}.19b Wilson vibration of the quinoid and benzoid rings are calculated at 1415 and 1398 cm^{-1} and are assigned to the bands at 1418 (Raman) and 1411(IR) cm^{-1} respectively. Infrared band at 1495 cm^{-1}[45] can be seen as 19a Wilson C–H stretching vibration and while the band at 1377 cm^{-1} is related with CC stretching + CH bending vibration of the quinoid ring (14).

C–H Bending Vibrations

In p-disubstituted benzenes, normal vibrations 3, 9a, 18b and 18a are classified as modes having C–H in-plane bending character. The vibration 3 appears generally as a weak absorption in the infrared spectra and in the PANI PNB form spectra we have a absorption dip at 1315 cm^{-1} which can be assigned to the modes calculated at 1327,

Table 2.
Calculated and Observed Modes of PANI PNB

Cal. freq.	Obs. Freq.		Assignment (% PED), δ = 0	Cal. freq	Obs. Freq.		Assignment (% PED), δ = π
	Raman	IR			Raman	IR	
3087	–	–	ν[C–H]$(99)	3087	–	–	ν[C–H]$(99)
3086	–	–	ν[C–H]$(99)	3086	–	–	ν[C–H]$(99)
3085	–	–	ν[C–H]$(99)	3085	–	–	ν[C–H]$(99)
3084	–	–	ν[C–H]$(99)	3084	–	–	ν[C–H]$(99)
3072	–	–	ν[C–H]*(99)	3072	–	–	ν[C–H]*(99)
3069	–	–	ν[C–H]*(99)	3070	–	–	ν[C–H]*(99)
3068	–	–	ν[C–H]* (99)	3068	–	–	ν[C–H]* (99)
3065	–	–	ν[C–H]* (99)	3065	–	–	ν[C–H]* (99)
1623	1612[a]	–	ν[C=C]*(29) + ν[C–C]*(27) + φ[H–C=C]*(12) + ν[C–N](10) + φ[C–C=C]*(9) + φ [H–C–C]*(9)	1623	1612[a]	–	ν[C=C]* (30) + ν[C–C]* (27) + φ[H–C=C]*(12) + ν[C–N](10) + φ[C– C=C]*(9) + φ[H–C–C]* (9)
1580	1579[a]	1570[a]	ν[C=C]$(33) + ν[C–C]$(20) + φ[C–C–H]$(18) + ν[N=C](13)	1583	1579[a]	1570[φ]	ν[C=C]$ (30) + ν[C–C]$ (19) + φ[C–C–H]$(17) + ν[N=C](15)
1562	–	1570[a]	ν[C–C]*(41) + ν[C=C]*(28) + φ[H–C=C]*(14) + φ[C–C=C]*(8) + φ[H–C–C]*(6)	1562	–	1570[a]	ν[C–C]*(40) + ν[C=C]*(29) + φ[H–C=C]*(13) + φ[C–C=C]*(8) + φ[H–C–C]* (6)
1558	1555[a]	–	ν[C=C]*(48) + ν[C–C]*(41) + φ[H–C=C]*(5)	1558	1555[a]	–	ν[C=C]*(48) + ν[C–C]* (41) + φ[H–C=C]*(5)
1545	1555[a]	–	ν[C=C]* (24) + ν[C–N](19) + φ[H–C=C]*(18) + φ[H–C–C]*(15)	1539	1555[a]	–	ν[C=C]* (25) + φ[H–C=C]*(21) + φ[H–C–C]*(18) + ν[C–N](16)
1496		1495[c]	ν[C=C]$(53) + ν[C–C]$(26) + φ[C–C–H]$(18) + φ[H–C=C]$(6)	1496		1495[c]	ν[C=C]$(53) + ν[C–C]$(26) + φ[C–C–H]$(15) + φ[H–C=C]$(6)
1487	1480[a]	1480[a]	ν[N=C](31) + ν[C–C]$(20) + φ[C–C–H]$(20) + φ[H–C=C]$(9)	1493	1480[a]	1480[a]	ν[N=C](38) + ν[C–C]$(21) + φ[C–C–H]$(20) + φ[H–C=C]$(9)
1415	1418[a]	–	ν[C–C]$(76) + φ[N=C–C](10) + φ[C–C=C]$(6) + φ[H–C=C]$(6)	1417	1418[a]	–	ν[C–C]$(71) + φ[N=C–C](10) + φ[C–C=C]$(6) + φ[H–C=C]$(6)
1398	–	1411[a]	φ[H–C=C]*(20) + ν[C=C]*(19) + ν[C–C]*(15) + φ[H–C–C]*(14) + ν[C–C]$(11)	1393	–	1411[a]	φ[H–C=C]*(24) + ν[C=C]* (23) + ν[C–C]*(17) + φ[H–C–C]*(17) + ν[C–C]$(6)
1381	–	1377[a]	ν[C–C]$(41) + φ[C–C–H]$(17) + φ[N=C–C](12) + φ[H–C=C]*(6) + ν[C=C]*(5) + φ[H–C=C]$(5)	1385	–	1377[a]	ν[C–C]$(52) + φ[C–C–H]$(21) + φ[N=C–C](16) + φ[H–C=C]$ (6)
1371	–	1377[a]	ν[N=C](45) + ν[C–N](22) + ν[C=C]$(16)	1367	–	1377[a]	ν[N=C](41) + ν[C–N](17) + ν[C=C]$(15) + φ[H–C=C]*(9) + a[H–C–C]*(6)
1334		–	φ[H–C=C]*(36) + φ[H–C–C]*(18) + ν[C–N](9) + φ[C–C–H]$(8) + φ[H–C=C]$(6) + ν[C=C]*(5) + φ[C–C=C]*(5)	1327		–	φ[H–C=C]*(48) + φ[H–C–C]*(31) + ν[C=C]*(6)
1324		–	φ[C–C–H]$(58) + φ[H–C=C]$(30)	1324	1320[b]	–	φ[C–C–H]$(61) + φ[H–C=C]$(31)
1317		1315[a]	φ[H–C=C]*(21) + φ[H–C–C]*(20) + φ[C–C–H]$(14) + ν[C–N](13) + φ[H–C=C]$(11) + ν[N=C](10)	1323	1320[b]	1315[a]	ν[C–N](27) + φ[C–C–H]$(21) + φ[H–C=C]$(17) + ν[N=C](16) + ν[C–C]*(6)
1217	1215[a]	1211[a]	ν[N=C](23) + ν[C–N](16) + φ[C–C=C]*(14) + φ[H–C=C]$(12) + φ[C–C–H]$(10) + φ[H–C–C]*(8) + φ[H–C=C]*(7)	1219	1215[a]	1211[a]	ν[C–N](23) + ν[N=C](19.72) + φ[C–C=C]*(17) + φ[H–C–C]*(11) + φ[H–C=C]*(10) + ν[C=C]$(7)
1214	1215[a]	1211[a]	ν[C–N](26) + ν[N=C](16) + φ[H–C=C]*(15) + ν[C–C]*(11) + φ[H–C–C]*(10) + [C=C]*(7) + ν[C=C]$(6)	1205	1215[a]	1211[a]	φ[H–C=C]*(17) + ν[C–N](17) + ν[N=C](16) + φ[H–C–C]*(12) + φ[H–C=C]$(8) + ν[C–C]*(8) + φ[C–C–H]$(7)
1176	–	1164[c]	φ[H–C=C]*(32) + ν[C=C]*(26) + φ[H–C–C]*(20) + ν[C–C]*(7) + ν[C–N](7)	1175	–	1164[c]	φ[H–C=C]*(29) + ν[C=C]*(26) + φ[H–C–C]*(18) + ν[C–N](8) + ν[C–C]*(8)
1154	1157[a]	1157[a]	φ[C–C–H]$(45) + φ[H–C=C]$(29) + ν[C=C]$(20)	1154	1157[a]	1157[a]	φ[C–C–H]$(45) + φ[H–C=C]$(29) + ν[C=C]$(19) + ν[C–C]$(5)
1087	–	1098[a]	φ[C–C–H]$(25) + ν[C=C]$(24) + φ[H–C=C]$(22) + ν[C–C]$(11) + φ[N=C–C](11)	1088	–	1098[a]	φ[C–C–H]$(26) + ν[C=C]$(25) + φ[H–C=C]$(23) + φ[N=C–C](12) + ν[C–C]$(11)

(Continues)

Table 2.
(Continued)

Cal. freq.	Obs. Freq.		Assignment (% PED), δ = 0	Cal. freq	Obs. Freq.		Assignment (% PED), δ = π
	Raman	IR			Raman	IR	
1008	–	1006[a]	φ[C–C=C]*(32) + ν[C–C]*(25) + ν[C=C]*(17) + φ[H–C–C]*(14) + φ[H–C=C]*(12)	1008	–	1006[a]	φ[C–C=C]*(32) + ν[C–C]*(25) + ν[C=C]*(17) + φ[H–C–C]*(14) + φ[H–C=C]*(12)
978	–	965[c]	ω[C–H](78) + ω[C–N](9)	976	–	965[c]	ω[C–H](77) + ω[C–N](11)
966	–	965[c]	ω[C–H](83) + t[C=C]$(7)	969	–	965[c]	ω[C–H](81) + t[C=C]$(8) + ω[N=C](6)
953	–	954[a]	ν[C–C]$(58) + φ[C–C=C]$(20.52) + φ[C–C–H]$(17)	951	–	954[a]	ν[C–C]$(54) + φ[C–C=C]$(21) + φ[C–C–H]$(16)
948	–	954[a]	ω[C–H](82) + t[C=C]$(13)	948	–	954[a]	ω[C–H](82) + t[C=C]$(13)
938	–	931[c]	ω[C–H](83) + t[C–C]*(8) + t[C=C]*(8)	939	–	931[c]	ω[C–H](83) + t[C–C]*(8) + t[C=C]*(8)
866	–	–	ν[C–C]*(30) + ν[C=C]*(25) + ν[C–C]$(17) + ν[C–N](12) + φ[C–C=C]*(7)	861	–	–	ν[C–C]*(30) + ν[C=C]*(26) + ν[C–N](13) + φ[C–C=C]*(8)
842	848[a]	844[a]	ω[C–H](55) + ω[C–N](17) + φ[C–N=C](11) + ω[N=C](6)	841	848[a]	844[a]	ω[C–H](33) + ν[C–C]$(30) + ω[C–N](14) + ν[C=C]$(7)
807	790[b]		ω[C–H](45) + φ[C–C=C]*(8) + ν[C–N](7) + ω[N=C](7) + φ[C–C=C]S(6)	823	–	–	ω[C–H](32) + ν[C–C]$(22) + φ[C–C=C]*(8) + ω[C–N](8) + ν[C–N](7)
790	790[b]	–	ν[C–C]$(47) – ν[N=C](9) + ν[C=C]$(6) + φ[C–C=C]*(6) + φ[C–N=C](5)	802	790[b]	–	ω[C–H](67) + ω[N=C](12) + τ[C–C]$(6) + φ[C–N=C](5)
776		–	ω[C–H](38) + φ[C–C=C]$(10) + ν[C–N](10) + φ[C–C=C]*(9) + φ[C–C–C]$(7) + φ[N=C–C](5)	761		–	ω[C–H](85) + τ[C–C]*(8) + τ[C=C]*(7)
761	–	775[a]	ω[C–H](85) + τ[C–C]*(7) + τ[C=C]*(7)	752	–	775[a]	ω[C–H](86) + τ[C–C]$(14)
752		750[b]	ω[C–H](86) + τ[C–C]$(14)	718	–	–	ν[C–N](20) + ω[C–H](13) + φ[C–N=C](13) + φ[C–C=C]*(12) + ν[C–C]$(10) + ω[N=C](6)
658	–	–	ω[C–H](27) + φ[C–N=C](21) + ν[C–N](9) + ν[N=C](7) + φ[C–C=C]$(6)	708		697[b]	φ[C–C=C]$(15) + φ[C–C–C]$(12) + ν[N=C](10) + φ[C–N=C](10) + ω[C–H](10) + φ[N=C–C](9) + ω[C–N](8) +
657	–	–	φ[C–N=C](30) + ω[C–N](22) + ω[N=C](14) + φ[C–C=C]*(5) + τ[C=C]S(5) + τ[C–C]*(5)	629	–	–	φ[N=C–C](61) + τ[C–N](7) + ν[C–C]$(7) + τ[N=C](6) + φ[C–C–N](6) + φ[C=C–N](6)
649	–	–	φ[N=C–C](51) + φ[C=C–N](12) + φ[C–C–N](12) + τ[C–N](7) + ν[C–C]$(6) + τ[N=C](6)	618	–	–	ω[C–N](24) + φ[C–N=C](18) + φ[C–C=C]*(13) + ω[C–H](8) + τ[C–C]*(7) + τ[C–C]*(6)
614	–	624[c]	φ[C–C=C]*(69) + φ[H–C–C]*(11) + φ[H–C=C]*(7)	617	–	624[c]	φ[C–C=C]*(66) + φ[H–C–C]*(10) + φ[H–C=C]*(6) + ν[C=C]*(6) +
544	–	–	φ[N=C–C](35) + φ[C–C–N](11) + φ[C=C–N](10) + τ[C–N](10) + τ[N=C](10) + ν[C–C]$(9) + ν[C=C]$(6)	593	–	–	φ[C–N=C](19) + ω[N=C](17) + ω[C–N](11) + ω[C–H](9) + τ[C=C]$(9) + φ[C–C–C]$(7) + φ[N=C–C](5)
543	–	542[b]	ω[C–N](28) + ω[N=C](27) + τ[C=C]$(12) + τ[C–C]*(9) + τ[C=C]*(8)	577	–	–	φ[N=C–C](25) + φ[C=C–N](18) + φ[C–C–N](18) + τ[N=C](9) + τ[C–N](9) + ν[C–C]$(7) + φ[C–C=C]*(6)
519	518[b]	–	φ[C–C=C]$(66) + φ[H–C=C]$(14) + φ[C–C–H]$(9) + 6) ν[C–C]$(9)	519	518[b]	–	φ[C–C=C]$(66) + φ[H–C=C]$(14) + φ[C–C–H]$(9) + ν[C–C]$(9)
498		498[b]	ω[C–n](36) + ν[C–N](9) + ν[N=C](8) + ω[C–H](7) + φ[C–N=C](6.52)	505		498[b]	ω[N=C](24) + ω[C–N](23) + τ[C–C]*(12) + τ[C=C]*(11) + φ[C–N=C](8) + φ[C–C=C]*(7) + ω[C–H](6)
482	–	498[b]	φ[C–C=C]*(27) + φ[C–C–C]$(14) + φ[N=C–C](11) + φ[C–C=C]$(7) + ν[C–N](6) + τ[C=C]$(5)	498	–	498[b]	ω[C–N](35) + ω[N=C](27) + τ[C=C]$(18) + ω[C–H](7) + τ[C–C]$(7)

(*Continues*)

Table 2.
(Continued)

Cal. freq.	Obs. Freq.		Assignment (% PED), $\delta = 0$	Cal. freq	Obs. Freq.		Assignment (% PED), $\delta = \pi$
	Raman	IR			Raman	IR	
442	438[b]	–	ω[N=C](53) + ω[C–N](24) + ω[C–H](7) + τ[C–C]$(5)	379	374[b]	–	ω[C–N](20) + φ[C–C–C]$(17) + φ[N=C–C](13) + φ[C–C=C]$(8) + τ[C–C]$(7) + φ[C–N=C](6) + ν[N=C](5)
406	–	410[b]	τ[C–C]$(14) + φ[C–N=C](13) + τ[C=C]$ (12) + τ[C–C]*(11) + τ[C=C]*(10) + φ[C–C–C]$(7) + φ[N=C–C](6)	374	374[b]	–	ω[N=C](37) + φ[C–C=C]*(22) + ν[C–N](12) + τ[C–C]$(10)
358	359[b]	–	φ[C=C–N](31) + φ[C–C–N](31) + φ[N=C–C](29)	286		–	φ[N=C–C](31) + φ[C=C–N](24) + φ[C–C–N](24)
284	–	–	τ[C=C]$(55) + τ[C–C]$(24) + ω[C–H](17)	283		–	τ[C=C]$(52) + τ[C–C]$(22) + ω[C–H](15)
272	–	–	τ[C–C]*(42) + τ[C=C]*(38) + ω[C–H](15)	275		–	τ[C–C]*(33) + τ[C=C]*(29) + ω[C–H](11) + φ[N=C–C](7) + τ[C–N](6)
224		–	τ[C–C]$(27) + τ[C–C]*(16) + τ[C=C]*(15) + ω[N=C](13) + ω[C–N](11) + ω[C–H](11)	264		–	τ[C–C]$(37) + φ[C–N=C](11) + τ[C=C]$ (8) + ω[C–H](8) + ω[N=C](7) + φ[C–C–C]$(6)
202	–	–	τ[C–C]$(11) + τ[C–C]*(10) + φ[C–C=C]*(9.52) + τ[C=C]*(9) + ν[C–N](8) + φ[C–C–C]$(8) + φ[N=C–C](6) + ν[N=C](6) + ω[C–H](5)	263		–	φ[C–C–N](23) + φ[C=C–N](21) + φ[N=C–C](17) + τ[C–C]*(11) + τ[C=C]*(8) + τ[N=C](8)
200	212[b]	–	φ[C–N=C](42) + τ[C–C]$(16) + ω[N=C](15) + ω[C–N](7) + τ[C–C]*(7) + τ[C=C]*(7)	259		–	τ[C–C]*(26) + τ[C=C]*(26) + φ[C–N=C](10) + ω[C–H](8) + ω[C–N](6) + ω[N=C](6) + φ[C–C=C]*(5)
190	–	–	φ[N=C–C](31) + φ[C=C–N](27) + φ[C–C–N](26)	116	–	–	φ[C–N=C](30) + τ[C–C]$(26) + ω[C–N](16) + τ[C–C]*(6) + τ[C=C]*(6)
91	–	–	τ [C–N](46) + τ [N=C](35) + φ[N=C–C](7)	111	–	–	φ[C–N=C](28) + ω[N=C](20) + τ[C–C]*(13) + τ[C=C]*(12) + τ[C–C]$(12)
66	–	–	τ[C–C]$(41) + τ[C–C]*(22) + τ[C=C]*(21) + ω[C–H](12)	64	–	–	τ[C–N](87) + φ[N=C–C](10)
65	–	–	τ[N=C](55) + τ[C–N](44)	60	–	–	τ [C–N](83)

Note: 1. [a]Ref.-[30], [b]Ref.[29], [c]Ref.-[45]; 2. All freq. are in cm^{-1}; 3. ν, φ, ω and τ denote stretch, angle bend, wag and torsion respectively; 4 * and $ stands for benzoid and quinoid ring respectively.

1324 and 1317 cm^{-1}. The mode calculated at 1327 cm^{-1} is associated with the C–H bending mode of benzene ring, at 1324 cm^{-1} is associated with C–H bending mode of the quinoid ring and the mode at 1317 cm^{-1} is a mix C–H bending mode of benzoid and quinoid ring.

Mode calculated at 1176 cm^{-1} is associated with C–H bending mode of benzoid ring and normal mode calculated with 1154 cm^{-1} is with quinoid ring and these modes corresponds to the Wilson vibration 9a of p-disubstituted benzenes which are assigned to Raman peaks at 1164 and 1157 cm^{-1} respectively. The 18b vibration appears at 1098 cm^{-1} in the IR spectra and is assigned to the calculated normal mode at 1087 cm^{-1}. The mode calculated at 1008 cm^{-1} is a mixed mode of CCC bend and C–H bending of the benzene ring and is assigned to the IR weak band at 1006 cm^{-1}, can be seen as 18a Wilson vibrational mode.

C–N Stretching Vibrations

In the Raman as well as in IR spectra the most intense band is at 1480 cm^{-1}, this band is not seen in the reduced form of PANI and its oligomers and hence we can conclude that this band should be characteristic of oxidized form. Consequently, this band is assigned to C=N stretching mode. In the p-disubstituted benzenes the C–X (X=N in the PANI) stretching vibrational mode are no. 13 and 7a. In the spectra of PANI PNB form we have a band at about 1211 (IR) and 1215 cm^{-1} (Raman) that can be assigned to the calculated modes at 1217 and 1214 cm^{-1} respectively that have N=C & N–C stretching vibrations as the main contribution in their PED.

C–H Out-of-Plane Vibrations

In the p-disubstituted benzenes normal vibration 17a, 10a, 10b and 11 are classified as C–H out of plane vibrations. The IR dip at 965 cm^{-1} is assigned to the calculated modes at 978 & 966 cm^{-1} and can be seen as 17a vibration of the quinoid and benzoid ring respectively. In the IR spectra we have a dip around 950 and 931 cm^{-1} and these dips can be seen as 10b vibration of the quinoid and benzoid rings and are assigned to the calculated modes at 948 and 938 cm^{-1}. Bands at 845(IR), 790 cm^{-1} (Raman) are found to be associated with 10a vibration of the benzoid and quinoid ring respectively and the IR bands at 775 and 750 cm^{-1} are with Wilson vibration no. 11.

Radial Skeletal Vibrations

Radial skeletal vibrations of the p-disubstituted benzenes are the Wilson no. 1, 12, 6a & 6b. Wilson no. 1 and 12 both vibrations are C–N stretch sensitive and appear in mix. After examining the PED of the modes calculated at 776 (assigned to the IR band at 775 cm^{-1}) & 708 cm^{-1} are completely mixed modes having radial skeletal vibrations of the benzoid and quinoid rings and can be seen as vibration 1 & vibration 12. The mode calculated at 614 and 519 cm^{-1} (assigned to the bands at 624(IR) and 518(Raman) cm^{-1} respectively) are associated with the radial skeletal vibrations of the benzoid (Wilson no. 6b) and quinoid ring (Wilson no. 6a) respectively.

Out-of-Plane Skeletal Vibrations

Wilson no. 4, 16a & 16b corresponds to the out-of-plane skeletal vibrations in p-disubstituted benzenes. Normal mode calculated at 718 cm^{-1} has C–C=C out-of-plane bend of the benzene ring and C–N stretch vibrations as the major component in the PED and can be seen as Wilson vibration 4. In PANI PNB form, IR dip at 498 cm^{-1} is assigned to the calculated mode at 498 and 482 cm^{-1}. Calculated normal mode at 498 cm^{-1} is purely associated with the out-of-plane skeletal vibrations of the quinoid ring (Wilson no.16b) and the mode calculated at 482 cm^{-1} is associated with out-of-plane skeletal vibrations of both benzoid and quinoid rings (Wilson no.16b). The band at 410 cm^{-1} is assigned to the calculated normal mode at 406 cm^{-1} and corresponds to 16a Wilson no. and is associated mainly with out-of-plane skeletal vibrations of the quinoid and benzoid rings.

C–N in-Plane Bending Vibrations

In p-disubstituted benzenes C–X (for PANI X=N) in-plane bending normal vibrations are 9b and 15. The Raman peak at 359 cm^{-1} is assigned to the calculated normal mode at 358 cm^{-1} and is associated with C–N in-plane bending vibrations and can be seen as 9b Wilson vibration. Calculated normal mode at 190 cm^{-1} is also C–N in-plane bending vibration and is assigned as Wilson vibration 15.

C–N Out-of-Plane Vibrations

Raman peak at 374 cm^{-1} is assigned to the calculated mode at zone boundary at 379 and 374 cm^{-1}, these modes are associated with out-of-plane C–N bend and can be seen as Wilson no. 5 vibration. Raman peak at 212 cm^{-1} is assigned to the modes calculated at 224, 202 & 200 cm^{-1}. These modes are associated with out-of-plane C–N vibrations and can be seen as 17b in-phase out-of-plane C–N vibration.

Dispersion Curves

In general, the IR absorption spectra and Raman spectra from polymeric systems are complex and can not be unravelled without the full knowledge of the dispersion curves. Besides providing knowledge of density-of-states, dispersion curves give information on the extent to the coupling of a mode along the chain in the ordered state. Dispersion curves and frequency distribution function are important for an understanding of thermodynamical and elastic

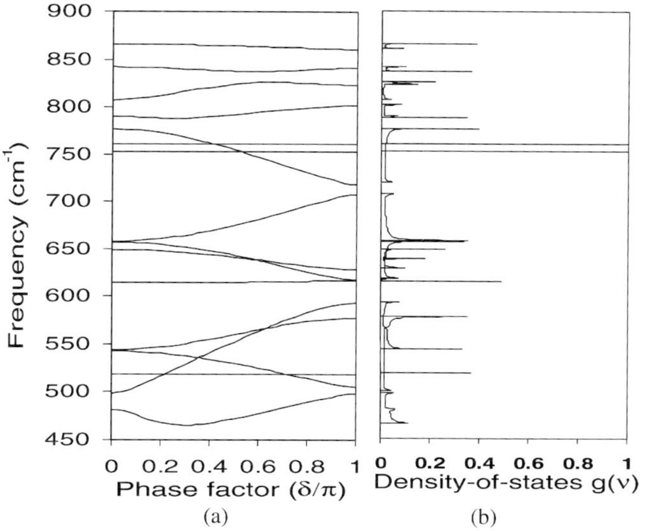

Figure 2.
(a) Dispersion curves and (b) Density of states of PANI PNB form (900–450 cm^{-1}).

properties of solids. Also a study of these is necessary to appreciate the origin of both symmetry independent and symmetry dependent spectral features. The dispersion curves of pernigraniline are shown in Figure 2(a) for frequencies 450–900 cm^{-1} and in Figure 3(a) 0–450 cm^{-1} for frequencies. The dispersion curves above 900 cm^{-1} are not shown because they have negligible dispersion.

The modes of PANI PNB have some special features like crossing. All such modes showing crossover are listed in Table 3 along with the PED and the δ values at which these features occur. To ascertain whether it is a crossing or a repulsion calcu-

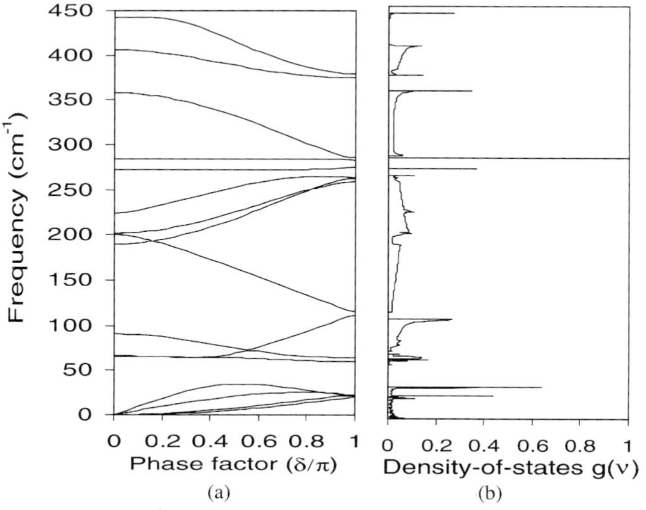

Figure 3.
(a) Dispersion curves and (b) Density of states of PANI PNB from 0–450 cm^{-1}.

Table 3.
Crossing between the pair of modes of PANI PNB form

Freq.(δ = 0)	Before Crossing				After Crossing		
	δ#/Π	δ*/Π	Freq.	PED	δ*/Π	Freq.	PED
i							
776	0.419	0.40	762	ν[C-C]$ (15) + ν[C-N] (12) + φ[C-C=C]*(11) + φ[C-C=C]$(10) + φ[C-C-C]$ (8) + φ[C-N=C] (6) + φ[N=C-C] (6) + ν[N=C] (6)	0.45	761	ω[C-H] (84) + τ[C-C]* (7) + τ[C=C]*(7)
761	0.419	0.40	761	ω[C-H] (83) + τ[C=C]*(7) + τ[C-C]*(7)	0.45	758	ν[C-C]$(15) + ν[C-N] (12) + φ[C-C=C]*(11) + φ[C-C=C]$(10) + φ[C-C-C]$(8) + φ[C-N=C] (7)
ii							
761	0.532	0.50	755	ν[C-C]$ (15) + ν[C-N](12) + a[C-C=C]* (11) + a[C-C=C]$(9) + a[C-C-C]$(8) + a[C-N=C](7) + ν[N=C] (6) + a[N=C-C](6)	0.55	752	ω[C-H](86) + τ[C-C]$(14)
752	0.532	0.50	752	ω[C-H] (86) + τ[C-C]$ (14)	0.55	751	ν[C-C]$(14) + ν[C-N]*(12) + a[C-C=C]*(11) + a[C-C=C]$(9) + a[C-N=C](8) + a[C-C-C]$(8) + ν[N=C](6) + a[N=C-C] (6)
iii							
657	0.589	0.55	641	φ[C-N=C](24) + ω[C-N] (20) + ω[N=C](11) + φ[C-C=C]*(7) + ω[C-H] (6)	0.65	636	φ[N=C-C] (54) + φ[C-C-N](9) + φ[C=C-N] (8) + τ[C-N] (7) + ν[C-C]$(6) + τ[N=C](6)
649	0.589	0.55	640	φ[N=C-C] (52) + φ[C=C-N](10) + φ[C-C-N](9) + τ[C-N](6) + ν[C-C]$(6) + τ[N=C] (6)	0.65	635	φ[C-N=C] (23) + ω[C-N](21) + ω[N=C](11) + φ[C-C=C]* (8) + ω[C-H] (7)
iv							
544	0.629	0.60	565	φ[N=C-C] (30) + φ[C-C-N](16) + φ[C=C-N] (15) + τ[N=C](9) + τ[C-N](9) + ν[C-C]$(8) + ν[C=C]$(5)	0.65	569	ω[C-N](14) + φ[C-N=C](13) + ω[N=C](11) + φ[C-C=C]*(8) + ω[C-H] (8) + φ[C-C-C]$(7) + φ[N=C-C](6) + ν[C-N] (5)
543	0.629	0.60	563	ω[C-N](15) + φ[C-N=C] (12) + ω[N=C](10) + φ[C-C=C]*(9) + ω[C-H] (8) + φ[C-C-C]$(7) + φ[N=C-C](6) + ν[C-N](5)	0.65	567	φ[N=C-C] (29) + φ[C=C-N] (16) + φ[C-C-N] (15) + τ[N=C] (9) + τ[C-N](9) + ν[C-C]$(8) + ν[C=C]$ (5)
v							
543	0.357	0.35	536	ω[C-N] (27) + ω[N=C] (26) + τ[C=C]$(11) + τ[C-C]*(9) + τ[C=C]*(8)	0.40	541	ω[C-N] (16) + φ[C-C=C]* (11) + φ[C-N=C] (9) + ω[N=C] (8) + φ[C-C-C]$ (8) + ω[C-H] (6) + ν[C-N](6) + φ[N=C-C] (6)
519	0.357	0.35	535	ω[C-N] (17) + φ[C-C=C]*(10) + φ[C-N=C] (9) + ω[N=C] (8) + φ[C-C-C]$(8) + ν[C-N] (7) + ω[C-H] (6) + φ[N=C-C] (6)	0.40	534	ω[C-N] (27) + ω[N=C] (26) + τ[C=C]$ (11) + τ[C-C]*(9) + τ[C=C]* (8)
vi							
519	0.713	0.70	520	ω[C-N] (27) + ω[N=C] (26) + τ[C=C]$ (9) + τ[C-C]*(9) + τ[C=C]* (8)	0.75	519	φ[C-C=C]$ (66) + φ[H-C=C]$ (14) + φ[C-C-H]$ (9) + ν[C-C]$(9)
498	0.713	0.70	519	φ[C-C=C]$(66) + φ[H-C=C]$ (14) + φ[C-C-H]$ (9) + ν[C-C]$ (9)	0.75	519	ω[C-N] (27) + ω[N=C] (26) + τ[C-C]* (9) + τ[C=C]$ (8) + τ[C=C]* (8)
vii							
519	0.219	0.20	519	φ[C-C=C]$ (66) + φ[H-C=C]$(14) + φ[C-C-H]$(9) + ν[C-C]$(9)	0.25	523	ω[C-N](19) + φ[C-C=C]*(11) + φ[C-C-C]$(7) + φ[C-N=C] (7) + ν[C-N](7) + ω[N=C](7) + φ[N=C-C](6) + ω[C-H](6)
498	0.219	0.20	517	ω[C-N](20) + φ[C-C=C]*(10) + ν[C-N] (7) + φ[C-C-C]$ (7) + φ[C-N=C](7) + ω[N=C] (6) + φ[N=C-C] (6) + ω[C-H] (6)	0.25	519	φ[C-C=C]$ (66) + φ[H-C=C]$ (14) + φ[C-C-H]$(9) + ν[C-C]$ (9)
viii							
202	.826	0.80	250	τ[C-C]*(15) + τ[C=C]*(14) + τ[C-C]$ (11) + ω[C-H] (7) + φ[C-N=C](7) + φ[C-C=C]* (6)	0.85	254	φ[C-C-N] (26) + φ[N=C-C] (24) + φ[C=C-N] (23) + τ[N=C] (7)

(*Continues*)

Table 3.
(Continued)

Freq.($\delta = 0$)	Before Crossing				After Crossing		
	$\delta^{\#}/\Pi$	δ^{*}/Π	Freq.	PED	δ^{*}/Π	Freq.	PED
200	.826	0.80	249	φ[C=C-N] (26) + φ[N=C-C] (25) + a[C-C-N] (23) + τ[N=C](7)	0.85	253	τ[C-C]*(16) + τ[C=C]* (16) + τ[C-C]$ (10) + a[C-N=C] (8) + ω[C-H] (7) + a[C-C=C]* (5)
ix							
200	0.163	0.15	194	φ[C-N=C] (36) + τ[C-C]$ (18) + ω[N=C](15) + τ[C-C]*(9) + ω[C-N](8) + τ[C=C]*(8)	0.20	195	φ[N=C-C] (31) + φ[C=C-N] (26) + φ[C-C-N] (26)
189	0.163	0.15	193	φ[N=C-C] (31) + φ[C=C-N] (26) + φ[C-C-N] (26)	0.20	190	φ[C-N=C] (34) + τ[C-C]$ (18) + ω[N=C] (14) + τ[C-C]*(9) + ω[C-N](8) + τ[C=C]*(8)
x							
91	0.559	0.55	73	τ[C-N] (56) + τ[N=C] (22) + φ[N=C-C] (9) + φ[C=C-N] (5)	0.60	76	τ[C-C]$ (24) + φ[C-N=C] (22) + τ[C-C]* (13) + τ[C=C]*(12) + ω[N=C] (8) + ω[C-H] (7) + ω[C-N] (6)
66	0.559	0.55	72	τ[C-C]$ (26) + φ[C-N=C] (18) + τ[C-C]* (14) + τ[C=C]* (13) + ω[C-H] (7) + ω[N=C] (7)	0.60	71	τ[C-N] (60) + τ[N=C] (18) + φ[N=C-C] (10)
xi							
66	0.348	0.30	65	τ[N=C] (54) + τ[C-N] (40)	0.40	65	τ[C-C]$(35) +) τ[C-C]*(19) + τ[C=C]* (18) + ω[C-H] (10) + φ[C-N=C] (7)
65	0.348	0.30	64	τ[C-C]$ (37) + τ[C-C]*(20) + τ[C=C]*(19) + ω[C-H] (11)	0.40	64	τ[N=C] (57) + τ[C-N] (39)
xii							
66	0.196	0.15	65	τ[C-C]$ (39) + τ[C-C]*(22) + τ[C=C]* (20) + ω[C-H] (11)	0.25	65	τ[N=C] (53) + τ[C-N] (41)
65	0.196	0.15	65	τ[N=C] (53) + τ[C-N] (42)	0.25	65	τ[C-C]$ (38) + τ[C-C]* (21) + τ[C=C]* (19) + ω[C-H] (11)
xiii							
0.00	0.966	0.95	22	τ[C-C]$ (23) + τ[C-C]*(18) + φ[C-N=C](17) + τ[C=C]* (17) + ω[C-H] (9) + ω[N=C](8) + ω[C-N] (8)	0.97	22	τ[N=C] (49) + τ[C-N](36)
0.00	0.966	0.95	21	τ[N=C] (47) + τ[C-N](38)	0.97	22	τ[C-C]* (20) + τ[C-C]$(19) + τ[C=C]*(18) + φ[C-N=C](17) + ω[C-H] (9) + ω[C-N] (8) + ω[N=C] (8) +
xiv							
0.00	0.883	0.85	26	τ[C-C]$ (26) + φ[C-N=C] (18) + τ[C-C]* (16) + τ[C=C]* (14) + ω[N=C] (9) + ω[C-H] (9) + ω[C-N] (8)	0.90	25	τ[C-N] (46) + τ[N=C] (43)
0.00	0.883	0.85	25	τ[C-C]$ (26) + φ[C-N=C] (18) + τ[C-C]* (16) + τ[C=C]* (14) + ω[N=C] (9) + ω[C-H] (9) + ω[C-N] (8)	0.90	25	τ[C-C]$ (25) + φ[C-N=C] (17) + τ[C-C]* (16) + τ[C=C]* (15) + ω[C-H] (9) + ω[N=C] (9) + ω[C-N] (8)

Note: 1. δ# corresponds to crossing points; 2. δ^{*} corresponds to the points before/after crossing.

lations at very close intervals of $\delta = .001\ \pi$ have been performed and it was found that the modes cross-over. From symmetry considerations, it can be shown that when the approaching modes belong to different symmetry species then they can crossover. Since PANI PNB form has mirror plane of symmetry along the chain axis, hence crossing are permissible.[46] Therefore, no two dispersion curves both of which belong to the same symmetry species can cross because this would imply the existence of two modes of vibrations with the same symmetry species and same frequency. This is also obvious from the Table 3 in which we have shown the pair of modes, which cross, belongs to different symmetries, i.e. in plane and out-of-plane modes. Such cross-over occurs in the case of pair of modes at 776 & 761 cm^{-1}, 776 & 752 cm^{-1}, 657 &

649 cm^{-1}, 544 and 498 cm^{-1}, 543 & 498 cm^{-1}, 543 & 519 cm^{-1}, 519 & 498 cm^{-1}, 202 & 189 cm^{-1}, 91 & 66 cm^{-1}, 66 & 65 cm^{-1} and acoustic modes which cross each other once or twice. The intersection phenomenon of modes can be regarded as an inelastic collision of two phonons in energy momentum space.

Study of dispersion curves enables the calculation of frequency distribution function. The frequency distribution function (density-of-states) shows how the energy is distributed among the various branches of the normal modes. Figure 2(b) and 3(b) show the plots of density-of-states versus frequency as obtained from the dispersion curves. The peaks of the frequency distribution curves correspond to regions of high density-of-states (Von Hove type singularities).

As our calculations have been made for an isolated molecular chain, the interpretation of IR \ Raman spectral and theoretical calculations are subjected to certain limitations. A complete interpretation of the spectra requires calculations of a three dimensional system where interactions play an important role. The only additional correction in this work is to check the force field through the deuterated or isotopically substituted species of pernigraniline.

Conclusion

In this paper, a detailed investigation of the vibrational properties of PANI PNB form has been performed. The vibrations have been assigned based on para-disubstituted benzene derivatives which provides a better understanding of the spectra. Normal modes and set of force constants for the PANI PNB form are being obtained by Urey – Bradley force field and several new assignments are being reported. All the characteristic features of the dispersion curves such as regions of high density-of-states, cross-over and repulsion have been well interpreted from the vibrational dynamics of PANI PNB form. Similar studies on the lecuromeldine and emeraldine salts are in progress with the aim of determining the modifications induced by primary and secondary doping.

[1] L. A. Ray, G. E. Asturias, D. L. Kershur, A. F. Richter, A. G. MacDiarmid, A. J. Epstein, *Synth. Met.* **1989**, *29*, E141.
[2] K. G. Neoh, E. T. Kang, S. H. Khor, K. L. Tan, *Polym. Degrad. Stab.* **1990**, *27*, 107.
[3] K. L. Tan, B. T. G. Tan, E. T. Kang, K. G. Neoh, *J. Chem. Phys.* **1991**, *94*, 5382.
[4] H. Letheby, *J. Chem. Soc.* *1862*(15), 161.
[5] de Surville, M. Jozefowicz, L. T. Yu, J. Perichon, R. Buvet, *Electrochim. Acta* **1968**, *13*, 1451.
[6] A. F. Diaz, J. A. Logan, *J. Electroanal. Chem.* **1980**, *111*, 111.
[7] T. Kobayashi, H. Yoneyama, H. Tamura, *J. Electroanal. Chem.* **1984**, *177*, 281.
[8] E. M. Genies, C. Tsintavis, *J. Electroanal. Chem.* **1984**, *195*, 109.
[9] A. G. Macdiarmid, S. L. Mu, N. L. D. Somasiri, W. Wu, *Mol. Cryst. Liq. Cryst.* **1985**, *121*, 181.
[10] E. M. Genies, A. A. Syed, C. Tsintavis, *Mol. Cryst. Liq. Cryst.* **1985**, *121*, 187.
[11] W. R. Salaneck, I. Lundstrom, B. Liedberg, M. A. Hasan, R. Erlandsson, P. Konradsson, A. G. Macdiarmid, N. L. D. Somasiri, *Springer Ser. Solid State Sci.* **1985**, *63*, 218.
[12] L. W. Schacklette, M. Maxfield, S. Gould, J. F. Wolf, T. R. Jow, R. H. Baughman, *Synth. Met.* **1987**, *18*, 611.
[13] K. Okabayashi, F. Goto, K. Abe, T. Yoshida, *Synth. Met.* **1987**, *18*, 365.
[14] A. G. Macdiarmid, L. S. Yang, W. S. Haung, B. D. Humphrey, *Synth. Met.* **1987**, *18*, 611.
[15] E. M. Genies, P. Hany, Ch. Santier, *J. Appl. Electrochem.* **1988**, *18*, 751.
[16] W. Wang, G. Li, C. Li, F. Wang, *J. Appl. Electrochem.* **1988**, *18*, 751.
[17] E. W. Paul, A. J. Ricco, M. S. Wrighton, *J. Phys. Chem.* **1985**, *89*, 1441.
[18] A. Boyle, E. M. Genies, M. Lapkowski, *Proceedings of ICSM'88, Santa Fe/USA, Synth. Metals* **1989**, *63*, 211.
[19] T. Kobayashi, H. Yoneyama, H. Tamura, *J. Electroanal. Chem.* **1984**, *161*, 419.
[20] A. Kitani, J. Yano, K. Sasaki, *J. Electroanal. Chem.* **1986**, *209*, 227.
[21] B. Aurian-Blajeni, I. Taniguchi, J. O. M̀. Bochris, *J. Electroanal. Chem.* **1983**, 149, 291.
[22] A. G. Green, A. E. Woodhead, *J. Chem. Soc. Trans.* **1910**, *97*, 2388.
[23] A. G. Green, A. E. Woodhead, *J. Chem. Soc. Trans.* **1912**, *101*, 1117.
[24] A. Ray, A. F. Richter, A. G. Macdiarmid, A. J. Epstein, *Synth. Met.* **1989**, *29*, 151.
[25] A. Ray, G. E. Asturias, D. L. Kershner, A. F. Richter, A. G. Macdiarmid, A. J. Epstein, *Synth. Met.* **1989**, *29*, E141.; A. Ray, A. F. Richter, A. G. Macdiarmid, A. J. Epstein, *Synth. Met.* **1989**, 29, E151.
[26] T. Shimanouchi, *J. Chem. Phys.* **1949**, *17*, 245.
[27] H. G. Urey, C. A. Bradley, *Phys. Rev.* **1931**, *38*, 1969.

[28] S. Quillard, G. Louarn, S. Lefrant, A. G. Macdiarmid, *Physical Riew B* **1994**, *50*, 17,12496.
[29] M. Cochet, G. Louarn, S. Quillard, M. I. Boyer, J. P. Buisson, S. Lefrant, *J. Raman Spectrosc.* **2000**, *31*, 1029.
[30] M. I. Boyer, S. Quillard, E. Rebourt, G. Louarn, J. P. Buisson, A. Monkman, S. Lefrant, *J. Phys. Chem. B* **1998**, *102*, 782.
[31] W. Qian, N. G. Mirikin, S. Krimm, *Chem. Phys Lett.* **1999**, *315*, 125.
[32] G. Varsanyi, *Vibrational spectra of Benzene Derivatives*, Academic Press, New York **1969**, Chapter 3.
[33] F. R. Dollish, W. G. Fateley, F. F. Bentley, *Characteristic Raman, Frequencies of Organic Compounds*, Wiley, New York **1974**.
[34] N. B. Colthup, L. H. Daly, S. E. Wiberley, *Introduction to Infrared and Raman Spectroscopy*, Acedemic Press, Boston **1990**, Chapter 8.
[35] A. Stojiljkovic, D. H. Whiffen, *Spectrochim. Acta* **1958**, *12*, 47.
[36] A. Stojiljkovic, D. H. Whiffen, *Spectrochim. Acta* **1958**, *12*, 57.
[37] C. Garrigou-Lagrange, J. M. Lebas, M. L. Josien, *Spectrochim. Acta* **1958**, *12*, 305.
[38] D. H. Whiffen, *Spectrochim. Acta* **1955**, *7*, 253.
[39] R. R. Randle, D. H. Whiffen, *Molecular Spectroscopy*, Institute of Petroleum, London **1955**, p 111.
[40] E. B. Wilson, Jr., *J Chem Phys* **1939**, *7*, 1047.
[41] E. B. Wilson, Jr., *J Chem Phys* **1941**, *9*, 76.
[42] P. W. Higgs, *Proc Roy Soc London A* **1953**, *220*, 472.
[43] B. Mannfors, K. Palmo, S. Krimm, *J. Mol. Struct.* **2000**, *556*, 1.
[44] W. T. King, I. M. Mills, B. L. Crawford, *J. Chem. Phys.* **1957**, *27*, 455.
[45] H. Neugebauer, A. Neckel, N. S. Sariciftci, H. Kuzmany, *Synth. Metals* **1989**, *29*, E185.
[46] D. I. Bower, W. F. Maddams, *The vibrational spectroscopy of polymers*, Cambridge University Press, Cambridge **1989**.

DOI: 10.1002/masy.200850513

Development of Methods to Determine the Infrared-Optical Properties of Polymer Films

Gernot Oreski,[*1] Daniel Tscharnuter,[1] Gernot M. Wallner[2]*

Summary: In this paper, two methods for the determination of infrared optical properties of thick polymer films, based on FTIR spectroscopy, were implemented and used. Complex index of refraction data were generated for various ethylene copolymer films. Transmittance and reflectance spectra were measured in the mid infrared range using a gold-coated 100 mm-diameter integrating sphere. For the investigated films n and k values ranging from 1.3 to 1.6 and from 10^{-4} to 0.25 were determined, respectively. Regarding n, a good agreement was obtained for both methods, the transmittance/reflectance procedure (T/R method) used for transparent and semi-transparent regions, and Single Substractive Kramers-Kronig (SSKK) algorithm applied for non-transparent regions. The highest k values were determined for the CH_2 stretching vibration. The k values are dependent on comonomer content and film thickness. The combination of both methods allows for an accurate determination of n and k in the entire IR region relevant for solar application.

Keywords: complex index of refraction; ethylene copolymer films; FTIR spectroscopy; Kramers-Kronig; solar applications

Introduction

Polymeric materials are already in use for many different solar energy related components. Concerning solar thermal systems plastics play a decisive role as transparent cover and/or convection suppressing material. For such applications the infrared (IR) optical properties of plastics are of prime importance. Contrary to silica glass, which is almost completely opaque to the IR spectrum, the transparency of plastics to the IR spectrum is ranging from highly transparent to almost completely opaque depending on the molecular structure of the polymer. While for solar thermal systems with a black absorber an infrared radiation absorbing cover material is favorable, for solar thermal systems with a selective absorber a non-absorbing cover material is advantageous. For solar thermal applications, an accurate prediction of infrared optical properties as a function of film thickness and temperature is crucial.[1–3] While for low IR absorbing materials (e.g. polymers) often simple methods assuming a constant real part of the refractive index are used, for high IR absorbing materials (e.g. glass, ceramics) comprehensive methods allowing for the determination of both, the real and imaginary part of refractive index (n,k), are a prerequisite.[3,4]

To determine n and k in the infrared range, several methods are described in the literature.[3–14] Currently, mainly spectroscopic ellipsometry (SE) is used for the evaluation of n and k over a wide range of wavelengths.[5–10] By systematic combination of SE with Fourier transform spectroscopy (FT) and development of accurate calibration procedures, the range of application was extended from UV-visible to the infrared.[5] The ellipsometric data are

[1] Polymer Competence Center Leoben GmbH, Roseggerstraße 12, A-8700 Leoben, Austria
Fax: +43 3842 429 626;
E-mail: oreski@pccl.at
[2] Institute of Materials Science and Testing of Plastics, University of Leoben, Franz-Josef-Straße 18, A-8700 Leoben, Austria

usually recorded over the whole spectral range at various angles of incidence. An optical model, which assumes geometry and microstructure, is then applied to calculate the complex index of refraction.[6] A significant drawback of the SE procedure is the necessity to investigate thin films, which are coated on a substrate (e.g. silicon wafers[7] or gold-coated glass slides[8]), or prepared as freestanding films of 12–15 μm thickness.[9] However, commercial polymer films, currently used for greenhouse or transparent insulation applications, are thicker with values ranging from 50 to several hundred microns. Roodenko et al. described the difficulties, which can occur by investigating the infrared optical properties of thick films by FTIR-SE, and the effect on ellipsometric parameters.[10]

In the past, spectroscopic methods were used for determination of the refractive index.[4,11–14] These methods, often requiring special sample preparation techniques, were limited to spectral regions with certain optical properties (transparent, semi-transparent or non-transparent) or based on assumptions such as a constant real part of refractive index. In a recent study, an advanced spectroscopic method, the transmission fringe depth method (RTFD),[11] has been developed for the determination of n and k. However, this method, which is based on the measurement of directional and hemispherical transmittance and reflectance spectra, is applicable only to thin films where interference effects can be observed.

The overall objective of this paper is to implement and apply spectroscopic methods to determine the infrared optical properties of thick polymer films for solar energy applications in all regions of the spectral range (transparent, semi-transparent and non-transparent regions). While for the transparent and semitransparent regions a simple method based on hemispherical transmittance and reflectance spectra was applied, for the non-transparent regions of the IR spectra a more sophisticated algorithm based on the Kramers-Kronig (KK) relation requiring was implemented and used.[3,15]

Experimental Part

Polar ethylene copolymer film types with comonomer contents between 1.7 and 30 wt-% and three different comonomer types (acids, acrylates and acetates) were investigated (Table 1). As to the acid comonomers acrylic acid (AA) and methacrylic acid (MAA) were chosen. Regarding acrylate comonomers butylacrylate (BA), ethylacrylate (EA) and methylacrylate (MA) are currently in use for ethylene (E) copolymer materials. Furthermore, widely used ethylene vinylacetate copolymers (EVA) as well as a terpolymer material with two different comonomers (acrylic acid and butylacrylate (AA/BA) were investigated. In comparison, a non-polar linear low-density polyethylene (LLDPE) was investigated.

The material resins were supplied by Arkema (EBA), Basell (EAA/BA), BP (EMAA), Dow (EAA, LLDPE), DuPont

Table 1.
Investigated ethylene (E) copolymers (material type, comonomer unit, comonomer content and film thickness).

Material type	Comonomer	Content, wt%	Film thickness
EAA	acrylic acid	9.7	125 μm
EMAA	methacrylic acid	1.7, 8.7	110 μm, 90 μm
EBA	butylacrylate	7	125 μm
EEA	ethylacrylate	12	115 μm
EMA	methylacrylate	9	100 μm
EVA	vinylacetate	9, 18	70 μm, 125 μm
EAA/BA	acrylic acid	AA 4	125 μm
	butylacrylate	BA 7	
LLDPE			42 μm, 105 μm

(EMAA, EEA, EMA, EVA). Films with various thicknesses ranging from 40 to 150 μm were cast on a Rosendahl RO400 single screw extruder (Pischelsdorf, Austria) with a chill roll unit of SML Maschinengesellschaft mbH (Lenzing, Austria).

The transmittance (T) and reflectance (R) spectra over the mid-IR range from 4000 to 580 cm^{-1} (2.5 to 17.2 μm) were recorded using a Perkin Elmer Spectrum GX (Perkin Elmer; Überlingen, Germany) with an external 100 mm diameter, gold-coated integrating sphere.

Mathematical Modelling

To calculate the complex index of refraction (n-ik), at least two input data are necessary to derive the two unknown quantities n and k. For transparent and semitransparent regions of the spectral range of the investigated sample, the infrared optical properties can be calculated by measurements of spectral transmittance (T) and reflectance (R) of a single sample.[3] Therefore this method is called T/R method. It is assumed that samples are plane-parallel and sample properties are measured normal to the sample surface. With multiple reflections taken into accounts, T and R can be expressed by surface reflectance ρ and the internal transmittance τ[16]:

$$R = \rho\left[1 + \left(\frac{(1-\rho)^2\tau^2}{1-\rho^2\tau^2}\right)\right] \quad (1)$$

$$T = \tau\frac{(1-\rho)^2}{1-\rho^2\tau^2} \quad (2)$$

First, to determine the complex refractive index, the surface reflectance ρ and the internal transmittance τ values are calculated from R and T:

$$\rho = \frac{1}{2(2-R)}\left[C - \sqrt{C^2 - 4R(2-R)}\right] \quad (3)$$

with

$$C = 1 + 2R + T^2 - R^2 \quad (4)$$

$$\tau = \left(\frac{R}{\rho} - 1\right)\frac{1}{T} \quad (5)$$

Second step is the determination of k. The absorption coefficient a can be calculated from τ and the sample thickness l, k is then derived from a and the wavelength λ:

$$a = -\frac{\ln\tau}{l} \quad (6)$$

$$k = \frac{a\lambda}{4\pi} \quad (7)$$

Finally, the real part of the complex index of refraction is given by:

$$n = \frac{1+\rho}{1-\rho} + \sqrt{\frac{4\rho}{(1-\rho)^2} - k^2} \quad (8)$$

The T/R method is sensitive to n and k as long as T is larger than zero.

If the sample material is non-transparent and the experimental determination of two independent optical properties i.e. the reflectance and transmittance is not possible, the complex refractive index may be calculated from a single quantity such as the reflectance by a Kramers-Kronig integral. If the reflectance has been measured, the phase of the complex reflection coefficient is given by

$$\varphi(\nu) = -\frac{\nu}{\pi}P\int_0^\infty \frac{\ln(R(x))}{(x^2-\nu^2)}dx \quad (9)$$

Because in practice the measurement can only be carried out over a finite range of wave numbers, extrapolation procedures have been developed for the inaccessible wave numbers. However, the result of the integration tends to depend on the kind of extrapolation.[15] To reduce the ambiguity associated with this dependence, Ahrenkiel[15] introduced the Singly Subtractive Kramers-Kronig method (SSKK). The idea is to use the phase known at one wave number, which is obtained by reflectance and transmittance measurements, to improve the convergence of the Kramers- Kronig integral and therefore make it less dependent on the extrapolation. By using the known phase φ_0 at a wave number ν_0 in the

Kramers-Kronig integral, Ahrenkiel derived the subtractive Kramers-Kronig integral

$$\varphi(\nu) = \frac{\nu}{\nu_0}\varphi(\nu_0) + \frac{\nu(\nu_0^2 - \nu^2)}{\pi} P \int_0^\infty \frac{\ln(R(x))}{(x^2 - \nu_0^2)(x^2 - \nu^2)} dx \quad (10)$$

By using cubic splines to represent ln(R(x)) the integrand is a proper rational function on each interval defined by the spline nodes and therefore the integral can be simplified by partial fraction decomposition,

$$\frac{ax^3 + bx^2 + cx + d}{(x^2 - \nu_0^2)(x^2 - \nu^2)} = \frac{A}{x - \nu} + \frac{B}{x + \nu} + \frac{C}{x - \nu_0} + \frac{D}{x + \nu_0} \quad (11)$$

where A, B, C and D depend on ν, ν_0 and the spline parameters a, b, c and d. For each of the four terms the standard integral as well as the Cauchy Principal Value integral, which needs to be taken when either ν or ν_0 are in the interval, can be solved in terms of logarithms.

However, the SSKK method is sensitive to n in the entire spectral region, but insensitive to k in the transparent and semitransparent region.[3]

Results and Discussion

For greenhouse or transparent insulation wall applications it is crucial to absorb the infrared radiation as well as possible in order to keep the heat radiation low. Thus, high IR absorption is required in the range between 200 and 1500 cm^{-1}, in which a blackbody at 20 °C emits mostly. Figure 1 displays the transmittance and reflectance spectra over the medium IR range for selected ethylene copolymer films with film thicknesses ranging from 40 to 125 μm.

Significant absorptions in the investigated polar polyethylene copolymers can be attributed to CH_2 and CH_3 groups (between 3000 and 2800, at about 1470, 1380 and 720 cm^{-1}), to C=O double bonds (between 1750 and 1700 cm^{-1}) to C–O single bonds (between 1300 and 1000 cm^{-1}) and to hydroxyl (O–H) groups in acid copolymers (at about 950 cm^{-1}). Below 1000 cm^{-1} mainly molecular vibrations of the polymer chain contribute to the absorption. However, the molecular vibrations are rather weak for the investigated polymers. Close to the maximum of the emissive power function of a blackbody at 20 °C, the narrow absorption peak at 720 cm^{-1} of the CH_2 group can be found for all investigated polymers. CH_2 groups are rather weak dipoles with low absorbance in the relevant IR wavelength range. In contrast, C–O and O–H single bonds are stronger dipoles with a higher specific absorbance in the relevant range. However, in polar ethylene copolymers CH_2 groups are dominating.

The reflectance values of all investigated polymer films are rather constant over the mid-IR range, only little variations can be seen in the areas of high absorbance. The reflectance values are well below 10%.

Figure 2 shows the integrated transmittance (T), reflectance (R) and absorbance values (A), which were determined by averaging the spectral values, weighted by a blackbody emissive power function, P(ν,T) at a temperature of 20 °C, at each wave number.[11–14] The reflection values of the investigated polymer films show little difference and are ranging from 0.042 to 0.048, except for the 18% EVA film (125 μm), with a significant lower value of 0.031, and for the EAA 9.7% film (125 μm) with a significant higher value of 0.056. As expected, the absorption of the LLDPE films was lowest, with transmission values of 0.82 (42 μm film) and 0.74 (105 μm film). For EMAA 1.7% also low absorption was observed, which can be attributed to its low comonomer content. The ethylene acrylate copolymers (EBA; EMA; EEA) exhibit medium transparency in the IR region with absorption values ranging from 0.33 to 0.38, despite the different film thicknesses and

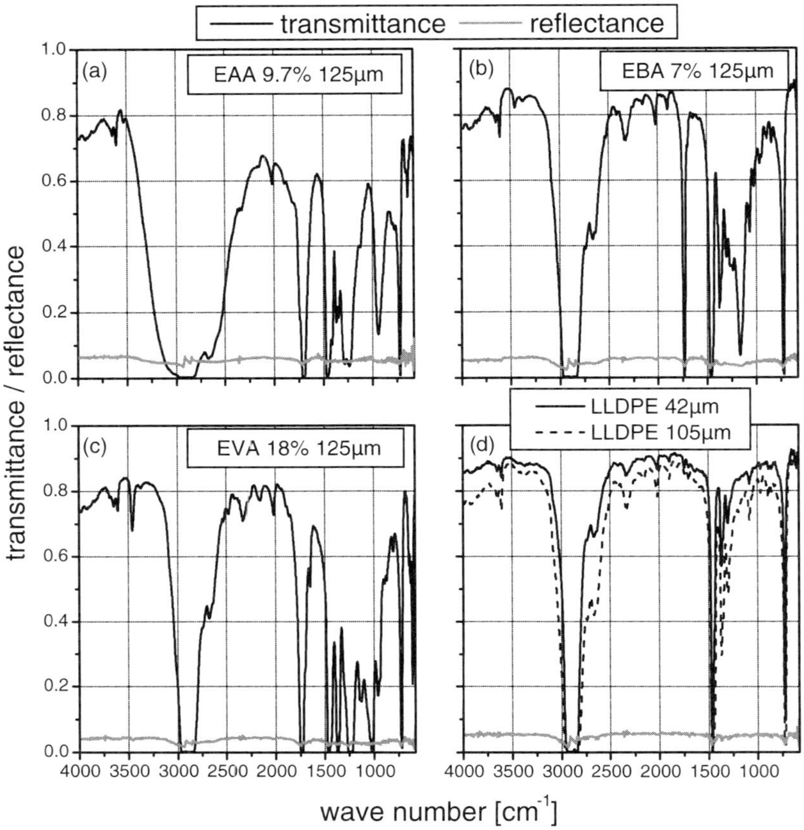

Figure 1.
Transmittance and reflectance spectra of (a) EAA 9.7% (125 μm film), (b) EBA 7% (125 μm film), (c) EVA 18% (125 μm film) and (d) LLDPE (42 and 105 μm film).

conomomer contents. In comparison, for the ethylene acid copolymers higher absorption values of about 0.41 for EMAA 8.7% (90 μm), 0.52 for EAA-BA 11% (125 μm) and EAA 9.7% (125 μm) were found. Due to the high amount of vinyl acetate groups EVA 18%, for these films the highest absorption values of about 0.6 were determined. As already reported in previous studies,[13,14] ethylene copolymers containing acid groups exhibit the highest infrared absorption values, provided that film thickness and comonomer content are identical. This can be attributed to material dependent differences in the IR spectra, especially in the region between 1300 and 900 cm^{-1}. Copolymers containing an acid group (EMAA, EAA, EAA-BA) show a medium peak at about 950 cm^{-1} (Figure 1a) caused by the carboxylic group of the polymer. This intense peak is located very next to the emittance maximum of a black body at 20 °C.[13,14]

Usually *T*, *R* and absorbance *A* are independent of film thickness in the transparent region of the spectral range. Banner et al.[17] defined the transparent region by an absorption coefficient $a < 10^0$ cm^{-1}. In the non-transparent region ($a > 10^3$ cm^{-1}), the internal transmittance becomes zero and reflectance and absorbance are almost surface properties. In the semitransparent region (absorption coefficient between 10^0 and 10^3 cm^{-1}) transmittance, reflectance and absorbance are strongly thickness dependent. In case of

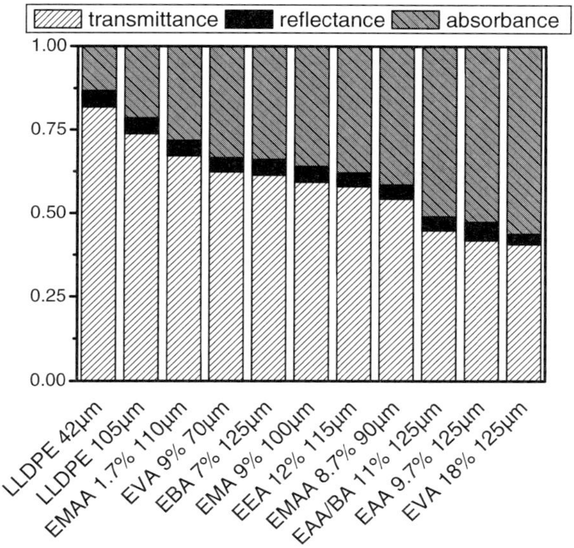

Figure 2.
Transmittance, reflectance and absorbance values, weighted with the emissive power function of a blackbody at 20 °C, for the investigated polymer films.

LLDPE (Figure 1(d)) it could be observed, that transmittance values show a higher dependence on film thickness than reflectance values, which are nearly identical.

In general, scattering in the infrared wavelength range of the material is neglected.[16] This assumption was confirmed by comparing hemispherical transmittance spectra measured with an integrating sphere to spectra measured directionally. The results for 125 µm thick EVA (18%) and EAA (9.7%) films are shown in Figure 3. Only small differences solely in areas of high transmittance can be seen.

In Figure 4 the validity regions of the T/R and the SSKK method is shown with reference to a transmittance spectrum of a 125 µm EAA 9.7% film. The threshold to distinguish the non-transparent from the semitransparent region was defined at a transmittance value of 0.01. Regarding the real part of the refractive index, the SSKK algorithm is sensitive over the whole spectral range and therefore the fastest method to determine *n* accurately. But, since the investigated ethylene copolymers are highly transparent in the IR range, also the T/R method is applicable nearly over the whole spectral range. Only small areas of total absorption have to be excluded. The determination of the imaginary part of the refractive index is more complicated. A combination of both methods (Figure 4)

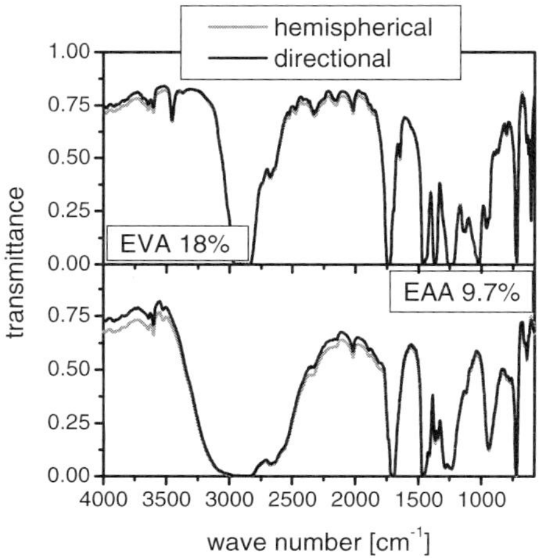

Figure 3.
Comparison of hemispherical (recorded with an integrating sphere) and directional transmittance spectra of 125 µm EVA 18% (top) and EAA 9.7% film (bottom).

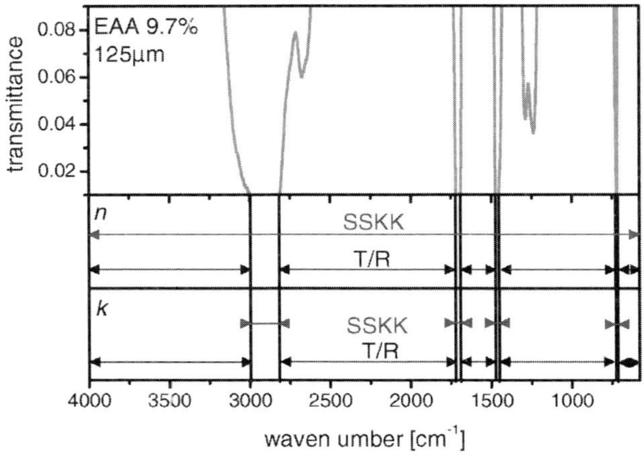

Figure 4.
Validity of the T/R and the SSKK method shown with reference to the transmittance spectra of 125 μm EAA 9.7% film.

has to be applied to calculate k over the whole spectral range.

The real part of the refractive index, determined by the SSKK algorithm, is shown in Figure 5 as a function of wave number for selected ethylene copolymer films. Whereas the highest values for *n* ranging from 1.5 and 1.6, were found for EAA 9.7%, EVA 18% exhibited the lowest values for *n* ranging from 1.3 and 1.4. The *n* values of the other ethylene copolymer films are lying between 1.4 and 1.5. The order of the n values is in agreement with the measured reflectance values.

For LLDPE it is observable, that slight differences in reflectance cause significant differences in *n* (Figure 6). On average, the 42 μm thick film showed a value of about 0.02 lower than the 105 μm thick film.

Figure 7 shows the real part of refractive index, calculated by the T/R and the SSKK

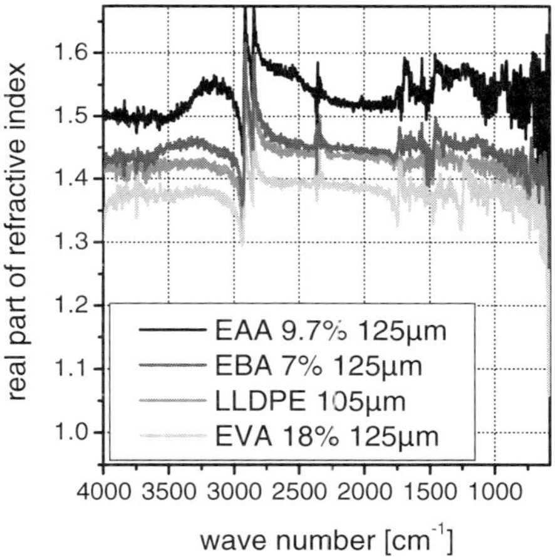

Figure 5.
Real part of refractive index calculated by the SSKK algorithm for selected ethylene copolymer films, as a function of wave number.

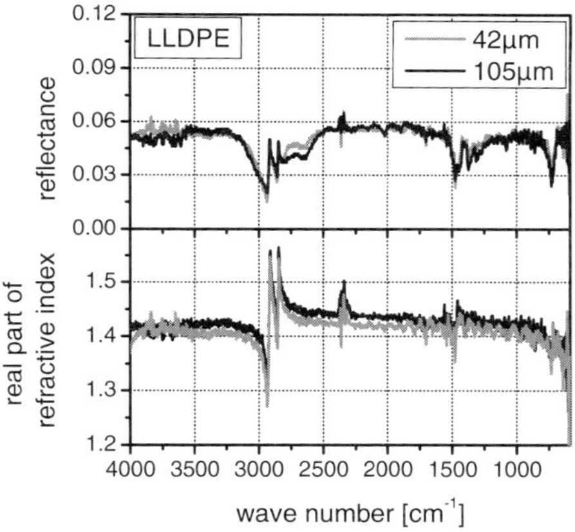

Figure 6.
Reflectance spectra and real part of refractive index, calculated by the SSKK algorithm for 42 and 105 μm thick LLDPE films as a function of wave number.

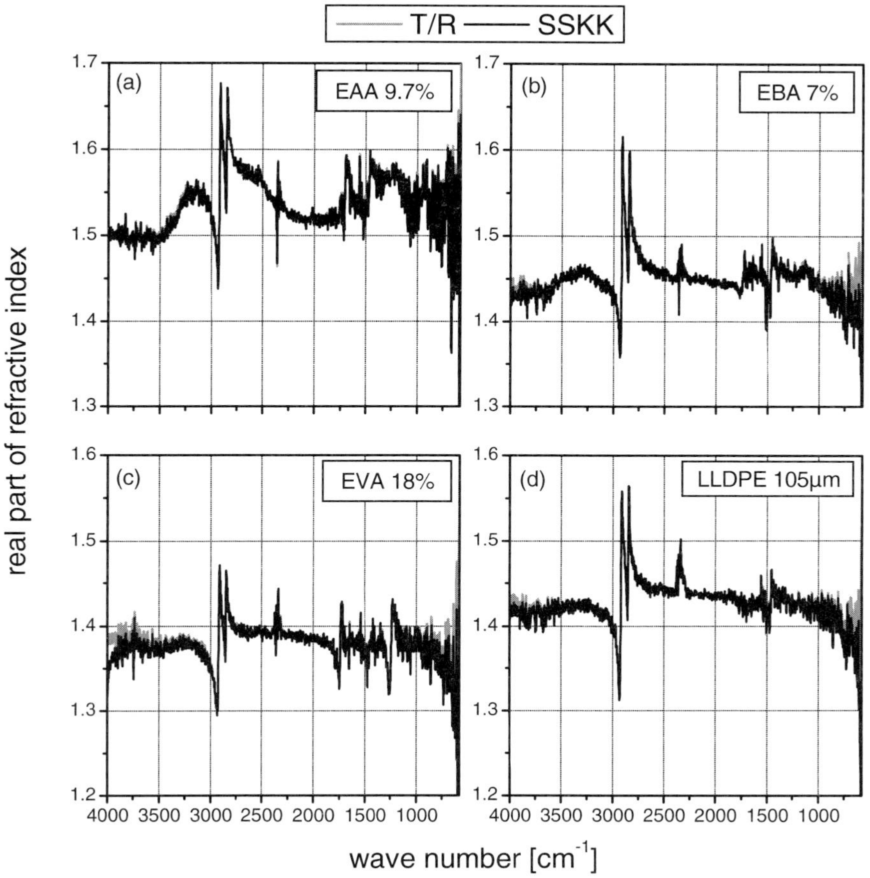

Figure 7.
Comparison of the T/R and the SSKK method for calculation of the real part of refractive index, shown for (a) EAA 9.7% (125 µm film), (b) EBA 7% (125 µm film), (c) EVA 18% (125 µm film) and (d) LLDPE (105 µm film).

method, as a function of wave number. In case of the T/R method, the non-valid values in non-transparent regions were excluded. In general, both methods are in good agreement regarding *n*. Slight differences can be seen only between 4000 and 3700 cm^{-1} and at wave numbers below 800 cm^{-1}.

Comparing both methods, the SSKK algorithm is more appropriate to calculate the real part of the refractive index, as it yields values for *n* without any limitations over the whole range of interest.

For the calculation of the imaginary part of the refractive index both methods have to be applied to obtain results over the whole spectral range. In Figure 8 the k values from 4000 to 580 cm^{-1} are shown for selected films. In the transparent region, the k is about 10^{-4}, increasing to 10^{-2} in the semitransparent region. In areas of total absorption, *k* values reach about 0.25 for the strongest absorption peak, (i.e. CH_2 stretching vibration between 3000 and 2800 cm^{-1}). Medium absorption bands (e.g. νC=O, νC–O, δCH_2, ρCH_2) exhibit *k* values up to 0.1. The area of change-over between the *k* values calculated by the T/R method and the values determined by the SSKK algorithm is not fully covered, especially at the absorption peaks below 1800 cm^{-1}. In some cases (e.g. ρCH_2 at

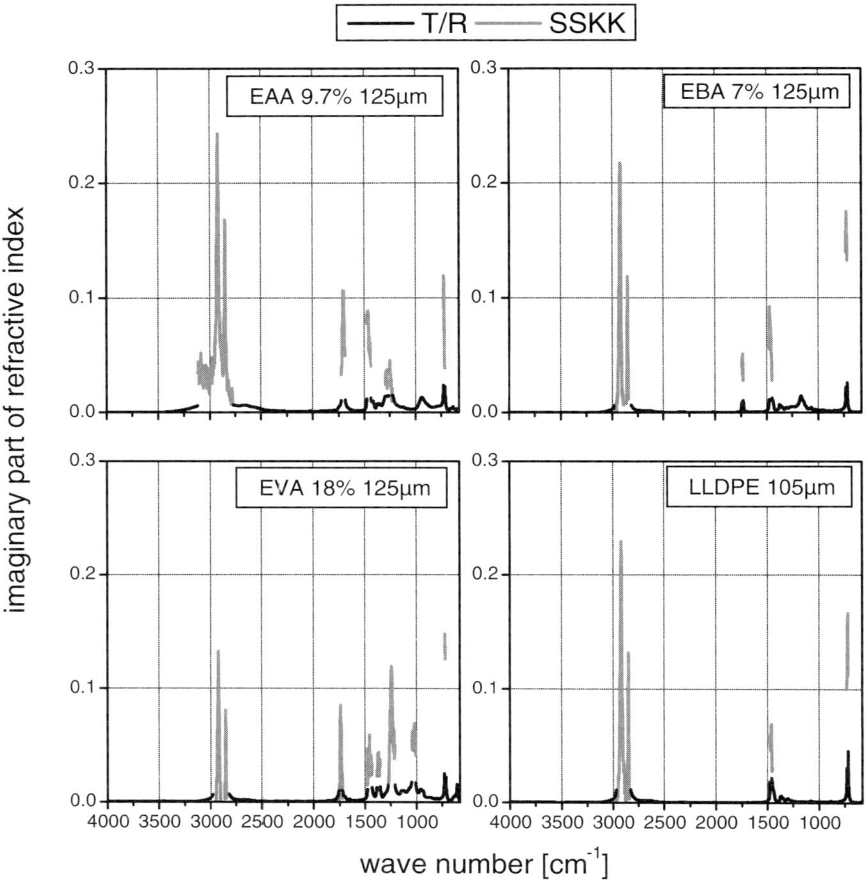

Figure 8.
Imaginary part of refractive index as a function of wave number, shown for (a) EAA 9.7% (125 μm film), (b) EBA 7% (125 μm film), (c) EVA 18% (125 μm film) and (d) LLDPE (105 μm film).

about 720 cm^{-1}) the values of k switch over one order of magnitude in one step. For a better definition of the change-over between both methods, higher resolution spectra should be measured.

Acknowledgements: This research work was performed at the Polymer Competence Center Leoben (PCCL) within the research project II-S.5 in cooperation with the Institute of Materials Science and Testing of Plastics at the University of Leoben. The PCCL is founded by the Austrian Government and the State Governments of Styria and Upper Austria within the K_{plus} program of the Austrian Ministry of Traffic, Innovation and Technology. The authors wish to express their gratitude to Arkema, Basell, DOW and DuPont for providing the materials.

[1] P. T. Tsilingiris, *Energy Conversion and Management* **2003**, *44*, 2839.
[2] G. Papadakis, D. Briassoulis, G. Scarascia Mugnozza, G. Vox, P. Feuilloley, J. A. Stoffers, *J. agri. Engng Res.* **2000**, *77*, 7.
[3] H. Mehling, *"Determination of infrared optical properties and phononic thermal conductivity of non-scattering inorganic and nonmetallic materials"*, PhD thesis, Universität Würzburg 1998.
[4] G. M. Wallner, W. Platzer, R. W. Lang, *Solar Energy* **2005**, *79*, 593.
[5] B. Drévillon, *Thin Solid Films* **1998**, *313–314*, 625.

[6] J. A. Wollam, C. Bungay, J. Hilfiker, T. Tiwald, *Nucl. Instrum. Methods Phys. Res., Sect. B* **2003**, *208*, 35.
[7] D. Tsankow, K. Hinrichs,, A. Röseler,, E. H. Korte, *phys. stat. Sol. (a)* **2001**, *188*, 1319.
[8] K. Hinrichs, M. Gensch, N. Nikonenko, J. Pionteck, K. J. Eichhorn, *Macromol. Symp.* **2005**, *230*, 26.
[9] M. Gioti, A. Laskarakis, S. Logothetidis, *Thin Solid Films* **2004**, *455–456*, 283.
[10] K. Roodenko, M. Gensch, H. M. Heise, U. Schade, N. Esser, K. Hinrichs, *Infrared Physics & Technology* **2006**, *49*, 39.
[11] Y. P. Zhang, X. S. Ge, *Solar Energy Materials and Solar Cells* **1995**, *37*, 379.
[12] G. Wallner, *"Kunststoffe für die transparente Wärmedämmung – Polymerphysikalische Einflüsse und Modellierung"*, PhD thesis, Montanuniversität Leoben 2000.
[13] G. Oreski, G. M. Wallner, *Solar Energy Materials & Solar Cells* **2006**, *90*, 1208.
[14] G. Oreski, G. M. Wallner, *Chemical Monthly* **2006**, *137*, 899.
[15] R. K. Ahrenkiel, *J. Opt. Soc. Am.* **1971**, *61*, 1651.
[16] R. Siegel, J. R. Howell, *"Thermal Radiation Heat Transfer"*, Hemisphere, Washington 1980.
[17] D. Banner, S. Klarsfeld, C. Langlais, *HighTemp – High Press* **1989**, *21*, 347.

 DOI: 10.1002/masy.200850514

Monitoring of the Sol-Gel Synthesis of Organic-inorganic Hybrids by FTIR Transmission, FTIR/ATR, NIR and Raman Spectroscopy

D. Fischer,[1] D. Pospiech,[1] U. Scheler,[1] R. Navarro,[2] M. Messori,[3] P. Fabbri[3]*

Summary: The sol-gel synthesis of organic-inorganic hybrids based on triethoxysilane-terminated poly(ethylene oxide) and tetraethylorthosilicate was monitored in-situ using three spectroscopic methods (FTIR/ATR, Raman, NIR). These spectroscopic methods allow in-situ monitoring of the evolution of hybrid materials starting from the modification of the polymer and the early steps of hydrolysis up to the network formation. By application of ^{29}Si solid-state NMR spectroscopy the assignment and quantification of the Raman bands to different end groups and different cross-linking states was made. The sol-gel reaction was also followed by in-line NIR spectroscopy. A multivariate data analysis was accomplished to obtain a conversion-time curve. Furthermore, we investigated spin-coated films on wafers using FTIR transmission spectroscopy.

Keywords: in-line spectroscopy; In-situ monitoring; organic-inorganic hybrids; poly(ethylene oxide); sol-gel

Introduction

Organic-inorganic hybrids based on polymers and silicon alcoxides represent an important class of materials because they allow the chemical and mechanical resistance of silica to combine with the properties of the polymer (e.g., flexibility of polyethers,[1] biocompatibility and hydrophilicity of poly(ethylene oxide),[2] abrasion resistance of silicones[3] and transparency of poly(methylmethacrylates)[4]). Thus, sol-gel chemistry with polymers has been developed as an important tool used to create new materials, in particular, coatings with improved scratch, chemical, UV and flame resistance, and the inclusion of special functions such as antibacterial behaviour, high transparency and barrier properties.[5] Advantages of sol-gel reactions in polymers technology are the low cost of the chemical educts, their versatility, preparation under mild conditions, and cheap and well-known technologies for application.[6]

Factors affecting the resulting silica network and, hence, the resulting properties of the materials are the type of metal alcoxide used, pH, temperature, time of reaction, reagent concentrations, catalyst nature and concentration, water/ethoxysilane-group molar ratio, aging temperature and aging time. The number of influencing parameters is rather high and the reaction, thus, not easy to control. Therefore, the main aim of our investigation was to monitor sol-gel reactions between tetraethylorthosilicate (TEOS) and telechelic poly(ethylene glycol) (PEO) by simultaneously applying FTIR/ATR, NIR, and Raman spectroscopy, in-situ in the reaction flask to find a procedure to control the sol-gel reaction for reproducible results of the resulting material properties.

[1] Leibniz Institute of Polymer Research Dresden, Hohe Str. 6, 01069 Dresden, Germany
Fax: (+49)-351-4658-565;
E-mail: fisch@ipfdd.de
[2] Institute of Polymer Science and Technology CSIC, Juan de la Cierva 3, 28006 Madrid, Spain
[3] University of Modena and Reggio Emilia, Department of Materials and Environmental Engineering, Via Vignolese 905/A, 41100 Modena, Italy

FTIR and Raman spectroscopy are often used to describe sol-gel reactions and to characterize thin films of the hybrid materials on wafers. Some research groups investigated the relevant wavenumber range between 400 and 1300 cm^{-1}, where the Si–O–Si and Si–O–R/Si–OH bands are located. Detailed information about vibration modes, assignments, interpretation of the bands and band deconvolution for FTIR- and Raman spectra are available in ref. [7–17]. Vibrational spectroscopic in-line monitoring has been shown as a powerful tool to understand reactions.[18–20] This method has already been applied to hybrid material sol-gel preparations using a setup different from that of the present study, with the Raman probe outside the reaction flask.[21,22] Furthermore, ^{29}Si solid-state NMR spectroscopy was used to characterize our final hybrids and for the correlation of NMR and Raman spectra. Assignments and interpretation of the NMR signals were taken from ref. [23–26].

Experimental Part

Materials

Tetraethylorthosilicate (TEOS, Sigma Aldrich), (3-isocyanatopropyl)-triethoxysilane (ICPTES, Fluka), absolute ethanol (Fluka) and conc. HCl (VWR) were used as received. Poly(oxyethylene), (poly(ethylene glycol), PEG, from Sigma Aldrich) with a molar mass of 1000 g/mol was dried in vacuum at 50 °C for 8 hours prior to use.

Synthesis

The PEO/silica hybrids were synthesized in a two-step synthesis as reported by Leonelli et al.[27] and Messori et al.[28]

In the first step, the reaction of the two OH-end groups of PEG 1000 (30 g; 0.03 mol) with

ICPTES (14.8 g; 0.06 mol) to triethoxysilane-terminated PEO (PEO-Si) (see **1**) was carried out at 120 °C in a three-necked flask equipped with ATR/FTIR silicon probe, condenser with CaCl2 trap, thermometer and magnetic stirrer under nitrogen and monitored by in-situ ATR-IR spectroscopy. An ATR/IR spectrum was taken every two minutes to observe the end of the reaction.

$$2\ O{=}C{=}N\text{-}(CH_2)_3\text{-}Si(OC_2H_5)_3 + HO\text{-}(CH_2CH_2O)_n\text{-}H$$

$$\longrightarrow (C_2H_5O)_3\text{-}Si\text{-}(CH_2)_3NH\text{-}CO\text{-}O\text{-}(CH_2CH_2O)_n\text{-}CO\text{-}NH(CH_2)_3\text{-}Si(OC_2H_5)_3$$

In the second step, the sol-gel synthesis of the triethoxysilane-modified PEO-Si obtained in step **1** with TEOS to a polymeric organic-inorganic hybrid (monomer ratio focused on a final polymer/SiO_2 ratio of 70/30 wt/wt) (see **2**) was performed and monitored. The reaction of PEO-Si (10.4 g; 0.006 mol) with TEOS (13.4 g; 0.064 mol) was carried out in ethanol with different solvent concentrations, the

Table 1.
Experimental conditions of the sol-gel reactions performed.

No.	$m_{(PEO\text{-}Si + TEOS)}/m_{EtOH}$ (g/g)	T (°C)	t_{gel} 1) (min)	SiO_2 content of dried hybrid (wt%)
1 (TEOS only)	1.0	70	100	93.9
2 (PEO-Si only)	1.0	70	–	10.2
3	1.0	70	106	
4-1	no ethanol	20	15	36.6
4-2	3.0	20	13	17.9
4-3	1.0	20	960	38.4
5-1	1.0	30	300	
5-2	0.7	30	480	
5-3	0.34	30	–	
6	1.0	40	120	36.4

1) time after which a gel formation was visibly observed.

stoichiometrically necessary amount of water (1 mol water/mol ethoxy) and HCl as catalyst (water/HCl molar ratio of 1/0.1 mol/mol) in a four-necked flask equipped with NIR, Raman and FTIR/ATR diamond probe, condenser with CaCl2 trap, thermometer and magnetic stirrer at different temperature (between 20 and 70 °C). FTIR/ATR and NIR spectra were taken every two minutes and Raman spectra were taken every four minutes. The performed experiments are summarized in Table 1.

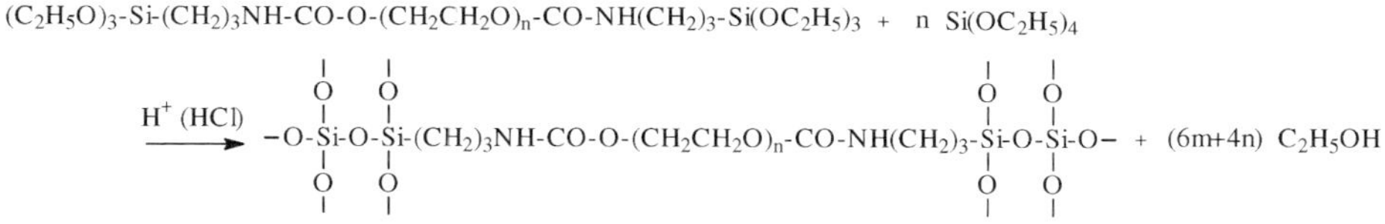

Ethanol was removed after the sol-gel reaction by distillation. The SiO_2 content of sample 4-2 is low because more ethanol was used than in the other samples and so Si as $Si(OH)_4$ was lost through evaporation during distillation. The resulting products were further dried for four days at 80 °C in a vacuum oven. The inorganic contents of the dried hybrids were determined gravimetrically by microwave treatment at 550 °C for 40 min.

Spectroscopic Techniques

The FTIR/ATR spectrometer ReactIR 4000 (Mettler Toledo), wavelength range between 4000 and 650 cm^{-1} (resolution 4 cm^{-1}, 100 scans per spectrum), equipped with silicon or diamond ATR probes, gold coated moveable tubes as lightpipe and MCT detector was used. The ATR crystal material was silicon in step 1 and diamond in step 2.

The Raman measurements were done with the Raman spectrometer HoloProbe785 (Kaiser Optical) which works in the range between 3500 and 150 cm^{-1} (resolution 4 cm^{-1}, 80 scans per spectrum). The Raman spectrometer is equipped with a 400 mW diode laser with an excitation frequency at 785 nm, a holographic notch filter and a CCD detector array and connected by optical fibers to a high temperature Raman probe.

NIR monitoring was performed by a diode array process spectrometer SentroProc (Sentronic) which works in the range from 10000 to 6000 cm^{-1} (resolution 4 nm, 1000 scans per spectrum) and a NIR transflectance probe (Solvias) connected by optical fibers to the spectrometer. Chemometric calculations were performed with the program GRAMS/AI/PLSplus/IQ (ThermoGalactic). Fitting of the conversion-time curve was done by polynominal 4-parameter fit using the software Origin 8.1.

All three optical probes dipped directly into the solution in the reaction flask.

FTIR transmission spectra were recorded with the spectrometer IFS66v (Bruker) under vacuum, equipped with a MCT detector, in the wavelength region from 600 to 4000 cm^{-1} (resolution 4 cm^{-1}, 200 scans per spectrum).

^{29}Si solid-state NMR of the dried organic-inorganic hybrids were acquired on a Bruker Avance 500 NMR spectrometer operating at a Larmor frequency of 99 MHz for ^{29}Si using a Bruker BL4 MAS probehead accepting 4 mm o.d. rotors with a sample spinning frequency of 10 kHz. Single pulse spectra have been acquired under continous-wave proton decoupling with an excitation pulse duration of 5 μs and a recycle delay of 20 s.

Thin Films

The sol obtained in experiment 5-3 was spin-coated after 3 and 7 hours reaction time on pre-cleaned silicon wafers using a Headway spin-coater (USA) at 2000 rpm for 90 seconds, resulting in film thicknesses of 300 nm and 780 nm, respectively, measured by ellipsometry. We used a

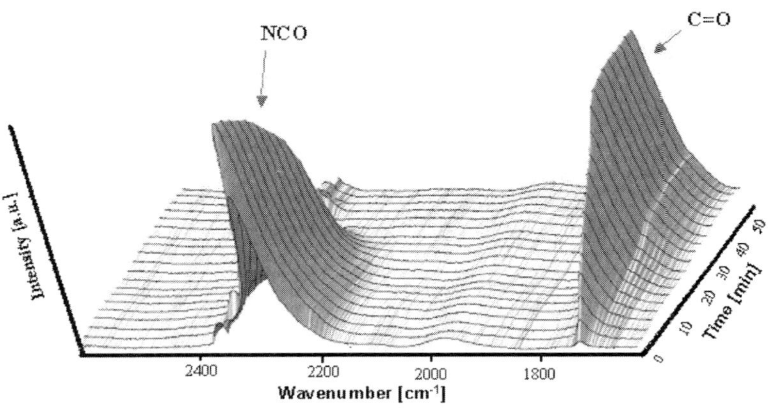

Figure 1.
In-line FTIR/ATR spectra (N=C=O/C=O region) of the reaction of PEO with (3-isocyanatoprcpyl)-triethoxysilane.

rotating analyzer type variable angle multi-wavelength ellipsometer M-44, for details see ref.[29]

Results and Discussion

The synthetic route to obtain the hybrid materials was carried out in two steps as already outlined. In the first step the polymer-analogous reaction of PEG with (3-isocyanatopropyl)-triethoxysilane leads to the functionalization of the polymer chains with two triethoxysilane groups. Thus, the polymer can participate in the following sol-gel process and will be incorporated into the inorganic network, finally forming the organic-inorganic hybrid. The reaction carried out in melt at 120 °C was followed by FTIR/ATR spectroscopy. Figure 1 shows the intensity changes of the isocyanate and the carbonyl bands.

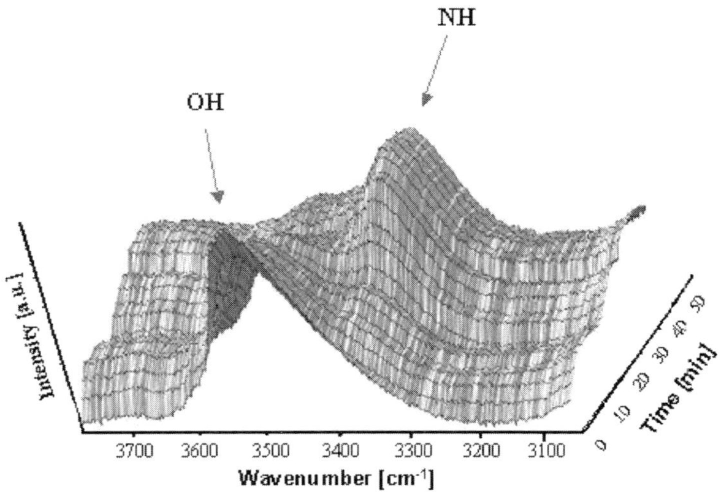

Figure 2.
In-line FTIR/ATR spectra (N–H/O–H region).of the reaction of PEO with (3-isocyanatopropyl)-triethoxysilane.

The intensity of the isocyanate band at 2270 cm^{-1} gradually reduces during the reaction, while the intensity of the urethane linkage formed at 1700 cm^{-1} increases. At the end of the reaction, the isocyanate band has completely vanished while the intensity of the modified polymer band has remained constant. The reaction was completed after 60 minutes at 120 °C. We could quantify the reaction and determine the conversion rate (not given here).

Figure 2 shows the shift of the maximum of the broad signal between 3200 and 3600 cm^{-1}. The intensity of the OH band of PEO is reduced with time while the NH band of the urethane group of the modified polymer are formed. Raman and NIR spectroscopy were not so suitable for the reaction of step1 like the ATR technique.

The triethoxysilane-terminated PEO (PEO-Si) obtained above (as shown in Figure 1 and 2) was used in the acid-catalyzed sol-gel reaction with TEOS in ethanol as solvent in second step. This reaction was carried out at different temperatures and ethanol concentrations at constant polymer/TEOS ratio, as outlined in Table 1. A molar ratio of ethoxysilane groups (TEOS + PEO-Si)/H2O of 1:1 was maintained. In all experiments, except 2 and 5-3, the gelation of the sol was observed (visible gelation point). After addition of TEOS, HCl and H_2O to the ethanol solution of PEO-Si the complete hydrolysis of TEOS and PEO-Si could be determined after 10 to 20 minutes. Figure 3 contains the Raman spectra of PEO-Si in ethanol, the reaction mixture after the addition of TEOS, water and HCl (hydrolysis), the spectrum after visible gelation in ethanol and the dried hybrid material. The spectrum after the addition of TEOS, water and HCl shows the disappearance of the TEOS bands at 930 cm^{-1} and 1291 cm^{-1} indicating the hydrolysis of TEOS. The visible gel in ethanol does not show the Si–O–Si cross-linking bands. Therefore, only a physically cross-linked state of the gel in ethanol at this reaction time can be assumed.

After removal of the solvent through distillation and annealing, these cross-linking bands can be observed by formation of the –Si–O–Si– bands at 840 and 1140 cm^{-1} and the internal Si–OH at 960 cm^{-1}. There is some controversy concerning the assignments of these bands in the literature. However, for our investigations, it is only important that we be able to differentiate the bands relevant for cross-linking or no cross-linking. It has to be stated that in the in-line measurements in

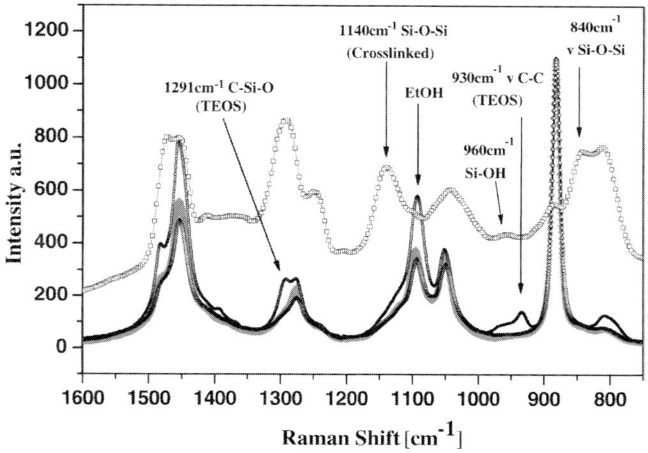

Figure 3.
Raman spectra in the range from 800 to 1600 cm^{-1} of the reaction mixture of experiment **4-3** (–: PEO-Si + EtOH; ,: PEO-Si + EtOH + TEOS; 8: ethanol-containing gel (after 960 min reaction time); #: dried hybrid 4-3.

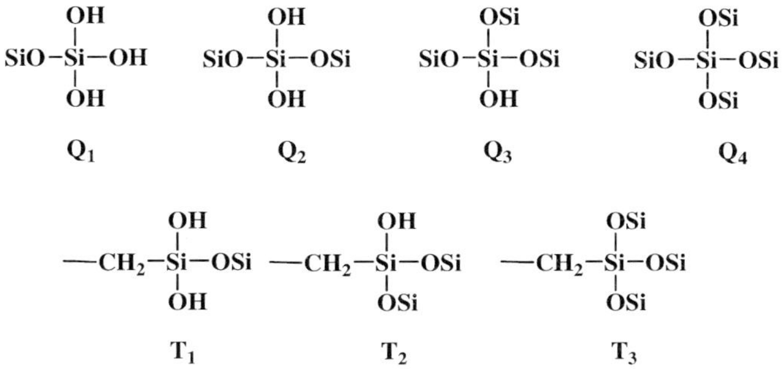

Scheme 1.
Types of Si atoms possibly contained in a PEO/SiO_2 network according to ref. [26].

ethanol solution under the conditions used in our study only the hydrolysis of TEOS can be observed spectroscopically.

The different times of visible gelation (physical network formation) are given in Table 1. They increase with decreasing concentration and reaction temperature, as expected.

We monitored the sol-gel reaction of PEO-Si with TEOS simultaneously with FTIR/ATR, NIR and Raman spectroscopy. The ATR-IR spectra, not given here, also show these cross-linking bands, so both applied spectroscopic methods demonstrate network formation by emerging of Si–O–Si bands.

Raman spectroscopy can also distinguish between different network formations and different end group modes in the range from 300 to 700 cm^{-1}. Here we can correlate Raman spectra with ^{29}Si solid-state NMR spectra of the dried samples. The types of Si atoms which can be expected according to reference[30,31] are given in Scheme 1. Q groups represent Si–O–Si linkages while T groups reflect covalent –C–Si–O– bonds. Figure 4 illustrates the ^{29}Si solid-state NMR spectrum of the dried PEO/SiO_2 hybrid sample 4-3.

The ^{29}Si-NMR spectrum of the final hybrid is dominated by Q_3 and Q_4 groups at -101 ppm and -110 ppm, respectively. The

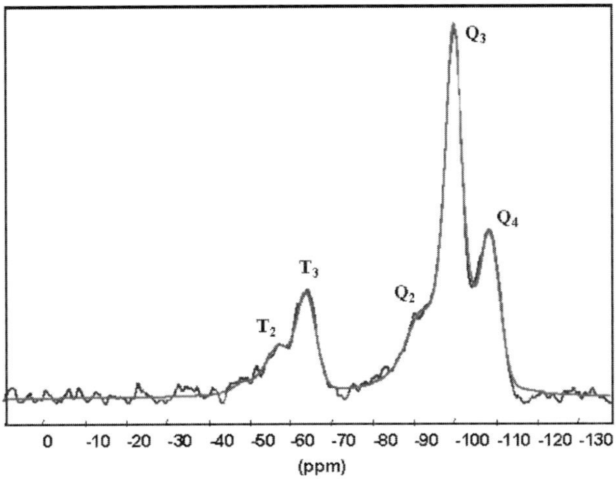

Figure 4.
^{29}Si solid-state NMR spectrum (direct polarization) of dried sample 4-3.

Table 2.
Quantitative analysis of T- and Q-groups in the organic-inorganicPEO/SiO_2 hybrids (after drying) as obtained from the ^{29}Si direct polarization spectra area. Intergration results were estimated from peaks derived via deconvolution using a Gaussian-Lorenzian model using DMFit.[32]

Sample	T_1	T_2	T_3	Q_2	Q_3	Q_4
	mol%	mol%	mol%	mol%	mol%	mol%
1 (TEOS only)	–	–	–	0	76	24
2 (PEO-Si only)	3	41	56	–	–	–
3	4	9	14	11	52	10
4-1	1	2	10	10	57	20
4-2	1	10	14	5	61	9
4-3	–	10	10	8	53	19
5-1	–	7	12	3	45	33
5-2	–	7	12	7	53	21
6	–	9	16	5	44	26

signal around −60 ppm originates from T_2 groups and that at −70 ppm from T_3 groups, respectively. Table 2 contains the calculated amounts of the groups after quantitative analysis.

The NMR spectra show the expected result that PEO-Si alone (**2**) cannot form Q-groups, while TEOS alone forms exclusively Q_3 and Q_4 groups. Mixtures of PEO-Si and TEOS give different ratios of T- and Q-groups, depending on the reaction conditions.

Figure 5 shows the Raman spectra of the dried hybrid (sample 4-3).

We could identify the following T and Q groups in the Raman spectra: Q_4 at 417 cm^{-1}, Q_3 at 488 cm^{-1},T_3 at 503 cm^{-1} T_2 at 532 cm^{-1}. The intensity ratio between Q and T bands in the Raman spectra is nearly the same as the ratios found by solid state NMR in Table 2.

The sol-gel process was also monitored in-situ using NIR spectroscopy. The NIR spectra of the reaction are illustrated in Figure 6.

The NIR spectra over reaction time show only small differences, mainly in intensities and in a shift of the baseline.

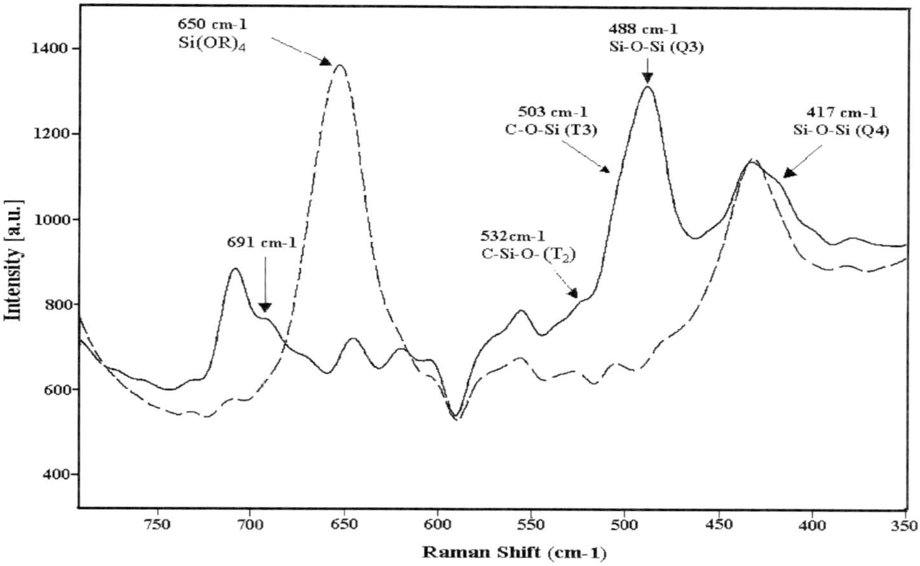

Figure 5.
Types of T and Q groups found in the Raman spectrum of the dried PEO/SiO2 hybrid (sample 4-3).

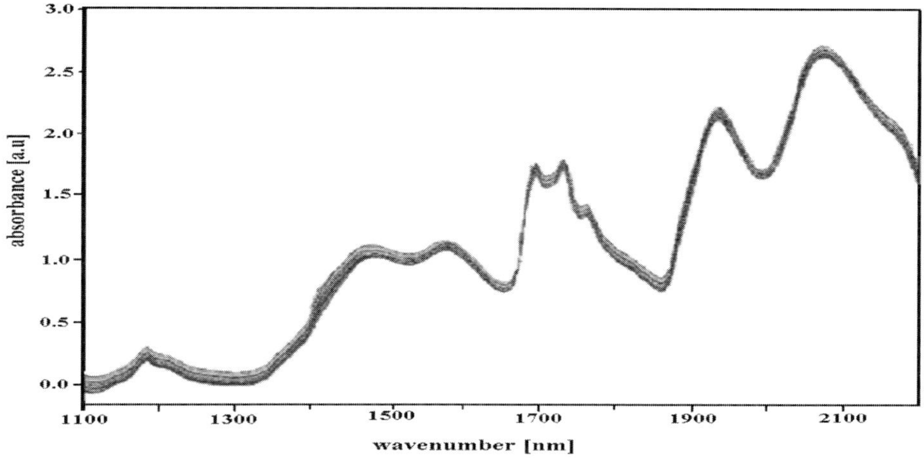

Figure 6.
In-line NIR spectra of the sol-gel reaction of PEO-Si/TEOS (sample 4-3).

A multivariate data analysis (chemometrics) is thus needed to extract the chemical information.

The partial least square regression method (PLS) was applied. The prediction model from the PLS method using two principle components allowed to track the conversion-time curve of the reaction by correlation of the complete spectra with the monitored time (t = 0 min; conversion = 0%, to t = 390 min; conversion = 100%) as regression parameter. We calculated the conversion curve until the visible gelation point, defining that the conversion at the end of the measurement is 100%. This value contains the changes in the spectra during the monitored time, and does not represent 100% cross-linking. Figure 7 displays the calculated and fitted conversion (time) curves expressing the reaction time in percent. The figure shows a typical conversion-time curve for a polycondensation reaction[33] which was expected from the chemistry of the system.

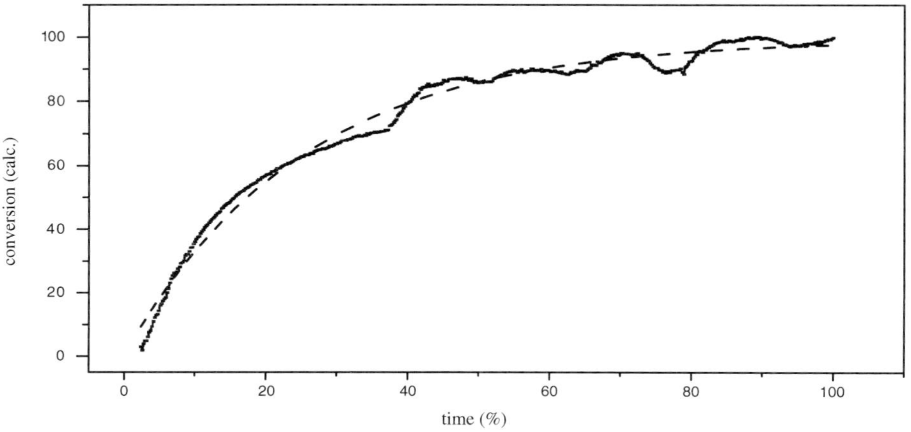

Figure 7.
Predicted and fitted conversion curve of the sol-gel reaction of PEO-Si/TEOS (sample 4-3) by PLS using NIR spectra (- calculated, --- fitted).

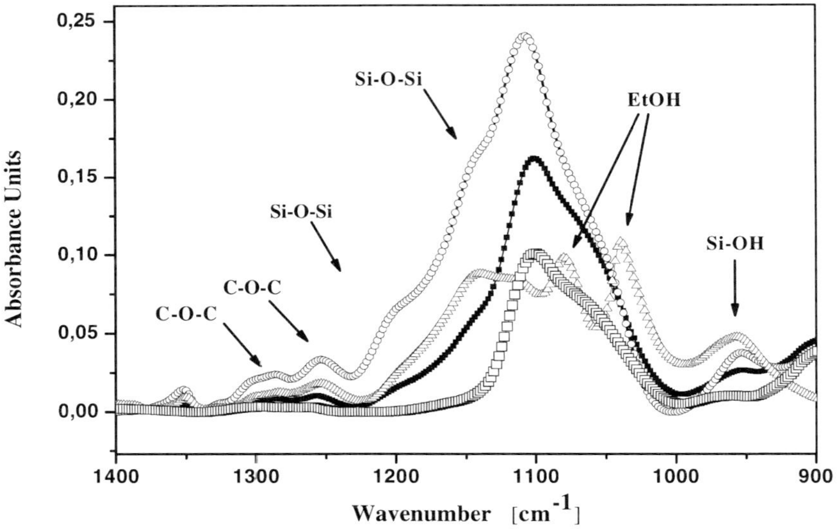

Figure 8.
FTIR-Transmission spectra of sample 5-3 (–300 nm film of the cross-linked hybrid on silicon wafer; 780 nm film of the cross-linked hybrid on silicon wafer; 8 ethanol-containing gel (partially cross-linked) 8: Reference wafer).

Besides in-line monitoring within the reaction vessel, the hybrid material as thin films were investigated. Thin films were prepared using the sol of sample 5-3 for spin-coating on Si-wafers after 3 and 7 hours reaction time (film thickness 300 and 780 nm, respectively). The samples were measured by FTIR in transmission technique. The spectra displayed in Figure 8 show the reference wafer, a partially cross-linked hybrid with solvent and two cross-linked hybrid films. One can clearly identify the –Si–O–Si–O– networking modes at 1140 and 1200 cm^{-1}, which are described in the references.[7–17] The detection of cross-linking bands in the thin films provides the opportunity to follow the cross-linking process in coatings.

Conclusions

Spectroscopic methods (ATR/IR, NIR, Raman) were used for the monitoring of the two-step sol-gel synthesis of organic-inorganic hybrid materials based on triethoxysilane-terminated poly(ethylene oxide) and tetraethylorthosilicate in ethanol catalyzed by HCl.

These spectroscopic methods allowed in-situ monitoring of the evolution of hybrid materials starting from the modification of the polymer (FTIR/ATR) and from the early steps of hydrolysis up to the physical network formation (NIR, Raman). A comparison with ^{29}Si solid-state NMR spectroscopic data allows an assignment and a quantification of the Raman bands to different end groups and to different cross-linking states. Additionally, we could also monitor the sol-gel reaction by in-line NIR spectroscopy until the visible gelation point. A conversion-time curve by multivariate analysis (PLS) was calculated. Thus, in-line NIR spectra can be a powerful and easy-to-handle method of following the reaction up to the visible gelation point. However, both Raman and NIR spectroscopy, need continued research to find the opportunities and limits for routine application as monitoring methods.

Furthermore, we could show with the investigation of spin-coated films on wafers by FTIR transmission spectroscopy that the cross-linking process in coatings can be investigated.

Acknowledgements: Participation in the EU NoE Nanofun-poly is gratefully acknowledged. Furthermore, the authors would like to thank R. Schulze for ellipsometric measurements and F. Näther for helpful technical assistance.

[1] I. Goodman, Ed., *Developments in Block Copolymers- 1*", Applied Science Publishers, London and New York **1982**, p. 261.
[2] J. M. Harris, S. Zalipsky, "*Poly(ethylene glycol). Chemistry and Biological Applications*", ACS Symp. Ser. 680, American Chemical Society, Washington, DC **1997**, p. 1.
[3] Z. Li, W. Han, D. Kozodaev, J. C. M. Brokken-Zijp, G. de With, P. C. Thüne, *Polymer* **2006**, *47*, 1150.
[4] K. F. Silveira, I. V. P. Yoshida, S. P. Nunes, *Polymer* **1995**, *36*, 1425.
[5] C. Sanchez, B. Julian, P. Belleville, M. Popall, *J. Mater. Chem.* **2005**, *15*, 3559.
[6] J. D. Wright, N. A. J. M. Sommerdijk, "*Sol-Gel materials. Chemistry and Applications*", Taylor & Francis Books Ltd, London **2001**, p. 15.
[7] N. Primeau, C. Vautey, M. Langlet, *Thin Solid Films* **1997**, *310*, 47.
[8] G. Orcel, J. Phalippou, L. L. Hench, *J. Non-Cryst. Solids* **1986**, *88*, 114.
[9] T. M. Parril, *J. Mater. Res.* **1992**, *7*, 2230.
[10] C. Vautey, M. Burgos, M. Langlet, *Thin Solid Films*, **1999**, *347*, 184.
[11] C. A. M. Mulder, A. A. J. M. Damen, *J. Non-Cryst. Solids* **1987**, *93*, 169.
[12] P. Innocenzi, *J. Non-Cryst. Solids* **2003**, *316*, 309.
[13] J. Gnado, P. Dhamelincourt, C. Pelegris, M. Traisnel, A. Le Maguer Mayot, *J. Non-Cryst. Solids* **1996**, *208*, 247.
[14] A. Fidalo, L. M. Ilharco, *J. Non-Cryst. Solids* **2001**, *283*, 144.
[15] B. Orel, R. Jese, U. L. Stangar, J. Grdadolnik, M. Puchberger, *J. Non-Cryst. Solids* **2005**, *351*, 530.
[16] K. Gigant, U. Posset, G. Schottner, L. Baia, W. Kiefer, J. Popp, *J. Sol-Gel Sci. Technol.* **2003**, *26*, 369.
[17] K. A. Mauritz, R. M. Warren, *Macromolecules*, **1989**, *22*, 1730.
[18] K. Sahre, U. Schulze, T. Hoffmann, M. Abd Elrehim, K. J. Eichhorn, D. Fischer, D. Pospiech, B. Voit, *J. Appl. Polym. Sci.* **2006**, *101*, 1374.
[19] K. Sahre, T. Hoffmann, D. Pospiech, D. Fischer, K. J. Eichhorn, B. Voit, *European Polymer Journal*, **2006**, *42*, 2292.
[20] D. Fischer, K. Sahre, M. Abdelrhim, B. Voit, V. B. Sadhu, J. Pionteck, H. Komber, J. Hutschenreuter, *Comptes Rendus Chimie*, **2006**, *9*, 1419.
[21] M. Gnyba, M. Keränen, A. Maaninen, J. Suhonen, M. Jedrzejewska-Szczerbska, B. B. Kosmowski, P. Wierzba, *Opto-Electronics Review* **2005**, *13*(1), 9.
[22] I.-G. Marino, P. P. Lottici, D. Bersani, R. Raschellà, A. Lorenzi, A. Montenero, *J. Non-Cryst. Solids* **2005**, *351*, 495.
[23] M. P. Besland, *J. Am. Chem. Soc.* **1991**, *113*, **1982**.
[24] G. Orgel, *J. Non-Cryst. Solids* **1986**, *88*, 114.
[25] J. W. de Haan, *J. Coll. Interf. Sci* **1986**, *110*, 581.
[26] G. E. J. Maciel, *J. Am. Chem. Soc.* **1980**, *102*, 7606.
[27] C. Leonelli, M. Messori, F. Pilati, P. Veronesi, *Macromol. Symp.* **2005**, *228*, 229.
[28] M. Messori, M. Toselli, F. Pilati, E. Fabbri, P. Fabbri, L. Pasquali, S. Nannarone, *Polymer* **2004**, *45*, 805.
[29] Y. Mikhlailova, V. Dutschk, C. Bellmann, K. Griundke, K.-J. Eichhorn, B. Voit, *Colloids Surf. A: Physicochem. Eng. Aspects* **2006**, *279*, 20.
[30] G. Engelhardt, H. Koller, ^{29}Si of NMR Inorganic Solids, *NMR Basic Principles and Progress 31*, **1993**, 1.
[31] J. Stebbins, M. in Duer, Eds., *Solid-State NMR Spectroscopy*, Blackwell Science Ltd., Oxford.
[32] D. Massiot, F. Fayon, M. Capron, I. King, S. Le Calve, B. Alonso, J. O. Durand, B. Bujoli, Z. H. Gan, G. Hoatson, *Magnetic Res. Chem.* **2002**, *40*, 70.
[33] P. J. Flory, "*Principles of Polymer Chemistry*", Cornell University Press, Ithaca, New York 1953.

 DOI: 10.1002/masy.200850515

ESR - Spin Probe Method in Studying Natural Rubber: an Effect of the Probe Incorporation Temperature

Mirna Didović,[1] *Damir Klepac,*[1] *Srećko Valić**[1,2]

Summary: Natural rubber (NR) latex containing 62 wt.% of solid phase was dried by evaporating the liquid phase at room temperature. 4-hydroxy-2,2,6,6-tetramethyl-piperidine-1-oxyl free radical, used as the spin probe, was incorporated into the NR matrix by swelling the samples in toluene probe solution at 4 °C, 24 °C and 37 °C. The probe incorporation temperatures were constant for each set of samples. When the solvent was slowly removed, all samples were annealed at 60 °C during 72 hours. Composite ESR spectra observed in the vicinity of T_{5mT} show a significant influence of the probe incorporation temperature on the spectral shape. More precisely, the amount of the broad component in a composite ESR spectrum increases with the increasing probe incorporation temperature. A simple model which describes the probe distribution inside the inhomogeneous NR matrix is proposed.

Keywords: ESR; natural rubber; probe-incorporation temperature; spin probe; swelling

Introduction

ESR spectroscopy has been extensively used in studying polymer systems during last decades.[1–4] Polymer systems usually do not possess paramagnetic centers and therefore cannot be directly studied by ESR spectroscopy. An introduction of such centers into the polymer matrix gives a possibility for the application of this method to the investigation of polymer materials. Most common paramagnetic centers that are used in studying polymers are free nitroxide radicals.[1–4] Nitroxide radicals can be introduced into a polymer either by mechanical incorporation of radicals into the polymer matrix or by chemical bonding of the radical on a polymer chain. The last method, known as spin labeling, usually requires longer experimental preparation and is rather limited due to the necessary functional groups of polymer chains and nitroxide radical that must react together resulting in a chemical bond. The simplest way to use the nitroxide radicals in studying polymer systems is to introduce them into the polymer matrix by diffusion. This method is known as spin probe method. The method implies that the polymer matrix and nitroxide radical must be soluble in the same solvent. In fact, the incorporation process of spin probe into a polymer matrix is regularly accomplished by swelling of polymer in the probe solution. This procedure is accompanied by diffusion of the probe molecules into the swollen polymer matrix and followed by slow solvent evaporation, which usually takes one or two days. During this time probe molecules immigrate from the solution to the polymer matrix and finally remain trapped in the free volume holes.

The main goal of this study is to investigate how the temperature of the probe incorporation process affects the shape of ESR spectra of a spin probe in the inhomogeneous polymer matrix. For that reason, we have chosen natural rubber as a typical inhomogeneous system.

[1] School of Medicine, University of Rijeka, Braće Branchetta 20, HR-51000 Rijeka, Croatia
Fax: (+385) 51 651135;
E-mail: valics@medri.hr
[2] Rudjer Bošković Institute, Bijenička 54, HR-10000 Zagreb, Croatia

Experimental Part

Sample Preparation

The natural rubber (NR) latex with a solid content of 62 wt.%, was kindly supplied by the Rubber Research Institute of India. Original latex was diluted with 0.6 wt.% ammonia solution in order to get 30 wt.% of the total solid content. The diluted latex was then placed in the aluminium square vessel, dimensions of 10 cm × 10 cm × 1 cm. The samples were left at room temperature for several days to remove the liquid phase by evaporation and finally a transparent NR film having the thickness of about 1 mm was obtained. The NR film was then dried in vacuum at 40 °C during 72 hours, until a constant weight was achieved.

The free nitroxide radical 4-hydroxy-2,2,6,6,-tetramethylpiperidine-1-oxyl (TEMPOL) was used as a spin probe for ESR measurements. The probe molecules were incorporated into NR samples by swelling the samples in probe solution. Toluene, known as a good solvent for both, the nitroxide radical and NR, was used as the swelling agent. NR samples, cut in small parts, were put in the toluene solution of TEMPOL. Three different temperatures of the probe incorporation were chosen and kept constant during the process: 4 °C, 24 °C and 37 °C. The solvent was slowly removed from the solution by evaporation. In order to avoid the presence of residual solvent all samples were annealed under the same conditions - in vacuum at 60 °C during 72 hours. This time period was long enough to reach the constant sample weights in a function of the annealing time, indicating that the residual solvent was completely removed. The total amount of the spin probe of 0.15 wt.% was high enough to give a good signal to noise ratio.

ESR Measurements

ESR measurements were performed on Varian E-109 spectrometer equipped with a Bruker ER 041 XG microwave bridge and a Bruker ER 4111 VT temperature unit. Spectra were recorded at −10 °C. This temperature is in the vicinity of T_{5mT}, the temperature related to the glass transition temperature[5] (T_g), defined as:

$$T_{5mT} - T_g = 52[2.9f\{\ln(1/f) + 1\} - 1]$$

where f is the ratio of the activation volume of a probe to the activation volume of a polymer segment undergoing relaxation. The samples were kept at the temperature of measurement at least 10 minutes before the beginning of accumulation. EW (EPR-Ware) Scientific Software Service program (EPR Data Acquisition System Version 2.41A) was used for data accumulation and manipulation.

Interpretation of Bimodal ESR Spectra

Figure 1a shows the model ESR spectrum of a spin probe in a polymer matrix measured well above the glass transition temperature. The spectrum consists of three well defined narrow lines characteristic of the fast molecular motions. On the contrary, when the segmental motions are frozen ($T < T_g$), ESR spectrum indicates slow dynamics of the spin probe determined mostly by the size of free volume holes. A typical spectrum characteristic of the slow probe dynamics, which consists of three overlapping broad lines, is presented in Figure 1b. In a particular case of an inhomogeneous system, the ESR spectrum measured in the vicinity of T_{5mT}, shows a

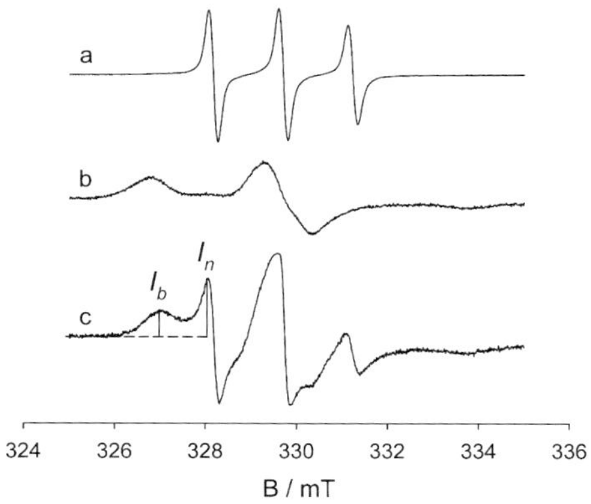

Figure 1.
Model ESR spectra of spin probe typical for the polymer: (a) well above T_g, (b) below T_g and (c) inhomogeneous system around T_{5mT}.

bimodal character. Such a composite spectrum consists of two components: the broad one being attributed to the slow motions and a narrow one that corresponds to the fast motions of the spin probe,[3] Figure 1c. The intensity ratio of the narrow and broad spectral line (I_n/I_b), as marked in Figure 1c, can be taken as a simple quantitative measure for the probe dynamics. The higher value of this ratio indicates a higher amount of the fast component present in a dynamically inhomogeneous system. The analysis of such composite spectra is often used in characterization of inhomogeneous systems, particularly elastomers, polymer blends, copolymers, block copolymers, interpenetrating networks and polymer-filler systems.[3,6–12]

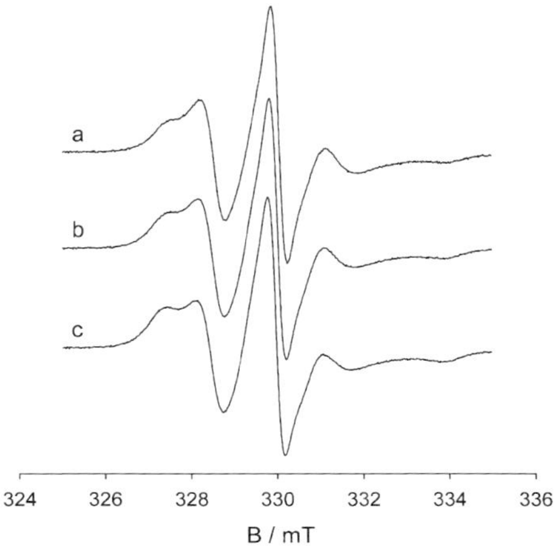

Figure 2.
ESR spectra of TEMPOL in NR measured at −10 °C. The probe incorporation temperature was: (a) 4 °C, (b) 24 °C and (c) 37 °C.

Swelling Experiment

To compare the results of ESR measurements with the swelling experiment, NR samples were swollen in toluene at temperatures applied for the probe incorporation. The degree of swelling was determined gravimetrically by measuring the mass of dry samples (before starting the swelling process) and the mass of swollen samples. The swelling time was limited to 50 minutes, since the NR was uncrosslinked. For a longer swelling time the structure of the gel fraction, which allows swelling of NR, was destroyed at 37 °C. The degree of swelling (α) was calculated as a difference in the mass of swollen and dry sample over the mass of dry sample. An average value of three measurements was taken in calculating α. It should be noted that the fraction of the sol component remained in the solution was not estimated.

Results and Discussion

Figure 2 represents ESR spectra of the spin probe diffusing through the NR matrix, measured at −10 °C. These spectra consist of two components arising from nitroxide radicals in two motional domains.[3] It is evident that the same spin probe incorporated in the same material, using the same solvent, but at different probe incorporation temperatures gives different spectral shapes. Spectrum observed with the sample in which the spin probe was incorporated at 4 °C has a more pronounced narrow component in comparison with the other two spectra when the probe was incorporated at higher temperatures. Moreover, the amount of the narrow component, as measured by the intensity ratio, decreases regularly with increasing the probe incorporation temperature. For the probe incorporation temperature of 4 °C, 24 °C and 37 °C, the corresponding ratios were 1.61, 1.35 and 1.15, respectively, see Table 1.

One could expect that the ESR spectra of spin probe should be the same when the temperature of the ESR measurement is kept constant, irrespective of the probe

Table 1.
The ratio of the narrow and broad ESR line intensities as a function of the probe incorporation temperature and temperature dependence of degree of swelling.

Temperature of probe incorporation and swelling/°C	I_n/I_b	Degree of swelling (α)
4	1.61	12.4
24	1.35	13.7
37	1.15	17.0

incorporation conditions. However, according to the results mentioned above, this is not the case. The spectra shown in Figure 2 clearly demonstrate that the probe incorporation temperature plays an important role in ESR spectroscopy of inhomogeneous polymer systems. In a special case of the composite spectra, characteristic of the temperature region of the phase transition, the amount of the broad and narrow component strongly depends on the temperature of probe incorporation. A higher amount of the narrow component for the lower value of the probe incorporation temperature is a consequence of higher probe mobility. Contrary, a lower amount of the narrow component, when the probe incorporation temperature is higher, indicates that the probe mobility is rather limited.

Such an interpretation of ESR results can be explained as follows:

At the lower value of the probe incorporation temperature the probe molecules enter in large free volume holes only and cannot penetrate into the high density regions of the NR matrix, Figure 3a. Therefore, in such case, they probe mostly the mobile parts of the matrix.

The degree of swelling (α) increases by increasing the temperature from 12.4 for 4 °C to 17.0 for 37 °C, Table 1. This allows the probe molecules to penetrate into the small holes, Figure 3b. Once the solvent is removed, the probe molecules remain trapped in these holes and consequently their motions are restricted.

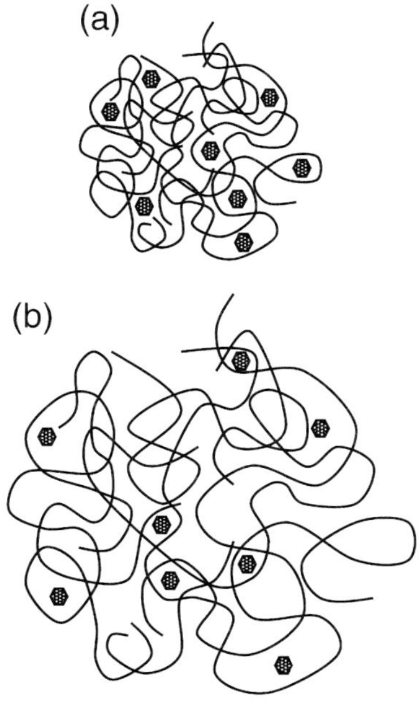

Figure 3.
Schematic representation of the probe incorporation process in NR at (a) 4 °C and (b) 37 °C.

Conclusion

On the basis of results of the present work it is clear that the probe incorporation temperature is an important factor when ESR spectroscopy - spin probe method is applied in studying inhomogeneous systems. Therefore, this temperature must be strictly controlled in order to avoid its unwished effect, which might lead to a possible wrong interpretation of ESR results.

[1] L. J. Berliner, Ed., *Spin labeling II: theory and application*, Academic Press, New York **1976**.
[2] G. G. Cameron, A. T. Bullock, *in*: *Developments in polymer characterisation*, Vol. 3, J. V. Dawkins, Ed., Applied Science Publishers, London **1982**, p. 107.
[3] Z. Veksli, M. Andreis, B. Rakvin, *Prog. Polym. Sci.* **2000**, *25*, 949.
[4] S. Schleidt, H. W. Spiess, G. Jeschke, *Colloid Polym. Sci.* **2006**, *284*, 1211.
[5] N. Kusumoto, S. Sano, N. Zaitsu, Y. Motozato, *Polymer.* **1976**, *17*, 448.
[6] T. Marinović, S. Valić, M. Andreis, Z. Veksli, *Polymer.* **1991**, *32*, 2519.
[7] S. Valić, M. Andreis, Z. Veksli, A. Charlesby, *Radiat. Phys. Chem.* **1991**, *37*, 257.
[8] A. L. Kovarskii, E. I. Kulish, Ya Placek, F. Szocs, S. V. Kolesov, K. S. Minsker, *Visokomol. Soedin. A* **1993**, *35*, 1669.
[9] Z. Hloušková, J. Tino, I. Chodák, *Eur. Polym. J.* **1994**, *30*, 175.
[10] J. Čulin, M. Andreis, Z. Veksli, A. Anzlovar, M. Žigon, *Eur. Polym. J.* **2005**, *41*, 1874.
[11] S. Schlick, R. D. Harvey, M. G. Alonso-Amigo, D. Klempner, *Macromolecules.* **1989**, *22*, 822.
[12] G. G. Cameron, M. Y. Qureshi, D. Stewart, R. Buscall, J. Nemecek, *Polymer.* **1994**, *36*, 3071.

 DOI: 10.1002/masy.200850516

Optical Spectra of Polygermane/Mesoporous Silica Nanocomposites

Nina Ostapenko,[*1] Nata Kozlova,[1] Shozo Suto,[2] Masato Nanjo,[3] Kunio Mochida[4]*

Summary: The monitoring of poly(di-n-hexylgermane) (PDHG) optical spectra in a variety of structures ranging from a bulk film to a nanosize polymer confined into a nanopore of SBA-15 was performed using the fluorescence and fluorescence excitation spectra in the temperature range from 5 to 240 K as well as the absorption and FTIR spectra at room temperature. The observed data were compared with those obtained for poly(di-n-hexylsilane). It was shown that PDHG film absorption and fluorescence spectra strongly depend on the polymer thickness and consist of a number of bands which were assigned to centers with different amount of trans- and gauche- conformers of the polymer chains. Conformations of the polymer chains found in a thin film and in a 10 nm pore are similar while differing from the conformations of a thick film. Optical spectra of the confined PDHG are blue-shifted relative to those of a thin film. The PDHG polymer chain conformation becomes disordered with the decrease of the polymer film thickness and the nanopore size from 10 to 6 nm.

Keywords: nanocomposites; optical spectra; poly(di-n-hexylgermane); SBA-15; films

Introduction

σ-conjugated polymers, polysilanes and polygermanes $[MR_2]_n$, (M stands for silicon or germanium and R is an organic molecule) have unique optoelectronic and electrophysical properties arising from delocalization of σ-electrons along the chain of M atoms. Significant interest to the investigations of these polymers is due to perspectives of their potential applications such as materials for photodiodes, photoresists, and materials for non-linear optics.[1]

The manufacture of the nanosize polymers opened an exciting new direction in the modification of the photophysical properties of the ultrathin polymer films. One of the reliable techniques of the fabrication of nanosize polymers is by filling polymers into the nanosize pores of the mesoporous silicas.

The optical spectra of nanocomposites based on poly(di-n-hexylsilane) PDHS and poly(methylphenylsilane) PMPS confined into mesoporous silicas were studied for the systems prepared by this method.[2–5] An essential influence of the space confinement on observed spectra was found to be a result of the competition between the polymer-surface and polymer-polymer interactions. It was shown that these interactions lead to new structures of nanosize polymer that are not observed in bulk films.[2–4]

In the family of polygermanes, poly(di-n-hexylgermane) PDHG has the structure similar to that of PDHS where the monomer unit $Si(C_6H_{13})_2$ is replaced by $Ge(C_6H_{13})_2$. Since the Ge–Ge bond length is longer than that of the Si–Si bond, intramolecular interaction is expected to be less pronounced compared to PDHS. This suggests that PDHG has more flexibility

[1] Institute of Physics, pr. Nauki 46, Kiev 03028, Ukraine
Fax: 38 (044) 525-15-89;
E-mail: ostap@iop.kiev.ua
[2] Tohoku University, 980-8578 Sendai, Japan
[3] Tottori University, 680-855 Japan
[4] Gakushuin University, 171-8588 Tokyo, Japan

near its backbone[6] and conformational changes in the polymer chain are more probable for PDHG than for PDHS. As a result the polymer-surface interaction as well as the optical spectra of PDHG film could change.

This paper presents a study of the monitoring of changes of the PDHG optical spectra varying the structure from that of a bulk film to a nanosize polymer. The nanocomposites PDHG/SBA-15 were fabricated by the incorporation of the polymer chains intro the pore of the mesoporous silicas of the SBA-15 type with pore diameter of 6 and 10 nm. PDHG films and composites were investigated using fluorescence (FL) and excitation spectra at (5–240) K as well as absorption and FTIR spectra at room temperature.

According to the analysis of the PDHG films absorption spectra published in literature,[7–9] a significant discrepancy exists between the results obtained by different authors. It was assumed that this discrepancy could be due to different conditions of the film preparation and could essentially depend on the polymer's molecular weight.[9] Our comprehensive study of the room temperature absorption and low-temperature (5–240 K) FL spectra of PDHG films and toluene solution has shown that the discrepancies observed between the reported absorption spectra are due to the different thickness of the polymer films. The nature of the absorption and fluorescence centers in PDHG films and the PDHG/SBA-15 nanocomposites is analyzed by comparing with the similar well-known data obtained on the previously studied PDHS films and PDHS/SBA-15 nanocomposites.[2–5]

Experimental Part

PDHG was synthesized as described in Ref.[8] Details of the preparation of the SBA-15 mesoporous silica was described in Ref.[10]

The parameters of the mesoporous system were calculated by the DFT method

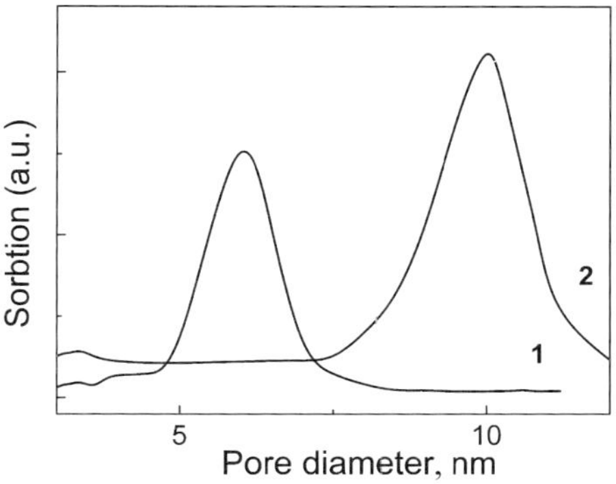

Figure 1.
Pore size distributions for the parent SBA-15 with the pore diameter of 6 nm (1) and 10 nm (2) derived from a nitrogen sorption isotherm (desorption branche, DFT method).

from the isotherm absorption branches. The resulting size distributions for pores are shown in Figure 1. The measurements showed quite a narrow scatter in the pore diameters.

In order to incorporate the PDHG polymer (Mw = 10600) into nanopores of SBA-15 with the pore diameter of 6 nm and 10 nm, the prepared silica matrixes were immersed in a 1% wt. solution of the polymer in toluene and slowly stirred in dark at 293 K for several hours and then kept in dark till evaporation of the solvent. Then the composite was twice washed in dark for approximately 2 h by stirring it in the fresh toluene to remove the polymer from the exterior surface. The PDHG films with different thickness were obtained by deposition on a quartz substrate by spin-coating.

The FL and excitation spectra were recorded by a DFS-12 and a Hitachi MPF-4 integrated with a helium cryostat in the temperature range from 5 to 240 K. FL is excited by a xenon lamp with the monochromator selected wavelength of 313 nm. The absorption spectra were recorded by KSVU-23 at room temperature. FTIR spectra of a polymer film and PDHG/SBA-15 composites were recorded by FTIR-66 Bruker instrument.

Results and Discussion

Absorption Spectra of PDHG Films

Figure 2 shows the room temperature absorption spectrum of PDHG thick polymer film and the result of the computer curve-fitting analysis. Insert depicts the structural formula of PDHG. It is seen that this spectrum consists of three bands with maxima at 373, 348 and 323 nm. It is worth mentioning that the band centered at 373 nm is two times narrower than the band centered at 323 nm. The spectrum is similar to the absorption spectrum reported in[9] for the PDHG with $M_w = 259660$. In Figure 3, the absorption spectra are show as a function of the film thickness. The film thickness varied from several micrometers to about 30 nm. As could be seen from Figure 3, a significant dependence of the absorption spectra on the film thickness is observed. In particular, when the film thickness decreases, the intensity of the short-wavelength absorption bands increases.

For example, in Figure 3, the spectrum depicted with the curve 2 shows the strongest band at 350 nm. It should be noted that this absorption spectrum is similar to the spectrum reported in[9] for the polymer with $M_w = 73940$. With further decreasing the film thickness, the short-wavelength band centered at 323 nm dominates the absorption spectrum. This spectrum coincides with the spectrum reported in[8] for polymers with the M_w within the range 7600–11400. Note that the band centered at 323 nm is similar to the absorption band observed in the spectrum of the polymer in a toluene solution (Figure 3, curve 4).

We would like to attract attention to the fact that the PDHG film absorption spectrum exhibits different patterns at room temperature, which, as known from the literature, were obtained by changing the polymer molecular weight. However, we have obtained the same spectral patterns for a polymer of a fixed molecular weight by changing the only thickness of the film. So, we have shown that the absorption spectra of the PDHG film strongly depend on the film thickness and consist of four bands with the maxima at 373, 348, 335 and

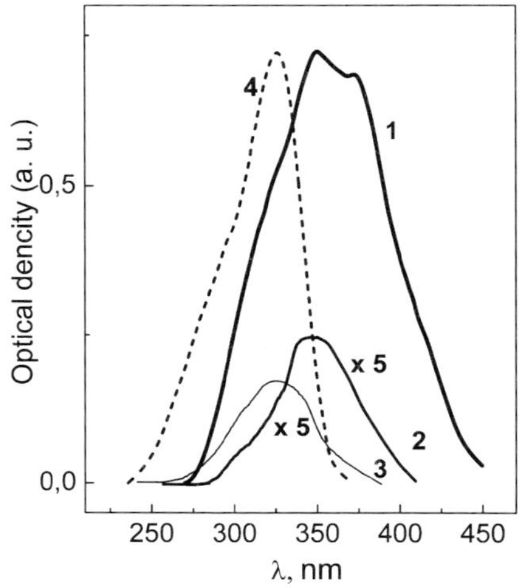

Figure 3.
Thickness dependence of the absorption spectra of PDHG films (293 K). Thickness of the films decreases from several micrometers (curve 1) down to 30 nm (curves 3). For a comparison the absorption spectrum of the PDHG toluene solution was measured (curve 4).

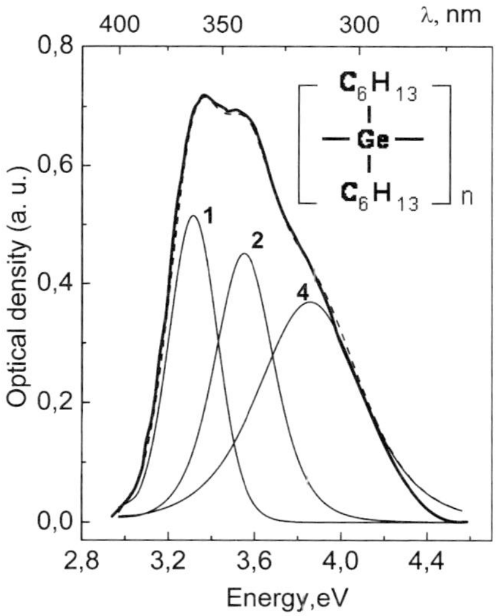

Figure 2.
A room temperature absorption spectrum of a thick PDHG film and the results of the computer curve-fitting analysis after base line substraction.

323 nm. Therefore, there are four types of absorption centers in PDHG film.

Like PDHS, PDHG is a thermochromic polymer. Its electron transition strongly depends on the conformation of the polymer chain. We have observed the thermochromic transition on heating the PDHG film from the room temperature to 323 K: the band at 373 nm is replaced by a broad band at 323 nm at this temperature. This temperature evolution of the absorption spectrum, reflecting the thermochromic transition, shows similarity to that, reported by Bukalov et al.[9] Since the Ge-Ge bond length is longer than that of the Si-Si bond, intramolecular interaction is expected to be less than in PDHS. This suggests that PDHG has more flexibility near the backbone and conformational changes in the polymer chain are more probable for PDHG than for PDHS. This fact may be responsible for the existence of four bands in the absorption spectra of the PDHG film instead of two bands observed in the spectra of PDHS. The presence of several bands in PDHG film absorption spectra at room temperature correlates well with the X-ray data[6] showing that PDHG is less ordered than PDHS. We suppose that these bands originate from the absorption centers with different amount of trans- and gauche-conformers of the polymer chain. For example, absorption centers of the type 1 (corresponding to the absorption band centered at 373 nm) may be associated with trans-conformation of the polymer chains, whereas the absorption centers of the type 4 (corresponding to the absoption band centered at 323 nm) may involve polymer chains with disordered gauche-conformation. The fact that the difference in the peak position between the absorption bands related to trans- and gauche conformers is the same for both polymers (about 50 nm) also confirms our interpretation of the PDHG film absorption spectra. It is known that the polymer chain in a polymer solution has a disordered conformation[11] In agreement with this is the fact that the maximum of the absorption band at 323 nm is close to a similar maximum in the polymer solution spectrum. The absorption centers of the type 2 and 3 (the corresponding absorption bands centered at 348 and 335 nm, respectively) may be associated with polymer chains with different ratio of trans- and gauche-conformers. In addition, since the polymer chain length is of the same order of magnitude as the polymer film thickness, this increases the probability to observe different conformations of the polymer chain.

Fluorescence Spectra of the PDHG Films

As seen from Figure 4, the FL spectra of a PDHG powder and of a thick film consist of two bands of different intensity ratio centered at 366 and 376 nm, the band at 376 nm being higher in the FL spectrum of a thick film, whereas the band at 366 nm is higher in the powder spectrum. In the FL spectrum of the polymer dissolved in 10^{-4} mol/l toluene a single band at 363 nm is observed. FL spectra of PDHG films are also strongly dependent on the film thickness. If the thickness of PDHG film obtained from a 1% polymer toluene solution decreases the FL spectrum reduces to a single band at 366 nm. If the film is obtained from a diluted solution in toluene,

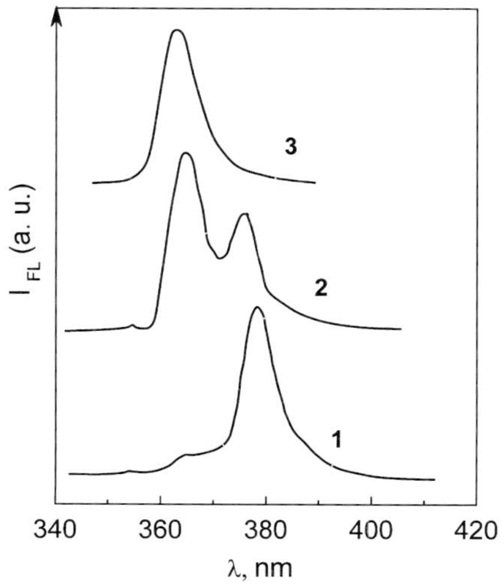

Figure 4.
Fluorescence spectra of the PDHG film, powder and solution at 5 K (curves 1–3, respectively).

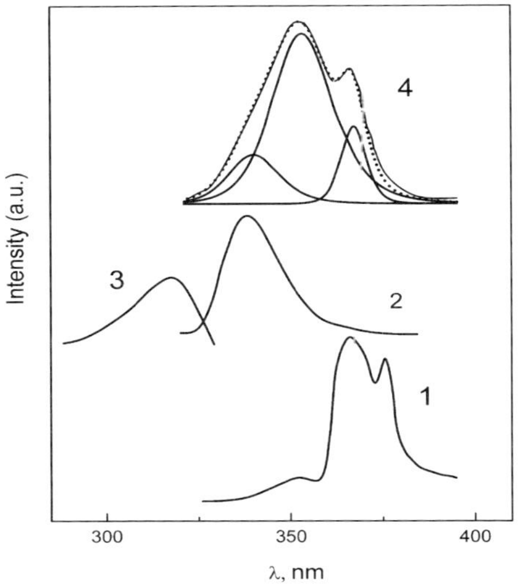

Figure 5.
Fluorescence spectra (T = 5 K, $\lambda_{exc} = 313$ nm): of PDHG thin film obtained from the diluted toluene solution (1), PDHG/SBA-15 composite with pore diameter of 6 nm (2) (the curve 3 corresponds the excitation spectrum of the composite); PDHS/SBA-15 composite with pore diameter of 10 nm and the result of the computer curve-fitting analysis (4).

an intensive blue band with a maximum at 352 nm appears in addition to the bands at 376 and 366 nm (Figure 5, curve 1). Since, in the PDHG film FL spectra recorded at 5 K, one can simultaneously observe several emission bands depending on the film thickness, this suggests that the emitting centers are spatially independent and located at different distances from the substrate, which results in a slow excitation energy transfer between them.

Analysis of the obtained data has shown that, in a whole, three bands centered at 352, 366 and 376 nm are observed in the FL spectrum of PDHG film. We suppose that these bands originate from the emission centers with different amount of trans-and gauche-conformers of the polymer chain. For example, we connected the band at 376 nm with the polymer chain having trans-conformation. The band at 366 nm was assigned to the polymer chains with different distribution of trans and gauche conformers over its length. Appearance of the shorter wavelength band in FL spectrum of PDHG film can indicate the shortening of the polymer segments having gauche conformation. So, when the PDHG film thickness decreases some polymer chains located close to the substrate become more disordered. Similar data was received for PDHS films. It is known that the FL spectrum (T = 5 K) of the thick PDHS film consists of a single band at 371 nm assigned to the trans-conformation of the polymer chain. A new blue band was observed in the room temperature optical spectra of PDHS film only when decreasing the film thickness to 7 nm.[12] The short-wavelength FL band at 343 nm and the corresponding absorption band at 316 nm were assigned to the gauche conformation of the Si backbone. Similar doublets were observed in the FL spectrum of the PMPS thin film[13] and in the diluted PMPS solution.[4] They were assigned to the polymer chains with different distributions of short and long segments over the backbone. Considering that the absorption and FL spectrum of PDHG films are more structured than the similar spectra of PDHS film, we can conclude that the PDHG films are less ordered than those of PDHS.

Low-Temperature Optical Spectra of Nanocomposite PDHG/SBA-15

The FL ($\lambda_{ex} = 313$ nm) and excitation spectra of PDHG incorporated in SBA-15 with pore diameter of 6 nm observed at 5 K are presented in Figure 5 (curve 2, 3). FL spectrum shows a single intensive broad band with a maximum at 338 nm. The excitation spectrum monitored for this FL band shows a single band with the maximum at 317 nm.

Figure 6 shows the temperature dependence of the FL spectra of the PDHG/SBA-15 composite. With increasing temperature, the intensive peak at 338 nm shows a red shift and a strong broadening. The intensity of this band decreases with increasing temperature.

As seen in Figure 5 (curve 2, 1), the FL spectra of the confined PDHG and the

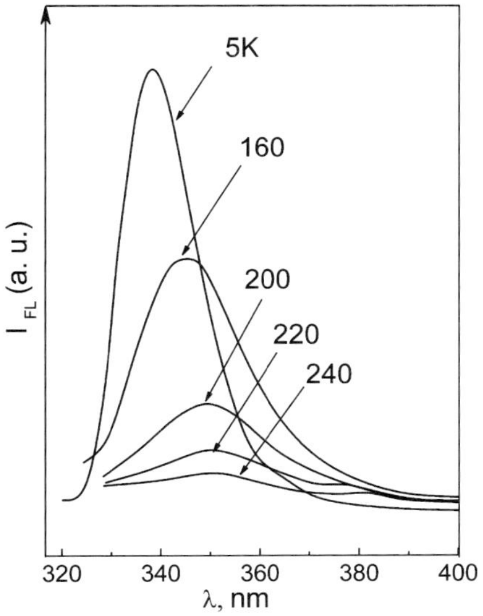

Figure 6.
Temperature dependence of the fluorescence spectra of PDHG/SBA-15 composite with the pore diameter of 6 nm.

PDHG thin film significantly differ. The FL spectrum of the PDHG confined in 6 nm pores displays a single band at 338 nm shifted by 14 nm to the short-wavelength side relative to the FL band in the spectrum of a thin film at 352 nm which has been assigned to the polymer chain having gauche-conformation. The PDHG monomer unit is 1.7 nm long. The simulation of the PDHS/ SBA-15 nanocomposite with the hexagonal pore with the diameter of 6 nm shows that this pore can hold three macromolecules in three alternate corners.[14] So, this significant blue shift is due to the reduction of the intermolecular interaction between the polymer chains. This confirms that the polymer chains are indeed incorporated into the pores of the SBA-15. Only a single band was observed also in the FL spectra of the PDHS/SBA-15 composite with the pore diameter of 6 nm.[2,3] Yet, the result obtained for the PDHS/ SBA-15 composite differs from those for the PDHG/SBA-15 composite in that significant aspect that the confined PDHS polymer chain has a trans–conformation[14] whereas the confined PDHG polymer chain has a gauche-conformation.

The temperature dependence of the band centered at 338 nm in FL spectrum of PDHG/SBA-15 is evidence in favour of our assumption. In the case of PDHS/ SBA-15 composite the band of the trans-form was replaced by the gauche- form band in the process of the thermochromic transition when the temperature was raised from 10 to 290 K,[5,14] whereas the temperature dependence of FL band at 338 nm in spectra of PDHG/SBA-15 is similar to those observed for the bulk film, i.e. the band broadens and its intensity decreases. We can conclude that a blue shift observed in the FL spectrum of this composite can indicate also the shortening of the polymer chains and their transformation into a disordered conformation.

As seen in Figure 5 (curve 4), the FL spectrum of PDHG/SBA-15 nanocomposite with a diameter pore of 10 nm becomes more structured exhibiting three bands instead of one. Note, that three bands in the composite spectrum (10 nm) are shifted to blue side with respect to those in a thin film (curve 1). The bands peaking at 339 and 354 nm, assigned to the polymer chains with more disordered conformations, are blue-shifted by 13 nm. The band at 368 nm, assigned to trans-conformers, has a smaller shift. So, we can conclude that the bands at 339 and 354 nm are connected with the transitions from the individual polymer chains, but the band at 368 nm is connected with the aggregates. Three bands were observed also in the spectrum of the PDHS/ SBA-15 composite with a pore diameter of 10 nm. The nature of these bands is well-known.[5] However, the data for PDHG composite differ from those for PDHS composite. In the case of PDHS composite, the band at 354 is assigned to the polymer chain with the trans-conformation,[5] whereas the similar band in the spectrum of PDHG composite is assigned to the polymer chain with both the trans- and gauche–conformers.

So, it can be assumed that these three bands could be attributed to different structural forms of the PDHG polymer chains coexisting in a restricted pore volume of SBA-15: the polymer chains of the gauche- conformers, the polymer chains with different number of trans- and gauche-conformers, and their aggregates, correspondingly.

FTIR Spectra of the Nanosize PDHG Incorporated into the Mesoporous Silica of SBA-15

Correlations between the changes in the electronic spectra and the structure of the side chains of the PDHG were studied with FTIR spectroscopy.

It is known that there are four major bands in the C–H stretching region (2850–3000 cm^{-1}).[15] The bands at 2851 cm^{-1} (a) and 2921 cm^{-1} (b) can be assigned to the symmetric CH_2^s and asymmetric CH_2^a stretching vibrations, respectively, while the 2871 cm^{-1} band and the 2956 cm^{-1} band (c) are assigned to the symmetric CH_3^s and the asymmetric CH_3^a stretching vibrations in the film, respectively. The two strongest bands around 2851 cm^{-1} and 2921 cm^{-1}, corresponding to the symmetric and anti-symmetric stretching modes of CH_2 groups, are conformation sensitive and their change is indicative of the modification of the trans/gauche ratio. They are shifted towards higher wavenumbers if either the content of gauche conformer or the disorder of the chain increases.[16]

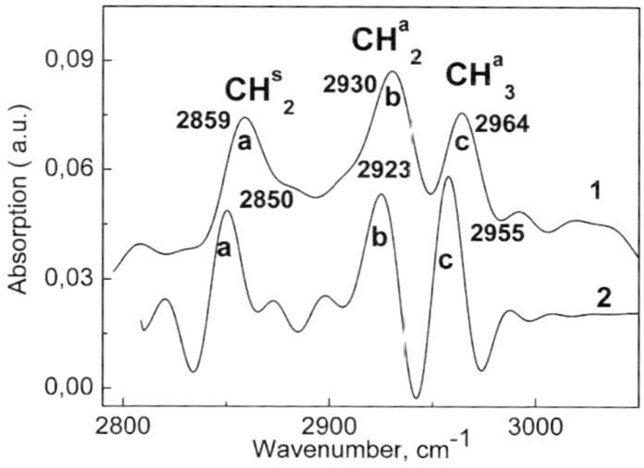

Figure 7.
FTIR spectra of PDHG in the CH stretching region for: the PDHG incorporated into SBA-15 with the diameter pore of 6 nm (1) and the bulk film on the quartz substrate (2).

Figure 7 shows the FTIR spectra of PDHG in the CH stretching region for: the PDHG incorporated into SBA-15 with diameter pore of 6 nm (1), and the bulk film on the quartz substrate (2). When PDHG is incorporated into SBA-15 with pore diameter of 6 nm, FTIR spectrum undergoes a remarkable change compared to the spectrum of the polymer film. Curve 1 in Figure 7 demonstrates these changes: both 2850 cm^{-1} (a) and 2955 cm^{-1} (c) bands experience high frequently shifts by 9 cm^{-1} from their bulk film positions, while the 2923 cm^{-1} band (b) moves by 7 cm^{-1}. An essential shift of the CH_2 and CH_3 symmetric and asymmetric stretching modes to the higher wave numbers in FTIR spectrum of PDHG/SBA-15 composite is evidence in favour of the availability of disordered conformers. These data coincide with the data for the FL spectrum of PDHG/SBA-15 composite with the pore diameter of 6 nm.

Conclusion

The optical spectra of PDHG film and PDHG/SBA-15 nanocomposites have been investigated. We have observed the same modifications of the absorption spectra of the PDHG film with changing the film thickness for the fixed molecular weight of the polymer as those which were previously reported for the case of increasing molecular weight. So, it has been shown that the absorption and fluorescence spectra of PDHG films depend strongly on the film thickness. The presence of certain bands in these spectra allowed us to conclude that there are some centers with different ratio of trans- and gauche-conformers in the polymer chain. The comparison of the observed data with those obtained for PDHS allowed a conclusion, that the polymer chains of the PDHG film are more disordered than the chains of PDHS.

The new nanosize PDHG/SBA-15 composites have been fabricated by introducing the semiconducting polymer poly(di-n-hexylgermane) into nanoporous silica with the pore diameters of 6 or 10 nm. It has been shown that their optical spectra depend on the size of pores and differ from those of the film. The PDHG polymer chain confined in nanopores of the mesoporous silica becomes more disordered with the decrease of the pore size from 10 to 6 nm. In the restricted pore volume of 10 nm there are three spatially separated polymer states having gauche- conformers, different amount of trans- and gauche-conformers, and their aggregates.

Acknowledgements: We gratefully acknowledge Dr. G. Telbiz for preparation of the composites and Dr. G. Dovbeshko for the investigations of FTIR spectra. The authors thank Dr. I. Goliney for useful discussions.

[1] J. Chojnovski, R. G. Jones, W. Ando, Eds., *Silicon-Containins Polymers: the Science and Technology of their Synthesis and Applications*, Kluwer Academic Publishers, Dordrecht, Netherlands **2000**.
[2] N. I. Ostapenko, G. Telbiz, V. I. Ilyin, S. Suto, A. Watanabe, *Chem. Phys. Lett.* **2004**, *383*, 456.
[3] N. I. Ostapenko, N. V. Kotova, G. Telbiz, S. Suto, A. Watanabe, *Fiz. Nizk. Temp.* **2004**, *30*, 658.
[4] N. I. Ostapenko, N. S. Kozlova, S. Suto, A. Watanabe, *Fiz. Nizk. Temp.* **2006**, *32*, 1363.
[5] A. Dementjev, V. Gulbinas, L. Valkunas, N. Ostapenko, S. Suto, A. Watanabe, *J. Phys. Chem.* **2007**, *111*, 4717.
[6] S. S. Patnaik, A. J. Greso, B. L. Farmer, *Polymer* **1992**, *33*, 5115.
[7] H. Tachibana, Y. Kawabata, A. Yamaquchi, Y. Moritomo, S. Koshihara, Y. Tokura, *Phys. Rev. B* **1992**, *45*, 8752.
[8] K. Mochida, H. Chiba, *J. Organometallic Chemistry* **1994**, *473*, 45.
[9] S. S. Bukalow, L. A. Leites, I. V. Krylova, M. P. Egorov, *J. Organomet. Chem.* **2001**, *636*, 164.
[10] D. Zhao, Q. Huo, J. Fenq, B. F. Chmelka, G. D. Stucky, *J. Am. Chem. Soe.* **1998**, *120*, 6024.
[11] R. D. Miller, R. Sooriyakumaran, *J. Polymer Sci. : part A: Polym. Chem.* **1987**, *25*, 111.
[12] M. M. Despotopoulou, R. D. Miller, J. F. Rabolt, C. W. Frank, *J. Poly. Sci., part B: Poly. Phys.* **1996**, *34*, 2335.
[13] Yu. A. Skryshevsky, *ФТТ* **2002**, *44*, 1705.
[14] N. I. Ostapenko, N. S. Kozlova, A. Dementjev, V. Gulbinas, L. Valkunas, S. Suto, A. Watanabe, *MCLC* **2007** (in press).
[15] J. Rabolt, D. Hofer, R. Miller, G. Fikes, *Macromolecules* **1986**, *19*, 611.
[16] B. Schrader, *Infrared and Raman Spectroscopy*, Weinheim, New York, Basel, Cambridge, Tokyo **1995**.

 DOI: 10.1002/masy.200850517

Assessment of Beam Damage in Polymers Caused by in situ ESEM Analysis using IR Spectroscopy

Armin Zankel,[*1] Boril Chernev,[1] Christian Brandl,[1] Peter Poelt,[1] Peter Wilhelm,[1] Michael Nase,[2,4] Beate Langer,[2] Wolfgang Grellmann,[2,3] Hans Joachim Baumann[4]*

Summary: The environmental scanning electron microscope (ESEM) enables in situ analyses of non-conducting samples such as polymers, thus allowing microscopic phenomena to be correlated to macroscopic measurement data. Unfortunately, irradiation of polymers with electrons always causes beam damage[1] and it is unclear whether this damage could influence the outcome of the experiments. The amount of beam damage in polymers is mainly determined by the electron dose, which is a function of the probe current, the irradiation time, the irradiated area and the type of imaging gas used. The beam damage during in situ tensile tests of peel films was assessed using Fourier transformed infrared spectroscopy (FTIR). The band at 965 cm^{-1} turned out to be significant for the estimation of beam damage in this material, which was verified by long-term measurements. The measurements were performed in an ESEM Quanta 600 FEG at parameters comparable to the prior in situ tensile tests. Additional measurements were performed in a Quanta 200 at parameters typical of in situ investigations. Again, the out-of-plane trans =C–H wag at 965 cm^{-1} turned out to be significant for beam damage and was used as an indicator for beam damage (dehydrogenation) for this type of material.

Keywords: electron beam irradiation; ESEM; FT-IR; peel test; PE/PB-1 blends

Introduction

The environmental scanning electron microscope (ESEM) enables in situ investigations of non-conducting specimens, thus allowing microscopic phenomena to be correlated with macroscopic measurement data. The principle of in situ tensile testing of polymers in the ESEM is described in[2] and results of in situ tensile tests of polyethylene/polybutene-1 (PE/PB-1) – peel systems in the ESEM have been published in[3].

Figure 1 shows a photograph of the setup of the tensile stage mounted on the microscope stage of an ESEM Quanta 600 FEG equipped with a Schottky emitter (FEI Company, Eindhoven, The Netherlands). T-peel tests[4] on polymer films were performed, at which the ends of the films were pulled apart with a constant clamp velocity of 1 mm/min using a motor. The T-geometry of the peel film was stabilized in a brass groove in order to emulate the characteristics of a classical T-peel test. During tensile testing, the cracks were imaged in situ with the microscope, with this type of setup enabling an excellent insight into the crack region during the recording of the whole force displacement curve (peel curve).

The imaging principle of the ESEM is fundamental for these tests as it allows the investigation of electrically non-conducting specimens without the necessity of a conductive coating.

This is achieved by a vacuum which is lower than in the conventional SEM,

[1] Institute for Electron Microscopy, Graz University of Technology, Steyrergasse 17, A-8010 Graz, Austria
E-mail: armin.zankel@felmi-zfe.at
[2] Polymer Service GmbH Merseburg, Germany
[3] Martin Luther University of Halle-Wittenberg, Centre of Engineering, Germany
[4] ORBITA-FILM GmbH, Germany

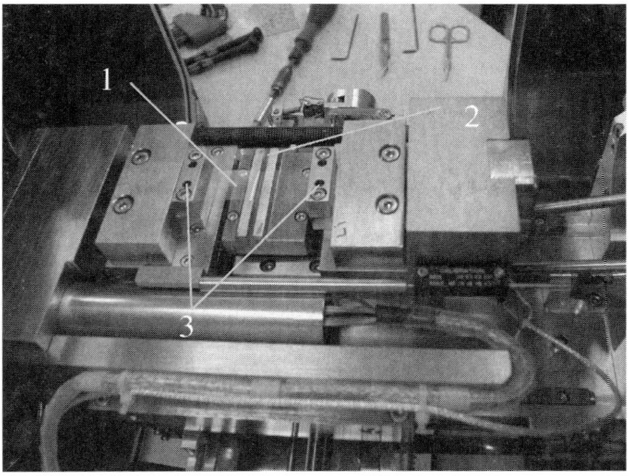

Figure 1.
Tensile stage mounted on the microscope stage of the ESEM Quanta 600 FEG. 1: peel film; 2: groove; 3: clamps.

where typically a vacuum of about 10^{-5} torr (1 torr = 133 Pa) is maintained.

A pressure between 0.08 and 1 torr can be chosen in the low vacuum mode of the ESEM,[5] thereby avoiding charging of the specimen surface.

The schematic in Figure 2 shows the interaction between the electrons accelerated to the detector and the gas molecules. The accelerated electrons collide with gas molecules, thus inducing ionization processes. These processes produce new electrons, called environmental secondary electrons,[6] and positive ions. This process is repeated many times between the surface of the specimen and the detector electrode, leading to a proportional cascade amplification of the original secondary electron signal.[7]

The ionization processes provide a double benefit for imaging. Firstly, the interesting signal is amplified thus enhancing the SNR, and secondly, positive ions are swept towards the surface of the specimen and neutralize the negative charges on the surface. This is the reason why ESEM allows stable imaging of non-conducting specimens.

However, it is well-known that imaging of polymers with electrons can cause beam damage, since the impinging electrons can

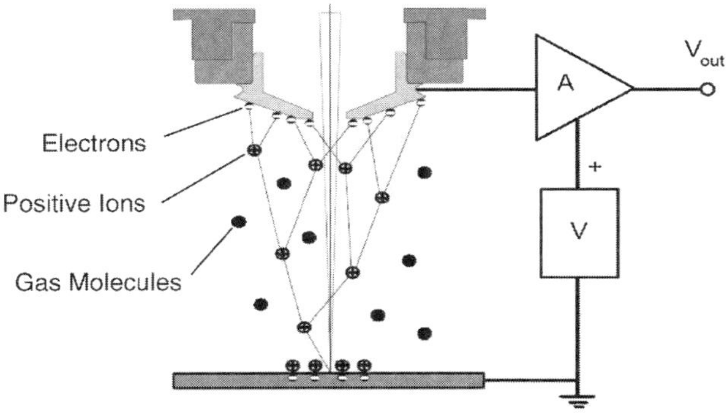

Figure 2.
Schematic of an environmental secondary electron detector.[7]

modify the material's morphology as well as its chemistry.[8,9]

Material and Methods

Peel films and peel systems are a growing field in the packaging industry. Examples of the wide use of these systems include food packaging, medical cutlery and multipurpose wrapping.[10]

Aside from the daily use of these systems in the household, they are indispensable in the field of medicine for objects like cutlery, catheters or prostheses, where quick and sterile handling without the use of a supporting tool is a necessity.[11]

It is not only a question of ease and hygiene in handling but also important for systematic working, to determine and present the fracture behavior of such films in a well designed manner. The following aspects are important for usage: Constant force during the tearing open, small peel forces and fracture at well designed regions of the wrap.

These properties can be determined by the material itself, the conditions during processing of the material and the parameters governing the production of the wrap.

Common peel film systems are based on a matrix-particle structure and thus consist of two phases. In our case the matrix was polyethylene (PE), with particles of 1-polybutene (PB-1) dispersed in it.

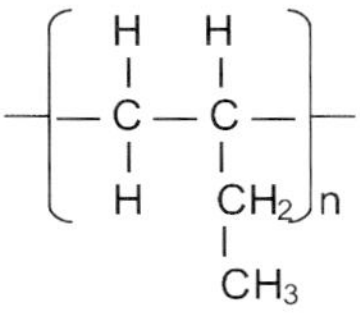

Figure 3.
1-Polybutene (PB-1).

PB-1 is produced by polymerization of 1-butene monomers using a stereospecific Ziegler-Natta catalyst, leading to a linear, isotactic, partially crystalline polymer. Figure 3 shows one unit of the polymer.

Its similar structure to polypropylene makes it compatible with this polymer, whereas it is not compatible with polyethylene. PB-1 copolymers are used in several types of PE foils as a matrix component in order to lower adhesion.[12]

Thus the peel effect is determined by the morphology of the material. Figure 4 shows the particle-matrix morphology. The brighter particles show lamellae, which are rather coarse compared to the lamellae of the PE matrix. Typically the particles have diameters of about 50 to 100 nm.

The build-up of a peel film is sketched in Figure 5. The material is sealed at a temperature of about 140 °C (seal pressure 175 kPa, seal time 2 s, cooling time in air 6.5 s). The seal area has a length (L) of 5.5 mm and a width (W) of 15 mm. The forces operate at the peel arms of the film. The fracture behavior is dependent on the angle between the arms of the peel film and

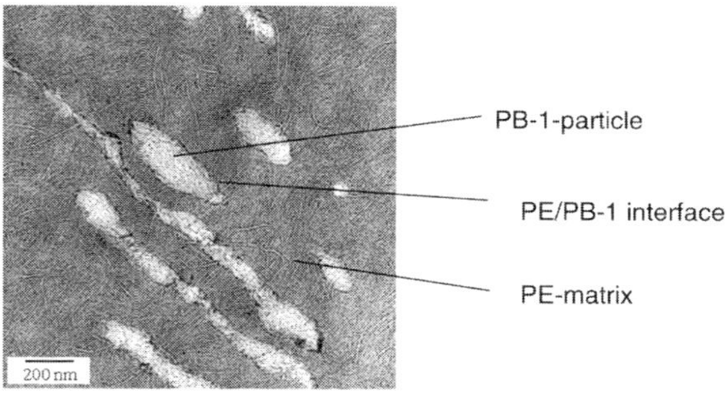

Figure 4.
TEM image of the cross-section of a peel film showing the size and distribution of the PB particles in the PE matrix.

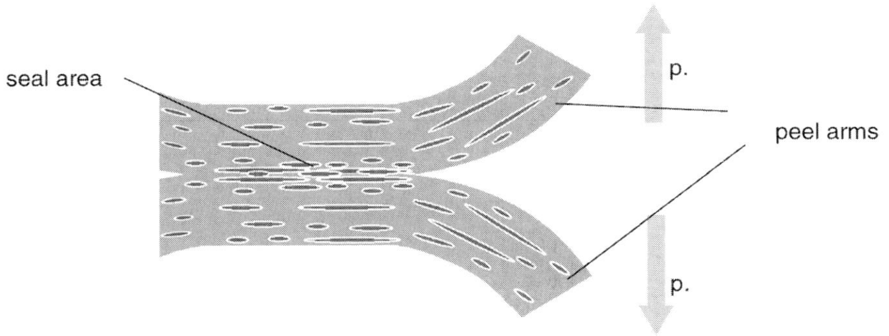

Figure 5.
Build-up of a peel film; p.: peel force.

can be investigated in a tensile testing machine equipped with an additional angle adjustment fixture. The angle geometry of 180° between the arms and 90° between the seal area and the arms is called T-peel test setup. A groove (Figure 1) was used to emulate this geometry for an in situ T-peel test.

Figure 6 shows four micrographs of the crack region of the film during a T-peel test correlated with the force displacement curve (peel curve). The points in the curve

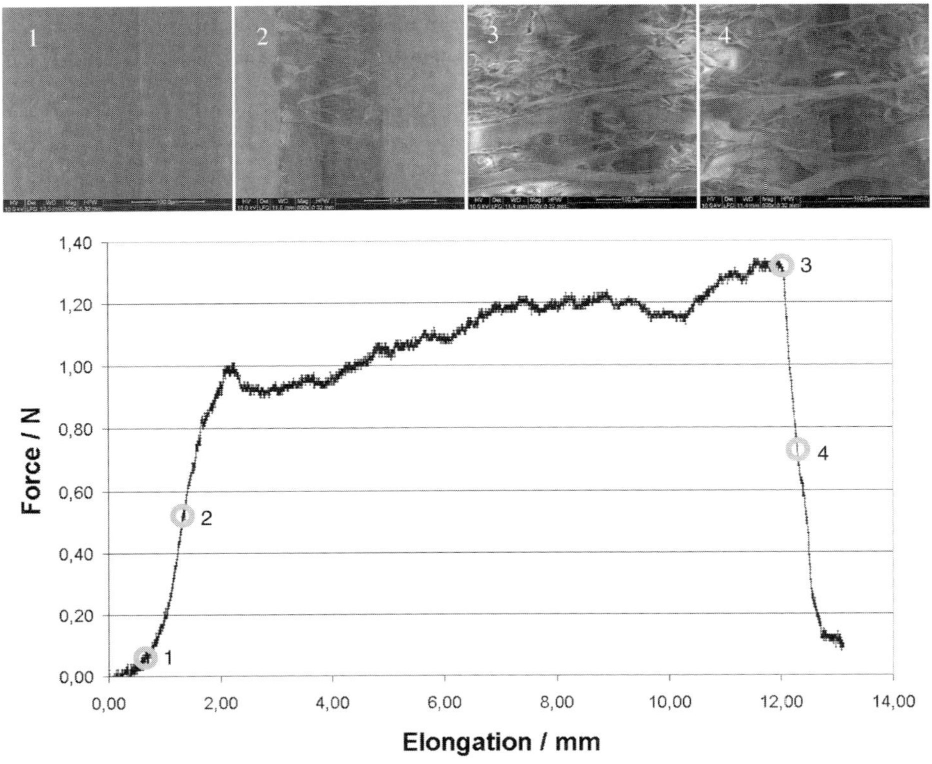

Figure 6.
Top: ESEM images recorded at the points marked in the force; bottom: elongation diagram of the corresponding peel test.

indicate the elongations at which the respective micrographs were recorded.

The correlation between the micrographs and the peel curve gives the opportunity to investigate the crack propagation through the material. Several types of fibrillation of the material can be found, depending on the amounts of polybutene in the matrix. The difference between interlaminar and intralaminar fracture[13,14] can be observed in situ.

But unfortunately, irradiation of polymers with electrons always causes damage to the polymers[1] and it is unclear whether this damage could influence the outcome of the experiments.

The time of exposure of the peel film was measured for assessing beam damage during in situ tensile tests using Fourier transformed infrared spectroscopy (FTIR). Subsequently, the peel films were irradiated for the same length of time with the same probe current.

The current was measured before and after the peel test and also during the irradiation experiment by use of a Faraday cup, which was mounted on the microscope stage, and an ammeter (Keithley 616 Digital Electrometer). A current of about 0.4 nA was measured.

In addition to the experiments in the ESEM Quanta 600 FEG (electron source: Schottky emitter), similar measurements were performed in an ESEM Quanta 200 (electron source: tungsten filament). The dependence of specimen degradation on some of the operating parameters of the microscope was also investigated.

FTIR measurements were performed in micro-ATR (attenuated total reflection) mode, using a Bruker Hyperion 3000 infrared microscope, coupled to a Bruker Equinox 55 FTIR spectrometer. The used 20× μ-ATR objective was equipped with a Ge crystal and an electronic pressure control system, assuring the reproducibility of the single measurements. The spectra were recorded in the spectral region 4000–500 cm^{-1}, spectral accumulation of 32 single scans with spectral resolution of 4 cm^{-1}. The evaluation of the spectra was carried out using the OPUS® software package (Bruker Optics GmbH).

Results

Investigation of Beam Damage in the ESEM Quanta 600 FEG

For the estimation of the beam damage with Fourier transformed infrared spectroscopy (FTIR) the polymer surfaces were irradiated in the ESEM using the same parameters as for the tensile tests, i.e. electron energy: 10 keV, probe current: 0.4 nA, pressure of the imaging gas (water vapor): 0.5 torr, field width of the irradiated area: 160 μm and 320 μm. The FTIR spectra of peel films with a polybutene content of 6% (PE + 6 wt-%PB-1) are depicted in Figure 7. Figure 7a shows the spectrum of an unirradiated film, while Figure 7b and 7c depict the spectra of irradiated films, with an area of a width of 160 μm irradiated for 15 s and an area with a width of 320 μm irradiated for 30 s, respectively. The colored region indicates the integration area for the calculation of the integral intensity.

The spectra demonstrate that nearly the same integral absorption was obtained for the two irradiated films. Since the integral absorption of the correlated band is low relative to other bands of polyethylene, the beam damage of the film at the parameters used can be regarded as very small.

The absorption was measured at a band of 965 cm^{-1}, because the respective spectra show no alteration at other wave numbers with changing electron doses. Consequently, the out-of-plane trans =C–H wag at 965 cm^{-1} was used as an indicator for beam damage (dehydrogenation) in this type of material.

This was verified by long-term measurements, where longer irradiation times correlated with a stronger absorption at 965 cm^{-1}.

Figure 8 presents three spectra, which show an increase in beam damage by a growth of the respective IR band. The first spectrum is that of an unirradiated film (PE + 6 wt-%PB-1), the second that of a

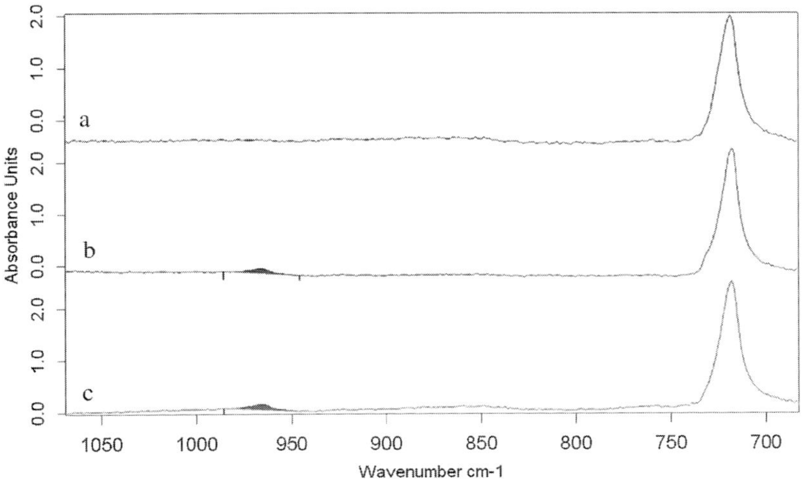

Figure 7.
Infrared spectra of the surface of a peel film (PE + 6 wt-%PB-1); a: not irradiated; b: irradiation of an area with a width of 160 μm for 15 s; c: irradiation of an area of a width of 320 μm for 30 s (the y-axis was scaled with respect to the height of the band at 720 cm^{-1}).

film irradiated for 30 s and the third that of a film irradiated for 10 minutes. A strong increase in the marked absorption band with increasing irradiation time can be observed.

Based on these results, the irradiation damage of peel films was more extensively investigated in an ESEM Quanta 200 with a tungsten filament at parameters typical of in situ investigations.

Investigation of Beam Damage in the ESEM Quanta 200

Specimen degradation caused by the electron irradiation was investigated as a

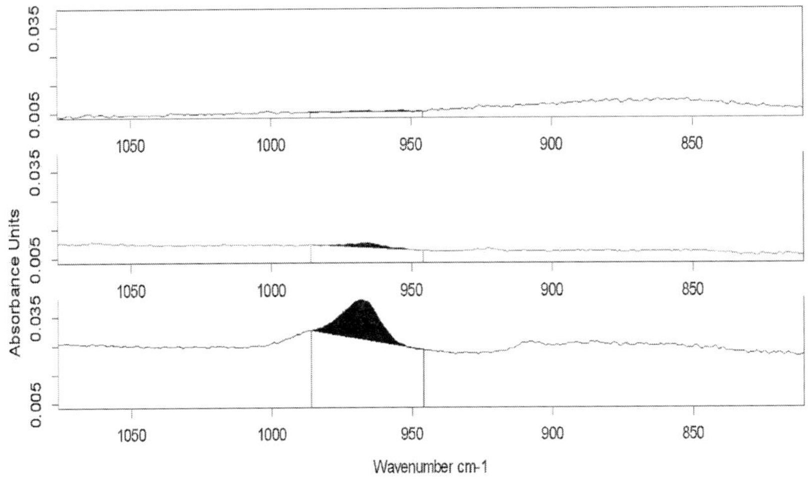

Figure 8.
Infrared spectra of the surfaces of peel films (PE + 6 wt-%PB-1); a: unirradiated; b: irradiation for 30 s; c: irradiation for 10 min (the y-axis was scaled with respect to the height of the band at 965 cm^{-1}).

function of irradiation time, pressure in the microscope chamber and electron energy for peel films with a polybutene content of 3% (PE + 3 wt-% PB-1) and 15% (PE + 15 wt-% PB-1).

Variation of Pressure

The pressure of the imaging gas (water vapor) was varied, with all the other parameters kept constant (electron energy: 10 keV, irradiation time: 30 s, probe current: about 0.04 nA). A varying gas pressure in the microscope chamber also causes a variation in the electron dose. Part of the electrons are scattered out of the beam by the interactions with the gas atoms or molecules (skirt effect[9]), reducing the number of electrons and thus the electron dose at the impact point of the residual focused beam.

The simulations in Figure 9 show this skirt effect for pressures of 0.4, 0.7 and 1 torr, with water vapor as the imaging gas. At 0.4 torr 82% of the electrons are unscattered, followed by 72% at 0.7 torr and 62% at 1 torr.

The infrared spectra in Figure 10 prove by the absence of the absorption peak at 965 cm^{-1}, that no measurable irradiation damage occurred, not even at the lowest gas pressure.

Thus, in contrast to the experiments in the Quanta 600 FEG, no irradiation damage was detected. This can be explained by the much lower probe current and thus lower electron dose than in the ESEM Quanta 600 experiments as well as the lower brightness and larger diameter of beams generated by a tungsten filament in comparison to a Schottky emitter. But this demonstrates that, at least for

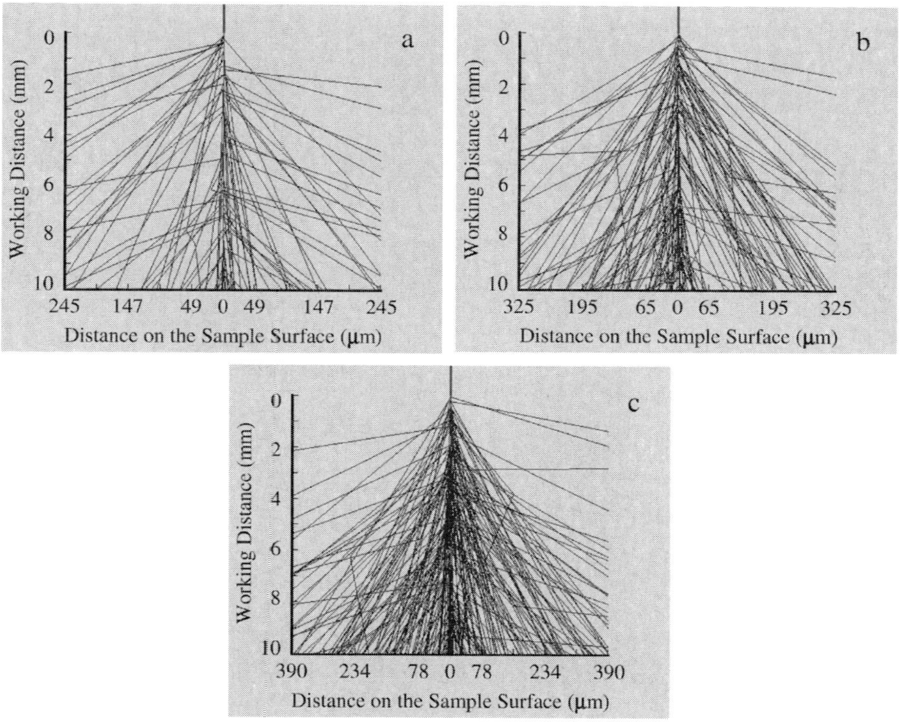

Figure 9.

Simulations of the skirt effect at various pressures in an ESEM using the program "Electron Flight Simulator™" (Small World); parameters: gas: water vapor, working distance 10 mm, electron energy: 10 keV pressure: a: 0.4 torr, b: 0.7 torr, c: 1.0 torr.

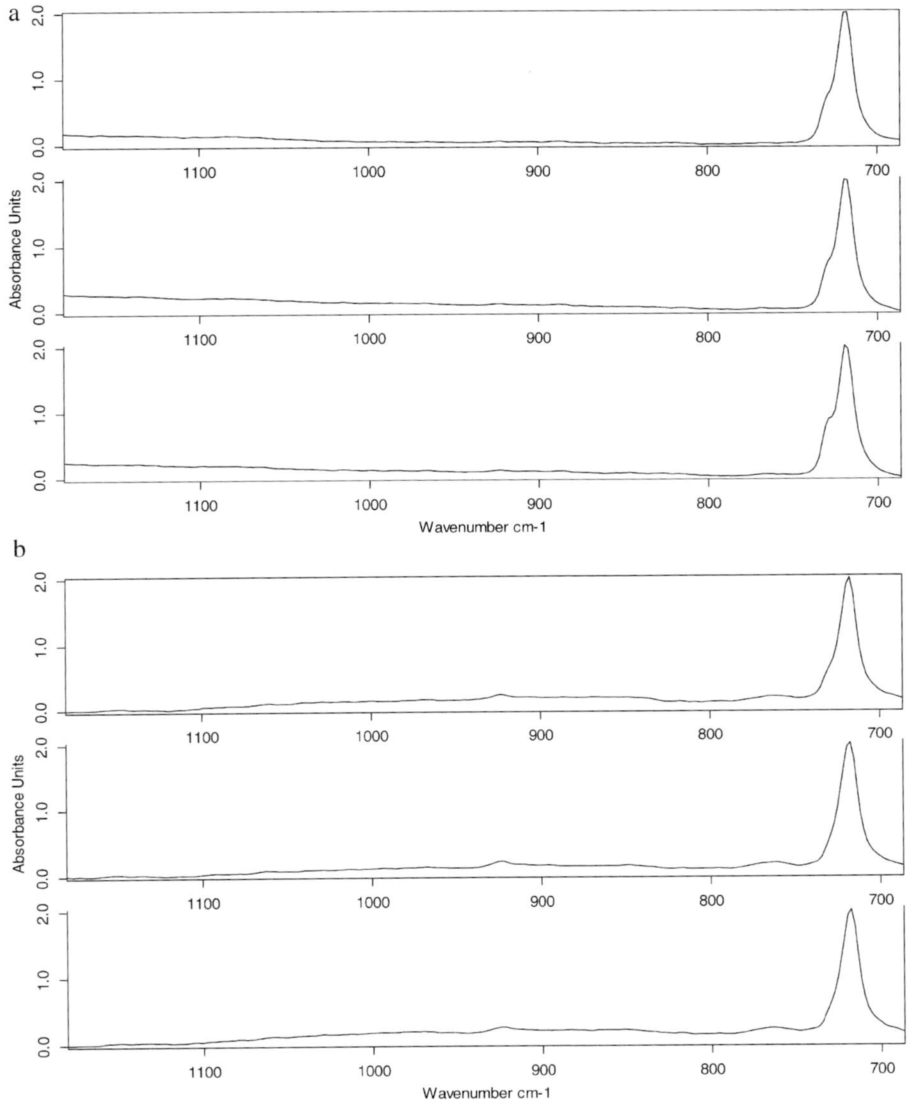

Figure 10.
Infrared spectra of the surfaces of peel films (a: PE + 3 wt-%PB-1, b: PE + 15 wt-%PB-1), with the pressure in the microscope chamber varied: from top to bottom: 0.4, 0.7, 1.0 torr.

this type of material, a threshold for the electron dose exists below which no measurable material degradation can be observed by IR.

Variation of Irradiation Time
However, a strong dependence on the irradiation times and the electron energy was found at higher doses (Figure 11, Figure 12).

Figure 11a shows five spectra of peel films with a polybutene content of 3% (PE + 3 wt-%PB-1), with the beam damage plotted as a function of time.

The difference in the height of the absorption peaks in Figure 11b and 11c

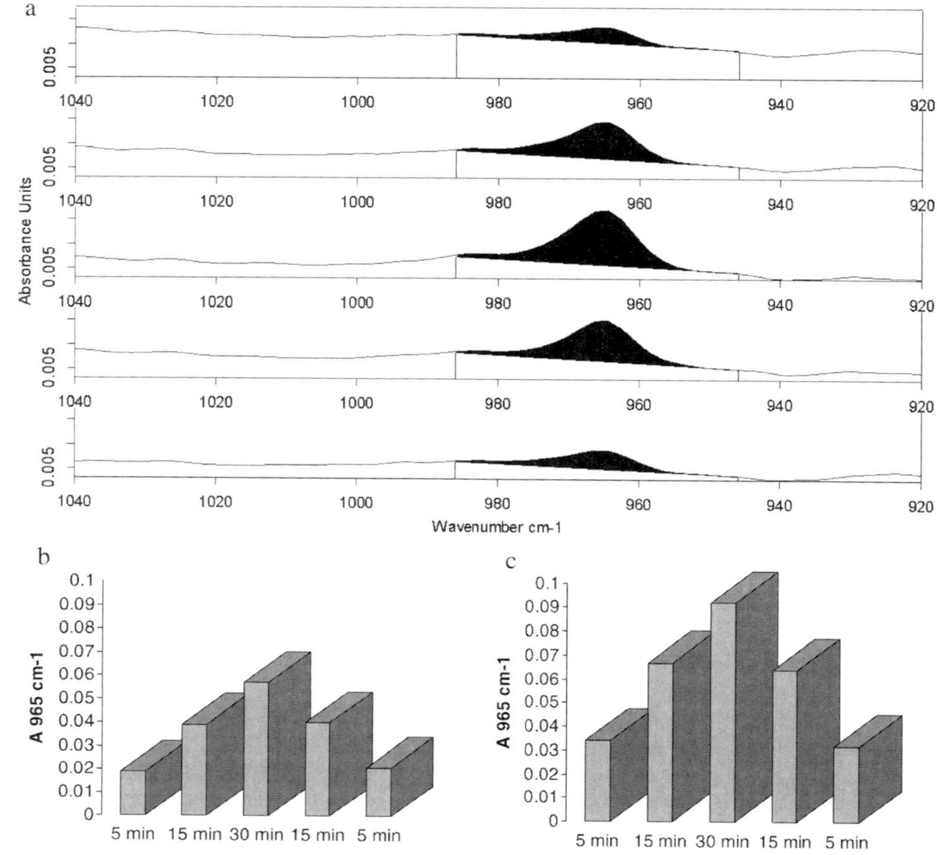

Figure 11.

a: Infrared spectra with the corresponding integration regions for determining integral absorption (PE + 3 wt-%PB-1; irradiation time: 5, 15, 30, 15, 5 min; probe current. 0.021 nA, y-axis: arbitrary units, same scaling for all five spectra); b: integral absorption as a function of the irradiation time (PE + 3 wt-%PB-1; 5, 15, 30, 15, 5 min; probe current: 0.021 nA); c: integral absorption as a function of the irradiation time (PE + 15 wt-%PB-1; 5, 15, 30, 15, 5 min; probe current: 0.032 nA).

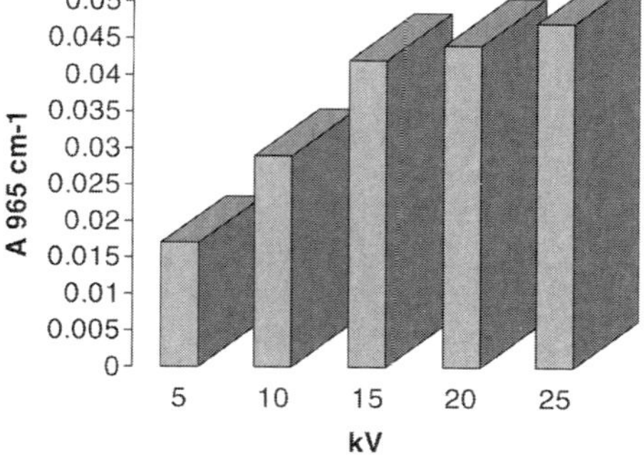

Figure 12.

Dependence of material degradation on the electron energy.

can be explained by the difference in the probe currents (Figure 11b 0.021 nA, Figure 11c 0.032 nA) at which the two investigations were performed. The electron energy was 10 keV in all cases.

Variation of Electron Energy

First measurements of the beam damage at different electron energies confirm that the peak at 965 cm^{-1} in the IR spectrum is significant for the evaluation of beam damage in this type of polymer even at higher energies. The integral absorption in dependence on the electron energy is shown in Figure 12.

It is obvious that beam damage increases monotonically with the electron energy, showing saturation at higher energies.

Conclusion

Fourier transformed infrared spectroscopy (FTIR) was used for the estimation of the beam damage caused by the electron irradiation of peel films during investigations in the environmental scanning electron microscope (ESEM). The films consisted of a polyethylene/polybutene-1 blend. Several parameters were varied to figure out at which parameters the investigation of this material can be performed with minimum beam damage. No measurable degradation of the material was observed for low electron doses.

The absorption band at 965 cm^{-1} turned out to be significant for the estimation of beam damage in this material. This was verified by long-term measurements with high electron doses, which were performed in an ESEM Quanta 600 FEG (electron source: Schottky emitter) under conditions comparable to those used in the peel tests.

Additional measurements in an ESEM Quanta 200 (electron source: tungsten filament) were also performed under conditions typical for in situ analyses. Once more the 965 cm^{-1} out-of-plane trans =C–H wag, due to dehydrogenation, was identified as a marker for beam damage.

[1] R. F. Egerton, P. Li, M. Malac, *Micron* **2004**, *35*, 399.
[2] A. Zankel, P. Poelt, M. Gahleitner, E. Ingolic, *Scanning* **2007**, *29*, 261.
[3] M. Nase, A. Zankel, B. Langer, H. Baumann, W. Grellmann, *Scanning* **2007**, *29*, 50.
[4] W. Grellmann, B. Langer, M. Nase, H. Baumann, International Conference "Werkstoffprüfung 2006 – Fortschritte der Kennwertermittlung für Forschung und Praxis", Conference Proceedings, **2006**, 439.
[5] D. J. Stokes, *Science, Technology and Education of Microscopy* **2006**, 564, http://www.formatex.org/microscopy1/content.htm.
[6] A. L. Fletcher, B. L. Thiel, A. M. Donald, *J. Phys. D: Appl. Phys.* **1997**, *30*, 2249.
[7] FEI, Philips Electron Optics, Eindhoven, The Netherlands,"Environmental Scanning Electron Microscopy, An Introduction to ESEM™", *Robert Johnson Associates* **1996**.
[8] L. C. Sawyer, D. T. Grubb, "*Polymer Microscopy*", 2nd ed., Chapman and Hall, New York **1996**.
[9] J. I. Goldstein, D. E. Newbury, P. Echlin, D. C. Joy, C. E. Lyman, E. Lifshin, L. Sawyer, J. R. Michael, "*Scanning Electron Microscopy and X-Ray Microanalysis*", 3rd ed., Kluwer Academic/Plenum Publishers, New York **2003**.
[10] W. Grellmann, "Mechanische Kunststoffprüfung", manual of the VDI-tutorial Nr. 356205, **2006**.
[11] P. Stober, H. Rist, *Kunststoffe* **2004**, *6*, 66.
[12] F. Schemm, F. Van de Vliet, J. Grasmeder, Polybuten (PB-1) – Faszinierendes Polyolefin, 7. Kunststofffrohrtage in Würzburg, Conference Proceedings **2001**, pp. 1ff.
[13] M. Nase, R. Androsch, B. Langer, H. Baumann, W. Grellmann, *J Polym Sci* **2008**, *107*, 3111.
[14] M. Nase, B. Langer, H. Baumann, W. Grellmann, International Conference "Polymerwerkstoffe 2006", Conference Proceedings, **2006**, 56.

 DOI: 10.1002/masy.200850518

Rheo-optical FT-IR Spectroscopy of LLDPE: Effect of Comonomer and Composite Materials

M. Plass,*[1] R. Streck,[1] J. Nieto,[2] H. W. Siesler[3]

Summary: In order to further evaluate the potential of FT-IR spectroscopic investigations on molecular processes during tensile testing experiments, the behavior of monolayer LLDPE films, made with ethylene-butene and ethylene-octene copolymers, was studied. Additionally, multilayer LLDPE films based on the same C4 and C8 copolymers were investigated. The stress-strain data obtained from the monolayer films indicate differences in the strain hardening region. It seems that the film samples PE 469-30-2 (C4-gas phase process) and PE 469-30-5 (C8 solution process) behave similarly whereas the strain hardening for the PE 469-30-3 (C4 solution process) requires lower stress values. The orientation function changes during the stretching of the films indicating that unfolding of the polymer chains occurs at lower strain for PE 469-30-5 (C8) than in the C4 materials. In the multilayer systems the Primplast 44 material (C8) shows a lower tendency for reorientation in the strain hardening region than the Coex 82 (C4) material. In this region of the stress-strain curve the lamellar structure is already transformed into the fibrillar arrangement. Regarding the orientation behavior of the material above 200% strain, a small increase in f_b was observed, which led to a decrease of f_c. In the octene product possibly the bulky side chains influence the unfolding significantly, producing a higher resistance to unfolding and alignment along the stretching direction. In part, this is potentially caused by the more perfect lamellae in the octene copolymer, which do not include the side chains, while the butene copolymer may have weaker lamellae because they contain a fraction of the side chains which create defects. Consequently, the octene copolymer requires higher stress values to be stretched and finally results in a lower stretchability of this material, as observed on an industrial scale during pallet wrapping tests. Based on the ratio of the structural absorbance parameters of the signals at 729 and 719 cm^{-1} changes in the crystallinity were studied. For the continuous stretching experiment, no monoclinic phase was detected even after Fourier self-deconvolution and peak fitting approaches. Literature data, however, describe that this crystalline transformation takes place as a result of mechanical deformation. Therefore, stepwise stretching experiments which allow an improvement of the spectral resolution to 1 cm^{-1} were carried out. In the deconvoluted spectra the monoclinic, orthorhombic and amorphous LLDPE modifications could be assigned. Ultimate stretchability and stretching force of the films, both monolayer and multilayer, was well correlated to the development of crystalline orientation in the films upon stretching. Other mechanical properties like Elmendorf tear and dart impact can also be better understood with these results.

Keywords: comonomers; FTIR; LLDPE; rheo-optical

[1] Analytical Sciences, Dow Deutschland Aktiengesellschaft, Industriestr. 1, 77834 Rheinmünster, Germany
Fax: (+49) 7227 91 3707;
E-mail: mplass@dow.com
[2] Polyolefins Research, Dow Chemical IBERICA S.L. Autovia Tarragona-Salou S/N, 43006 Tarragona, Spain
[3] Department of Physical Chemistry, University of Duisburg-Essen, Schützenbahn 70, 45117 Essen, Germany

Introduction

The end use properties of a polymer are determined by the complex interconnection of molecular structure, morphology and processing parameters. Thus, the mechanical properties of a polymer will be influenced by the processing conditions. By selection of the extrusion conditions morphological parameters such as crystallinity and orientation may be dramatically influenced which can affect the transparency and other properties of the end-product. Recently the effect of type of short chain branching on the mechanical properties of LLDPE has been investigated [1]. Based on their data the authors concluded that the understanding of the branching effect on mechanical properties is still not fully understood and requires further studies to understand the physical process of deformation.

Rheo-optical FT-IR investigations may allow to contribute to the understanding of those effects since the combination of tensile testing experiment and IR spectroscopy enables us to study changes of the crystallinity and orientation of polymer units during tensile testing and lead to a better understanding of the polymer deformation mechanism. The idea of continuous monitoring the structural transitions in polymers during their deformation was realized for the first time in the works of the Zhurov school[2]. In the early 80's an approach of a miniaturized tensile testing equipment which can be placed into the sample compartment of an FT-IR spectrometer was refined[3]. As a result of this experiment stress-strain curves and FT-IR polarization spectra of polymer films can be recorded simultaneously and provide information on the change of the orientation function and the crystallinity of the investigated polymer.

In the present paper we have applied this approach on mono- and multilayer polyethylene films, which were produced from different comonomers. Tensile testing experiments on butene and octene based PE films have shown that the type of comonomer influences the properties in the strain hardening region and the overall stretchability. The purpose of this work was to relate differences in tensile properties to changes in the orientation function or crystallinity.

Our approach to understand the molecular behavior during the stretching process by simultaneous stress-strain and infrared spectroscopic measurements could be described as follows:

- Investigation of the monolayer which was formed from butene and octene copolymers
- Investigation of multilayer films which also contained layers of the different copolymers.

Experimental Part

Samples

Monolayer films of three different linear low density polypropylenes (LLDPE's), two 1-butene (C4) copolymers and one 1-octene (C8) copolymer were used. Two of them are 1-butene (C4) copolymers produced by different processes which results in a significantly different molecular structure. Additionally an 1-octene (C8) copolymer made in the solution process was investigated. The molecular weight distribution and the comonomer distribution were analyzed by High-Temperature Gel Permeation Chromatography (HT-GPC) and by CRYSTAF, respectively. The basic descriptive parameters of these resins are given in Table 1 below.

Two sets of conditions were used to evaluate the films: First, mechanical properties of the original films were measured. Second, the films were stretched up to 200% strain using a Highlight stretcher, and selected properties were evaluated on the stretched film.

Additionally, two multi-layer co-extruded films of the A-B-C type were also included in the study. In this case A and B are identical and non-cling material, so mostly higher density LLDPE in the 0.916 to

Table 1.
Description of LLDPE resins used for the monolayer films.

Monolayer films	Process	Melt Flow Index [g/10min]	Density [g/cm³]	Comonomer
PE 469-30-2	Gas phase	2.6	0.917	C4
PE 469-30-3	Solution	2.0	0.918	C4
PE 469-30-5	Solution	2.3	0.918	C8

0.922 g/cm^3 density range. Layer C (typically 15% of the total thickness) serves as cling layer and is a lower density material in the 0.904–0.912 g/cm^3 density range. The first sample is labeled "Primplast 44" and consists of A and B layers made of a C8 LLDPE made in the solution process with 3.3 g/10 min melt index and 0.917 g/cm^3 density. The second film contains C4 LLDPE, labeled "Coex82".

The stretching performance of these films was measured using a Highlight stretcher, which reproduces conditions of an industrial pallet wrapper. The results of maximum stretchability were 260% for Primplast44 and 356% for Coex82. From DSC data it was determined that the crystallinity of both coextruded films is quite similar and can be estimated at 44 and 46%, for Primplast44 and Coex82, respectively[4].

FT-IR spectroscopy

The samples were analyzed using a self-built stretching device similar to the device which was mounted into the sample compartment of the FT-IR spectrometer. The setup follows the description given in [5]. The initial dimensions of the film are 10×15 mm^2. The thickness of the film was appr. 20 μm. Accurate values are given in Table 3. Using a linear velocity of 0.02 mm/min for the clamp movement, the film is stretched up to a stress of ca. 500%. During the continuous elongation of the film FT-IR polarization spectra were recorded with alternating rotation of a KRS5 polarizer using an IFS 66 FT-IR spectrometer (Bruker Optik GmbH, Ettlingen, Germany). 10 scans per spectrum were accumulated in the rapid-scan mode with a spectral resolution of 4 cm^{-1}.

Results and Discussion

Molecular Structure and Film Properties

A summary of the HT-GPC and CRYSTAF results is given in Table 2.

Butene LLDPE samples produced with the gas phase process are generally not as heterogeneous in comonomer distribution (CD) as the gas phase hexene LLDPE's. This is due to the much closer reactivity between butene and ethylene compared to hexene and ethylene. Therefore, the two butene samples have a relatively similar CD shape, in general terms. Still some more heterogeneity is seen for the gas phase sample, as observed by the higher Sigma (more heterogeneous) and the higher soluble fraction, compensated by the higher HDF peak temperature.

Table 2.
HT-GPC and CRYSTAF data on the monolayer films.

Resin	MWD given in g/mol measured by HT-GPC				Comonomer distribution measured by CRYSTAF					
	Mp	Mw	Mn	Mw/Mn	%HDF	HDF Tp in °C	%SF	Sigma	T_{median}	Copolym Tp in °C
PE 469-30-2	44076	84176	22580	3.73	17.3	80.2	9.5	14.9	57.8	61
PE 469-30-3	47987	88784	22962	3.87	17.1	78.4	5.2	13.1	62	65
PE 469-30-5	46238	84512	24001	3.52	18.4	81.4	6.6	14.1	62.2	63

HDF: High Density Fraction; SF: Soluble fraction at 30 °C; Sigma: Breadth of the Comonomer Distribution.

Table 3.
Characteristic parameters of the mono-layer films.

Sample	Stretch level	Thickness	Elmendorf Tear –CD		Dart Impact	Puncture	Yield Tensile	Ultim. Ten. Strength	Ultim. Elong.	Toughness
	%	micron	g	g/micron	g	J/cm^3	MPa	MPa	%	MJ/m^3
PE 469-30-2	0	18.2	314	17.3	55	7.17	5.5	39.6	401	66.2
	200		277		<50	5.41				
PE 469-30-3	0	21.1	281	13.3	55	7.51	5.9	38.4	330	56.1
	200		260		<50	5.9				
PE 469-30-5	0	17.8	526	29.6	140	17.06	5.8	36.3	335	52.1
	200		351		81	11.33				

The octene LLDPE is narrower in MWD, particularly on the low MW side. The CD is typical for the solution process.

Monolayer film properties, both for the unstretched and the stretched films, are reported in Table 3. Additionally, the stress-strain characteristics of the test in the highlight stretcher are listed in Table 4. The butene gas phase film is slightly more stretchable than the solution process grades. However, the octene grade stands out in terms of mechanical properties.

Stress-Strain Curves

Monolayer

For the mechanism of molecular reorganization different theoretical models are discussed in the literature [6,7]. Whereas one model assumes that the polymer actually melts under stress and then reorganizes, the second idea is based on the assumption that the crystallites themselves rotate as a response to the applied strain. Highly crystalline polyolefins like HDPE show a sharp peak in the yield region. This is due to the break-up of large lamellar crystals into smaller segments linked by newly formed tie-chains [8,9]. In contrast to HDPE, a less defined yield peak is expected in the stress-strain diagram for LLDPE films because the latter contain smaller crystals and lower crystallinity. For all monolayer films only diffuse yield points can be found at about 15 and 70% strain as can be seen in Figure 1. In the initial (elastic) region the monolayer samples behave similarly as expected from the very similar crystallinities. This is the case also for the yield and draw plateau regions. The differences in the stretchability become significant above strain values of ca. 150%. Surprisingly the samples PE 469-30-2 and PE 469-30-5, which are based on C4 (gas phase) and C8 (solution) comonomers, respectively, behave differently than the sample PE 469-30-3, which is also based on butene as comonomer in the solution process. In the strain hardening region the slope of the octene copolymer is higher, as expected, because octene is known to be more effective in building tie chains. More tie chains lead to an increase in the resistance of the sample to be stretched, therefore building up stress faster. Comparing the two butene copolymers, there is no clear reason why they should have the observed difference.

Table 4.
Tensile data of the raw material.

Sample	Highlight Performance Data		
	Max. Stretch %	Max. Str. Force N	Av. Str. Force N
PE 469-30-2	244	321	274
PE 469-30-3	227	340	312
PE 469-30-5	210	346	327

Multilayer

The differences caused by the use of butene and octene comonomers in LLDPE become more obvious in the multilayer films. Only a diffuse yield point can be observed for these films which are at similar strain amplitudes for both materials. In the strain hardening region Coex82 (C4) shows a lower slope than the Primplast 44 (C8) material. Again, this is expected from the higher tie-chain

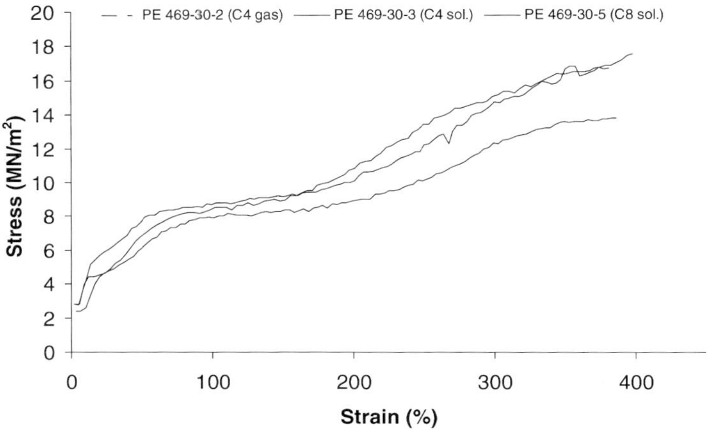

Figure 1.
Stress-strain curves of the monolayers

capability of the hexyl branches from 1-octene, compared to the ethyl branches from 1-butene.

Infrared Spectra

The infrared spectra of the multilayer samples (thickness: 18 μm) are given below (Figure 3). The infrared spectrum of semicrystalline polyethylene has been studied extensively by numerous authors and the detailed band assignment is described in [10–13]. In the spectrum below the $\nu(CH_2)$ bands are, even at this thickness, too intense and thus not appropriate for further analysis. However, the crystallinity sensitive bands in the out-of -plane deformation range between 750 and 680 cm^{-1} are accessible at appropriate absorbance intensity.

In the spectra the band doublet at 729 cm^{-1} and 719 cm^{-1} which can be assigned to the in- and out-of-plane CH_2 rocking vibrations of LLDPE, are the most important absorptions for the characterization of orientation. This band pattern is rather complex due to of orthorhombic and monocline crystalline and amorphous components [14]. For the analysis of the orientation behavior during tensile testing of LLDPE

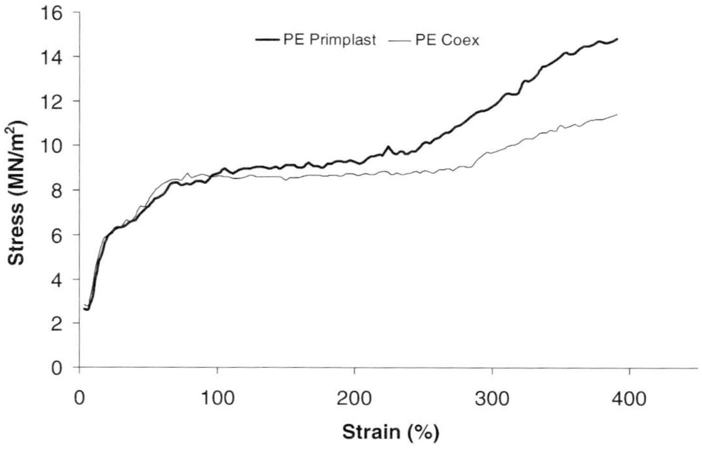

Figure 2.
Comparison of the stress-strain curves of the multilayer films

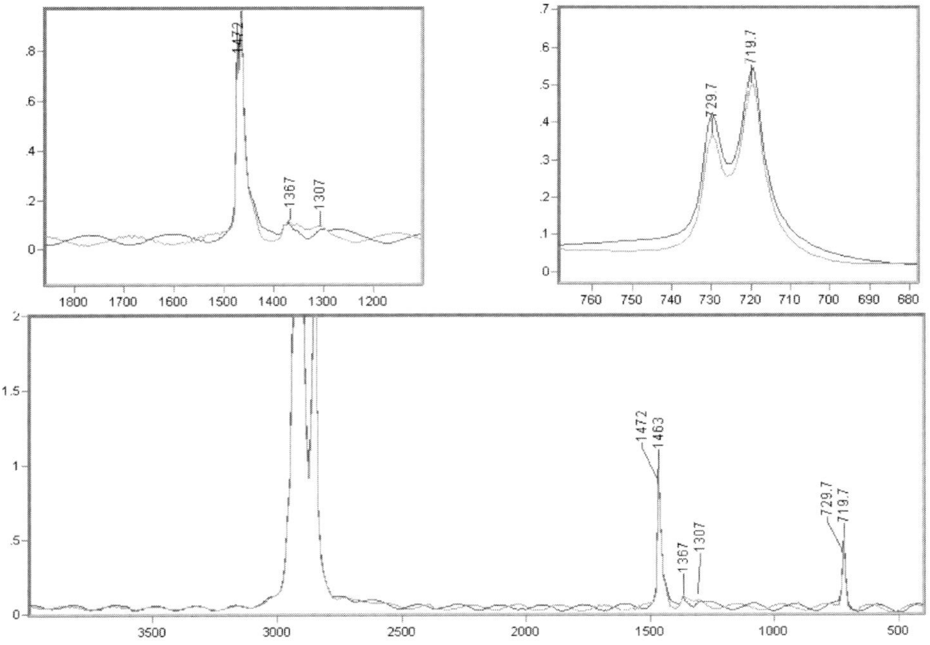

Figure 3.
Infrared spectra of Coex82 (red) and Primplast 44 (blue)

the assumption was made that the contribution of the amorphous component to the $719\,cm^{-1}$ band is relatively small. The validity of this assumption will be discussed based on the deconvoluted spectra in the later paragraph.

The calculation of the orientation functions for the crystallographic axes of the orthorhombic unit cell can be done using the dichroic ratio, R, of the signal at $729\ cm^{-1}$, which is polarized along the crystallographic a-axis and the signal at $719\ cm^{-1}$ is polarized along the crystallographic b-axis. Thus, the orientation functions for the three crystallographic axes a, b and c can be calculated according to the following equations:

$$f_a = \frac{R_{729} - 1}{R_{729} + 2}$$

and

$$f_b = \frac{R_{719} - 1}{R_{719} + 2}$$

The orientation function of the crystallographic c-axis is then obtained from the orientation functions f_a and f_b[15]:

$$f_a + f_b + f_c = 0$$

Orientation Function

According to Peterlin the application of stress results in tilting, orientation and ultimately fragmentation of the material which corresponds to the transformation of a chain folded lamellar structure to a fibrillar arrangement[8,9]. As a result of the molecular processes of the lamellar orientation and partial chain-unfolding, the orientation function of the crystallographic axes will change[9]. Onogi and Asada[16] have predicted for transformation of the lamellar unit that the f_b will increase whereas f_a will show up. The unfolding mechanism should result in a decrease of both orientation functions f_a and f_b.

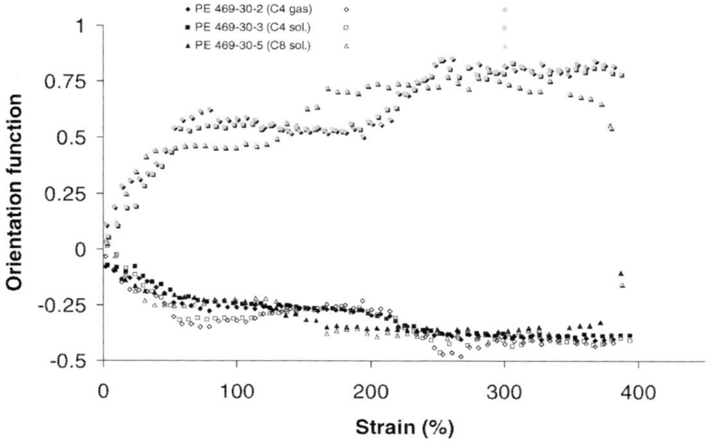

Figure 4.
Orientation function vs. strain of the monolayer LLDPE films (symbols indicate different LLDPE films, orientation functions: grey: f_c, black: f_b open: f_a)

The orientation function f_a, f_b and f_c for the crystallographic axes a, b and c of the orthorhombic structure are shown in Figure 4 and 5.

Monolayer

In terms of the orientation vs. strain behavior the curves in Figure 4 can generally be divided into four regions. At low strain amplitudes the elastic elongation below the yield point is observed which will not lead to substantial changes in the lamellar arrangement (<70% strain). In the following region the lamellar structures will be aligned towards the drawing direction. This will affect the orientation functions in such a way that orientation function f_a decreases significantly while there is no or only a small increase monitored in f_b and f_c.(strain: 70 to 140% for the C8 copolymer, 70 to 200% for the C4). Even f_c of the gas phase butene seems to decrease slightly.

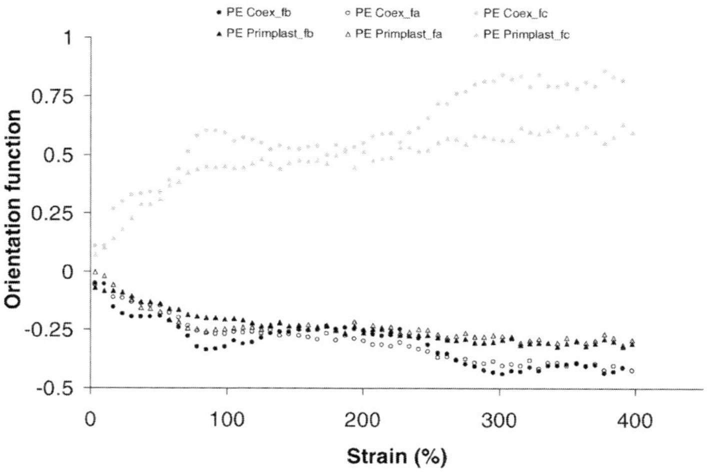

Figure 5.
Orientation function vs. strain curves of the crystallographic axis (symbols indicate different LLDPE films, orientation functions: grey: f_c, black: f_b open: f_a)

According to Siesler [3] this behavior can be interpreted as an effect of rotational motion of the c and b axis towards the direction of the applied stretch.

In the third region unfolding of the folded chains determine the orientation functions and will result in an increase of f_c while f_a and f_b decreases until all orientation functions reach a plateau, which is the fourth region. Further stretching of the films will cause a slight improvement in the alignment of the crystallographic axes and will finally end in the fracture of the material.

Discussing the differences of the behavior of the monolayer samples in the region above a strain of ca. 150% becomes important. As seen in Figure 4 at this stage the crystallographic axes are already aligned and the orientation function for the c axis approaches values of 0.5 whereas f_a and f_b approach values close to −0.25. Further stretching of the films results in an increase in the orientation function, f_c, for PE 469-30-5 (C8) at ca. 150% strain. In contrast, the C4 based PE films show the effect of strain hardening at slightly higher strain values (ca. 200%). Furthermore, the changes in the orientation along the c-axis are smaller for PE 469-30-5 (C8) than for the C4 copolymers. Obviously the unfolding of the chains can be better accomplished by the C8 comonomer. It seems that the larger side chains in the C8 copolymer force the unfolding of the chains (strain hardening) to start at a lower strain for the C8 copolymer. They also prevent the lamellae from aligning with the stretch direction as much as the lamellae in the C4 copolymers do.

Multilayer

The differences of the behavior of both samples in the region above a strain amplitude of 200% becomes important. As seen in Figure 5 at this stage the crystallographic axes are already perfectly aligned and any increase in the elongation needs to be assigned to further unfolding of the chains which can obviously be better accomplished in Coex 82 (butene based) than in Primplast 44 (octene based). The orientation functions f_c and f_b are affected by the unfolding of the chains. Obviously the hexyl side chains hinder the formation of all-trans conformers more significantly than the ethyl groups.

For the multilayer films the results discussed in the previous section on monolayer films can be confirmed. Clearly the Coex 82 (C4) can be oriented to a higher degree than the Primplast 44 (C8) film, and this is translated into the higher ultimate elongation or stretchability that can be achieved under industrial conditions (260% for the C8, 356% for the C4). Regarding the degree of orientation, the monolayer results and the results on the Primplast film show similar trends. However, it could not be observed that there are differences in the strain values which initiate the second phase of the orientation.

Crystallinity

Apart from the characterization of the changes in the orientation behavior during elongation, the structural absorbance of the band at 729 cm^{-1} should provide a tool for monitoring the distortion and destruction of the orthorhombic crystalline modification during deformation. However, in the absence of another thickness-relevant band in the spectrum, the intensity of the 719 cm^{-1} band was used as a fairly reliable reference signal. Hence the intensity ratio $A_{0(729)}/A_{0(719)}$ directly reflects any consequences of the mechanical treatment on the orthorhombic crystal lattice.

Keeping in mind that Figure 6 shows only trend changes due to the aforementioned approximation, some basic conclusions can be drawn.

From Figure 6 it can be concluded that the monolayer films behave similarly with reference to the changes in the crystallinity. The intensity ratio changes only at low strain amplitudes (<70%) until the diffuse yield point is formed. In all following phases described above the fraction of

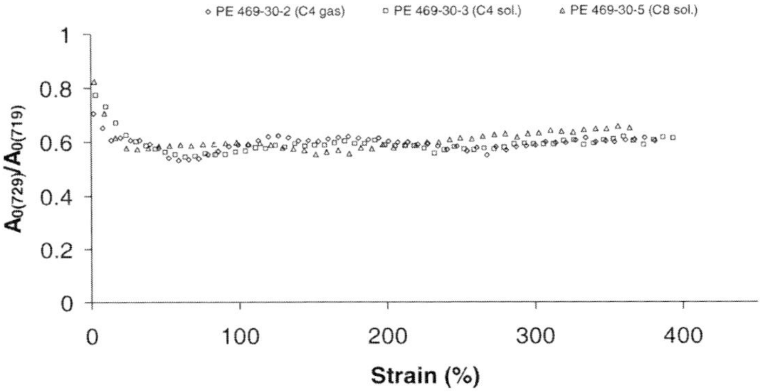

Figure 6.
Plot of the $A_{0(729)}/A_{0(719)}$ structural absorbance ratio vs. strain for the elongation of the PE monolayer films up to ca. 400% strain

orthorhombic crystalline material is constant. This observation is confirmed by Siesler for butene and octene LLDPE. As a significant difference to his study, the concentration of the crystalline polymer is higher in our films, and also the plateau region is already reached at low strain values [15].

It has been shown in the literature that the structural changes during elongation are superimposed by a partial transformation from the orthorhombic to a monoclinic crystal phase [17]. Due to its packing in the unit cell the monoclinic structure exhibits only one absorption band at 717 cm^{-1} for the CH_2 rocking mode. We have specifically looked for the identification of this signal, which is overlapped by the absorptions of the orthorhombic modification. Using Fourier self-deconvolution techniques, derivative methods and band fitting procedures in order to enhance the spectral resolution we have not been successful in analyzing this signal. Reasons for this result might be due to the low resolution (4 cm^{-1}) which was used in this experiment to using the rapid scan mode.

Therefore, in a second test the multilayer samples were stretched in a discontinuous mode (steps of 40% elongation) using a spectral resolution of 1 cm^{-1}. The spectra were corrected by subtraction of the overlapping sinusoidal fringing pattern and transferred to a band fitting program (Jandel Scientific). This program creates a fitting file based on the number of signals found in the second derivative of the spectra (Figure 7).

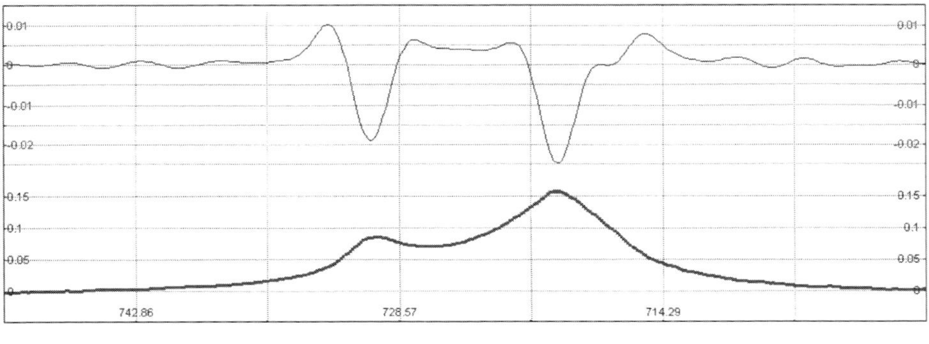

Figure 7.
Absorbance spectrum of Coex 82 at a strain of 440% and its 2nd derivative

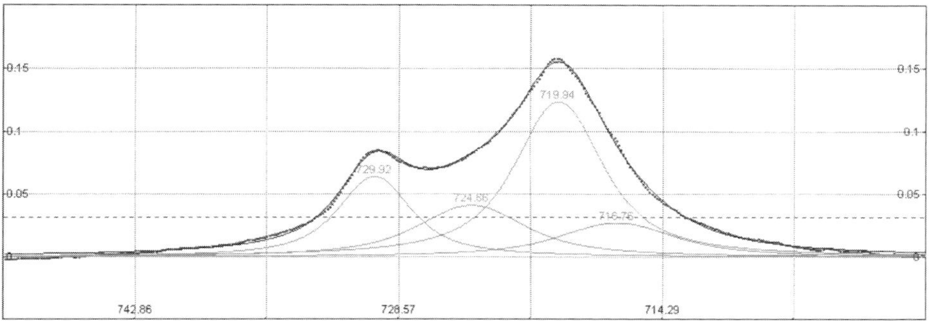

Figure 8.
Absorbance spectrum of Coex 44 at a strain of 440% overlaid by the four separate functions.

From the behavior of the 2nd derivative it becomes obvious that the band profile can be fitted with at least four separate functions. Two of them can be assigned to the two crystal axes of the orthorhombic LLDPE whereas the band at 717 cm^{-1} is due to the monoclinic modification. The broad overlaying band at 724 cm^{-1} can be assigned as amorphous PE. The results of the fitting procedure are shown in Figure 8.

If the relative changes of the intensities of the four bands with respect to its contribution to the total area are studied in dependence of the applied strain, then Figure 9 is obtained. Clearly the changes observed in the plot do not follow a trend. Hence even with this refined methodology we cannot verify the effect of a crystalline phase transition.

Relationships with Final Film Properties

As shown in Table 3 for the monolayer films, the C8 based film results in a lower maximum stretchability under industrial conditions, and it needs slightly higher force to be stretched. This is well in line with the observed stress-strain curve and the relationship between orientation functions and percentage of strain. A lower degree of orientation of the lamellae in the direction of stretch (machine direction MD) may contribute to a higher Elmendorf tear resistance in cross direction (CD). The orientation of the amorphous phase could be another morphological factor. The key role of the octene comonomer on the generation of tie chains is reflected in the

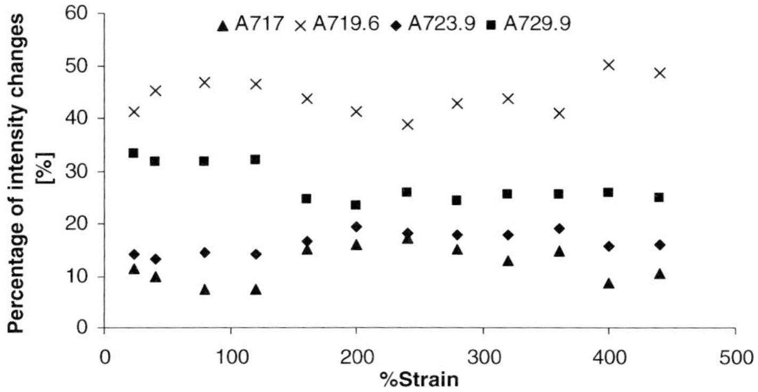

Figure 9.
Relative changes of the intensity of the fitted bands with respect to the applied strain as measure for the cryatlllinity changes during film stretching

excellent balance of tear and dart impact, both for the un-stretched and the stretched films. The hexyl side chains oppose more resistance to stretching, as shown by the lower crystalline orientation, earlier onset of lamellar unfolding and the higher slope of the strain-hardening region. The lamellar unfolding also releases a new number of tie chains that contribute to mechanical properties like tear and dart impact, and this starts to happen at a lower strain for the C8. At the same time, the fragmented lamellae from the C8 may be more effective in anchoring the stretched structure (providing toughness), as they do not contain defects, like the ethyl branches in the case of the C4 copolymers. It is possible that the weaker lamellae in the C4 copolymers are one reason for the different orientation behavior as described above.

The multilayer films, as discussed above, again confirm that a lower crystalline orientation is achieved for the octene film at a given strain, which indicates a stronger resistance of the lamellae to become aligned in the stretching direction. This results in a higher stress and therefore a lower ultimate stretchability: 260% for the Primplast44 film, 365% for the Coex82 (C4) film.

Conclusions

In an effort to further evaluate the potential of FT-IR spectroscopic investigations of the molecular processes during tensile testing experiments, the behavior of monolayer LLDPE films, produced with butene and octene comonomers, was studied. Additionally, multilayer LLDPE films based on the same C4 and C8 copolymers were the subject of our investigation.

With regard to the stress-strain data the monolayer films indicate differences in the strain hardening region of the curves. It seems that the samples PE 469-30-2 (C4) and PE 469-30-5 (C8) behave similarly whereas the strain hardening for the PE 469-30-3 requires lower stress values. The orientation function changes during the stretching of the films; the unfolding of the polymer chains occurs at lower strain for PE 469-30-5 (C8) than in the C4 materials.

In the multilayer systems the Primplast 44 material (C8) shows a lower tendency for reorientation in the strain hardening region than the Coex 82 (C4) material. In this region of the stress-strain curve the lamellar structure is already transferred into the fibrillar arrangement. Regarding the orientation behavior of the material above 200% strain, a small increase in f_b was observed which reflects on f_c where a decrease in the orientation was found. In the octene product the bulky side chains influence the unfolding significantly, producing a higher resistance to unfolding and alignment along the stretching direction. Consequently, the octene copolymer requires higher stress values to be stretched and finally results in a lower stretchability of this material, as observed also on an industrial scale during pallet wrapping tests.

Based on the structural absorbance ratio of the bands at 729 and 719 cm^{-1} changes in the crystallinity were studied. For the continuous stretching experiment the monoclinic phase could not be identified even after the performance of Fourier self-deconvolution and peak fitting approaches. Literature data, however, suggest that this crystalline transformation takes place as a result of mechanical deformation. Therefore, discontinuous stretching experiments which improved the spectral resolution to 1 cm^{-1} were carried out. The deconvoluted spectra now allow the assignment of the monoclinic, orthorhombic and amorphous LLDPE modifications. Analyzing the changes of the intensity of these bands with the elongation does not prove the orthorhombic-monoclinic transformation in the LLDPE films.

Ultimate stretchability and stretching force of the films, both monolayer and multilayer, was well correlated to the development of the crystalline orientation in the films upon stretching. Other mechanical properties like Elmendorf tear and dart impact can also be better understood with these results.

[1] M. Sukhadia, R. K. Krishnaswamy, M. J. Lamborn, S. W. Wharry, C. C. Tso, P. J. DesLauriers, T. Mansfield, F. L. Beyer, *Polymer* **2005**, *46*, 8819.
[2] V. R. Regel, P. Gupta, G. L. Wilkes, A. I. Slutsker, E. E. Tomashevskii, *Kinetical Nature of the Strenght of Solids*, Nauka, Moscow **1974**.
[3] H. W. Siesler, *Habilitation thesis*, Univ. Essen **1988**.
[4] Calculation is based on the melt enthalpy which was related to the enthalpy of the pure crystalline material (ΔHref = 290J/g).
[5] H. W. Siesler, K. Holland-Moritz, *Infrared and Raman Spectroscopy of Polymers*, Marcel Dekker, New York **1980**.
[6] J. D. Hofmann, *Polymer* **1983**, *24*, 3.
[7] J. D. Hofmann, *Polymer* **1982**, *23*, 656.
[8] A. Peterlin, *J. Mat. Sci.* **1971**, *6*, 470.
[9] A. Peterlin, *Colloid & Polym. Sci.* **1975**, *253*, 809.
[10] R. Zbinden, *Infrared Spectroscopy of High Polymers*, Academic Press, New York **1964**.
[11] J. Dechant, *Ultrarotspektroskopische Untersuchungen an Polymeren*, Akademie Verlag, Berlin **1972**.
[12] S. Krimm, *Adv. Polym. Sci.* **1960**, *2*, 51.
[13] H. Tadokoro, *Structure of Crystalline Polymers*, Wiley-Interscience, New York **1179**.
[14] R. G. Snyder, *J. Chem. Phys.* **1957**, *47*, 1316.
[15] H. W. Siesler, *Makromol. Chem.* **1989**, *190*, 2653.
[16] S. Onogi, T. Asada, *Progr. Polym. Sci. Jap.* Vol. 2, M. Imoto, S. Onogi, Eds., John Wiley New York **1979**, p. 261.
[17] H. W. Siesler, *Adv. Polym. Sci.* **1984**, *65*, **1984**, 1.

DOI: 10.1002/masy.200850519

Micro Raman Spectroscopy of Silica Nanoparticles Treated with Aminopropylsilanetriol

V. Volovšek,[1] K. Furić,[2] L. Bistričić,[3] M. Leskovac[1]*

Summary: Micro Raman spectroscopy was used to investigate the structures formed at the surface of amorphous pirogenic silica Aerosil 200. Nanoparticles were treated with 25% water solution of aminopropylsilanetriol and dried in laboratory atmosphere and under vacuum of about 5 mbar. Spectra of pure silica nanoparticles and treated nanoparticles were recorded in the frequency range 20–1700 cm^{-1} and 2500–3800 cm^{-1}. Comparative analysis reveals the existance of different structures formed at the surface of silica nanoparticles depending on drying conditions.

Keywords: aminopropylsilanetriol; hydrogen bonds; Raman spectroscopy; silica nanoparticles; vacuum

Introduction

Amorphous pirogenic silica Aerosil 200 is frequently used as a nanofiller for rubbers and different plastics. It reduces the sedimentation rate, strengthens the elastomers and disperses well in polymeric matrices. Silica nanoparticles have silanol groups at the surface and siloxane bonds inside the particles. The SiOH groups at the surface of the neighboring particles interact through hydrogen bonds creating three dimensional lattice like structure. In order to get hydrophobic silica and to prevent nanoparticles to interact among themselves their surface is frequently modified with organosilane compounds. Vibrational spectroscopies are powerful methods used in investigation of various structures formed at the surfaces of different inorganic particles. Shimizu et al.[1,2] used Raman scattering to study the interaction of 3-aminopropyltriethoxysilane on silica gel and DRIFT spectroscopy to investigate the effect of adsorbed water in the same samples.[3] In this work we studied the effect of the low pressure on the structures of aminopropylsilanetriol (APST) formed at the surfaces of silica nanoparticles. The APST molecules are stable only in the presence of water which prevents forming of siloxane (Si–O–Si) bridges. During the solidification of APST under vacuum, water molecules are being constantly removed from the surrounding as well as from the inside of the sample where they are created in the process of silanization. Therefore, we expect that the structures formed at the surface of nanoparticles will depend on the atmospheric conditions (underpressure) applied.

[1] Faculty of Chemical Engineering and Technology, University of Zagreb, Marulićev trg 19, 10000 Zagreb, Croatia
Fax: (+385) 01 4597135;
E-mail: volovsek@fkit.hr
[2] Ruđer Bošković Institute, Bijenička 54, 10000 Zagreb, Croatia
[3] Faculty of Electrical Engineering and Computing, University of Zagreb, Unska 3, 10000 Zagreb, Croatia

Experimental Part

Amorphous pirogenic silica Aerosil 200 was purchased from Degussa Company and was used without further purification. According to the specification of the producer the mean diameter of the investigated particles of pure Aerosil was 12 nm and specific surface was 200 m^2/g. 25% water solution of APST was purchased from ABCR Company. Silica nanoparticles were mixed with APST solution and the

WILEY InterScience®

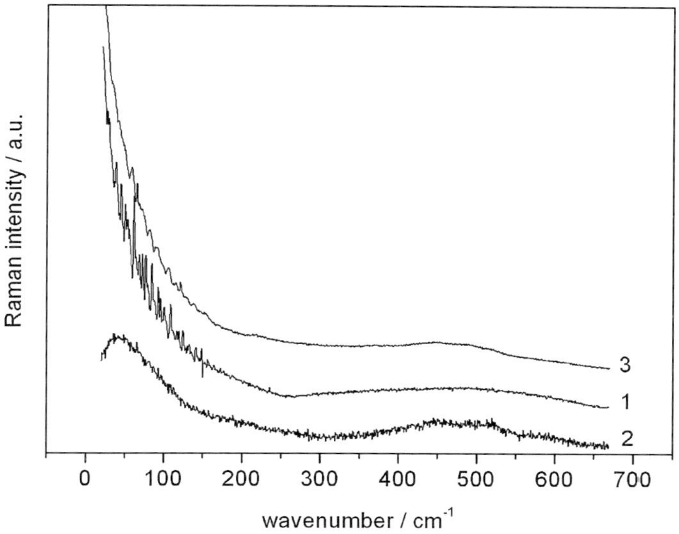

Figure 1.
Low frequency Raman spectra of 1 = pure nanoparticles, 2 = nanoparticles with APST dried in the air, 3 = nanoparticles with APST dried under vacuum.

part of the mixture was dried in the air, while the other part was placed in PVC vessel, connected to the vacuum pump and finally dried under low pressure of around 5 mbar. Raman spectra of pure nanoparticles (sample 1), nanoparticles with APST dried in the air (sample 2) and nanoparticles with APST dried under vacuum (sample 3) were then recorded on Horiba Jobin Yvon T64000 instrument equipped with the Olympus open microscope stage and CCD Symphony detector. Spectra were taken with notch filter in single mode in the frequency region from 700–1700 cm^{-1} and from 2500–3800 cm^{-1} with spectral resolution of 3.5 cm^{-1}. During the acquisition of spectra below 650 cm^{-1}, spectrometer was operating in triple subtractive mode. The 514.5 nm line of a Coherent INNOVA-400 argon ion laser was used for excitation with laser power of 20 mW at the sample place.

Results and Discussion

Comparison between Raman spectra of the two treated Aerosil samples reveals significant differences. In the low frequency region (Figure 1), the spectrum of nanoparticles treated in the laboratory atmosphere (sample 2) shows pronounced Bose band with maximum at 41 cm^{-1}. It represents acoustic modes of the sample which are becoming active in Raman spectrum due to the break-down of selection rules in amorphous materials. Its existence is obviously not related to the nanoparticles because it is missing in the spectra of the pure Aerosil (sample 1). Very similar low frequency spectra were recorded on APST polymer film[4] indicating the existence of medium range order.

In the spectral range from 700–1700 cm^{-1} (Figure 2) there are also few distinctions between the Raman spectra of the three samples. The noise level in the spectra 1 and 2 is much higher compared to the spectrum 3, although all three spectra were recorded under the same conditions. This noise is caused by high fluorescent background which was subtracted in order to compare all three spectra in the same figure. The probable source of this high fluorescence is in the larger amount of pore water incorporated in nanoparticles as well

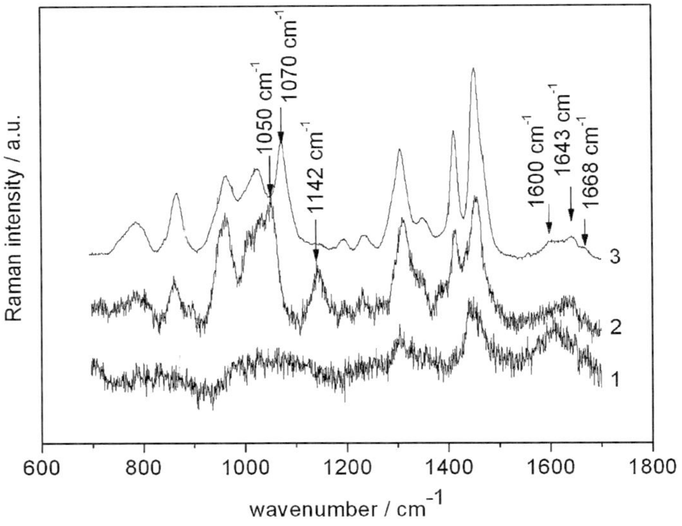

Figure 2.
Raman spectra of 1 = pure nanoparticles, 2 = nanoparticles with APST dried in the air, 3 = nanoparticles with APST dried under vacuum.

as in the mixture of nanoparticles and APST. When this mixture is subjected to high vacuum, water is extracted by rapid pumping, leaving only small amounts of water molecules forming strong hydrogen bonds with silica and APST.

Sample 2 generates bands at 1050 cm^{-1} and 1142 cm^{-1} which are missing in sample 3. Instead, in sample 3 the band at 1070 cm^{-1} is detected. Shimizu et al.[2] in their Raman scattering study of polyaminopropylsiloxane recognized the bands at 1053 cm^{-1} and 1074 cm^{-1} to be sensitive on the concentration of APST in aqueous solution and are useful as the indicators for investigation of the conformations of aminopropyl segments. Our DFT calculations of conformational stability and vibrations of aminopropylsilanol molecule ascribe these bands to mixed CH_2 and NH_2 rocking vibrations in *trans* and *gauche* conformations of aminopropyl chains, respectively.[6] Medium strong Raman band at 1142 cm^{-1} in the spectrum of sample 2 is characteristic for the in phase Si–O–Si vibrations in the rungs of the ladder structures[5] of polyaminopropylsiloxane. It indicates the creation of siloxane bonds among APST molecules and not between APST and substrate nanoparticles.

Another difference arises in the region of NH_2 scissoring vibrations around 1600 cm^{-1}. The APST molecule can form different kinds of intramolecular and intermolecular hydrogen bonds giving rise to slightly different vibrational frequencies. The DFT calculation of APST molecule helped us to assign the band at 1600 cm^{-1} to NH_3^+ bending vibration of hydrogen bonds of the type $SiO^-\cdots H{-}NH_2^+$ in the *gauche* conformation, and band at 1643 cm^{-1} to NH_2 scissoring mode[6] involved in the inter- and intramolecular hydrogen bonds of the type $SiO{-}H\cdots NH_2$. The third band at 1668 cm^{-1} might be attributed to OH bending vibration of water molecules involved in hydrogen bonds.

Shimizu et al.[3] investigated DRIFT spectra of 3-aminopropyltriethoxysilane layer modified onto the surface of silica gel. They also found differences in this part of the spectrum and they ascribed them to a marked variation in the environment of amino groups.

The broad Raman bands in the 2800–3000 cm^{-1} region (Figure 3) are assigned to

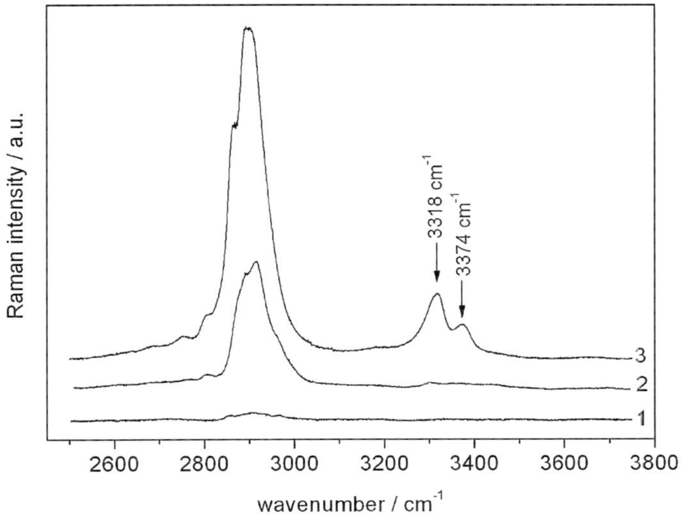

Figure 3.
High frequency Raman spectra of 1 = pure nanoparticles, 2 = nanoparticles with APST dried in the air, 3 = nanoparticles with APST dried under vacuum.

the CH_2 stretching vibrations of the propyl segment.[6] There are some differences in the shape of these bands that can be attributed to different environment of the CH_2 groups in the *trans* and *gauche* conformations of the aminopropyl chains. In the high frequency region the bands at 3318 cm^{-1} and 3374 cm^{-1} assigned to the NH_2 symmetric and asymmetric stretching have much higher intensities in the spectra of nanoparticles treated under vacuum. According to Björnström et al.[7] dehydration process of silica gel induced by thermal treatment is manifested by elongation of SiO–H bonds of the acidic silica groups and dramatic downshift of OH stretching frequencies from ~3600 cm^{-1} to ~3000 cm^{-1} depending on the number of water molecules involved. At sufficiently low water contents, water molecules are forming stronger hydrogen bonds with the silanol groups.

All the characteristic vibrations with suggested assignment are listed in Table 1.

Table 1.
Characteristic vibrations for different structures of polyAPST

ν_{exp} cm^{-1}	ν_{calc} cm^{-1}	Assignment	Structure
1050	1038	NH_2 rock	*trans*
1070	1064	NH_2 rock	*gauche*
1141	1142	Si–O–Si str	ladder
1600	1596	NH_3^+ bend	*trans, gauche*
1643	1619	NH_2 sciss	*trans, gauche*
1669		OH bend	
3318	3351	NH_2 str	*trans, gauche*
3374	3427	NH_2 str	*trans, gauche*

Conclusion

Present investigation of silica nanoparticles treated with aminopropylsilanetriol reveals significant influence of the vacuum drying on the structures formed at the surface of the nanoparticles. Both substances involved in the process have great affinity towards creating different hydrogen bonds between themselves as well as with the water molecules in their surroundings. Applied underpressure of about 5 mbar removes water molecules from the surrounding air, but also from the inside of the mixture. This condition favors the formation of stronger hydrogen bonds. When the underpressure is applied, disordered structure is detected in which propyl chains are predominantly in the *gauche* position and where zwitter-ionic structure of the type $SiO^{-}\cdots H{-}NH_2^{+}$ is created.

When the mixture is dried under normal pressure lot of water still remains in the sample forming droplets of pore water. Under these conditions the APST molecules are forming partially ordered ladder-like structures (Bose band, band at 1142 cm^{-1}) with aminopropyl chains in the *trans* positions.

[1] I. Shimizu, H. Okabayashi, K. Taga, A. Yoshino, E. Nishio, C. J. O'Connor, *Vib. Spectrosc.* **1997**, *14*, 125.
[2] I. Shimizu, H. Okabayashi, K. Taga, C. J. O'Connor, *Colloid. Polym. Sci.* **1997**, *275*, 555.
[3] I. Shimizu, H. Okabayashi, K. Taga, E. Nishio, C. J. O'Connor, *Vib. Spectrosc.* **1997**, *14*, 113.
[4] V. Volovšek, L. Bistričić, K. Furić, V. Dananić, I. Movre Šapić, *J. Phys.: Conference Series* **2006**, *28*, 135.
[5] V. Volovšek, L. Bistričić, V. Dananić, I. Movre Šapić, *J. Mol. Struct.* **2007**, 834-836, 414.
[6] L. Bistričić, V. Volovšek, V. Dananić, I. Movre Šapić, *Spectrochim. Acta Part A* **2006**, *64*, 327.
[7] J. Björnström, A. Martinelli, J. R. T. Johnson, A. Matic, I. Panas, *Chem. Phys. Lett.* **2003**, *380*, 165.

Thermal Degradation of Poly(ε-caprolactone), Poly(L-lactic acid) and their Blends with Poly(3-hydroxy-butyrate) Studied by TGA/FT-IR Spectroscopy

Christian Vogel, Heinz W. Siesler*

Summary: The thermal degradation behavior of poly(ε-caprolactone) (PCL) and poly(L-lactic acid) (PLA) have been studied in different environment. It was found that these polymers undergo completely different degradation processes in nitrogen and oxygen atmosphere. In oxygen environment PCL and PLA mainly decompose to CO_2, CO, water and short-chain acids. In nitrogen atmosphere PCL releases 5-hexenioc acid, CO_2, CO and ε-caprolactone, whereas PLA decomposes to acetaldehyde, CO_2, CO and lactide. The polymer blends of poly(3-hydroxybutyrate) (PHB) with PCL and PLA decompose similar to the individual homopolymers with crotonic acid as the initial decomposition product of PHB.

Keywords: poly(ε-caprolactone) (PCL); poly(3-hydroxybutyrate) (PHB); poly(L-lactic acid) (PLA); thermal degradation; thermogravimetric analysis (TGA)/Fourier-Transform infrared spectroscopy (FT-IR)

Introduction

Bioplastics[1] have lately gained increasing attention because their biosynthesis can alleviate the production problem of oil-based synthetic plastics resulting from possible shortage of oil. In contrast to synthetic plastics, bioplastics have the fundamental advantage of being biodegradable. The great advantage of bioplastics – the conservation of fossile resources and reduction of CO_2 emissions – make them one of the most important innovations for sustainable development.

In the present research the degradation processes of poly(ε-caprolactone) (PCL)[2] and poly(L-lactic acid) (PLA)[3,4] were studied. PCL is a semi-crystalline, linear, aliphatic polyester which can be produced from ε-caprolactone. It is a commercial biodegradable polymer with good mechanical properties and is frequently used as blend component.[2] PLA can be produced by polycondensation of lactic acid or by ring-opening polymerization of the cyclic lactide dimer. The monomeric raw material can be obtained from renewable agricultural sources such as corn. PLA is already used for several applications such as plastic bottles or coated papers for food and beverage packaging.

Polyhydroxyalkanoates (PHA)[5–7] are nature's high molecular weight thermoplastic polyesters which are occurring in the storage granules of a variety of bacteria. In the present study the most common member of the PHAs, poly(3-hydroxybutyrate) (PHB), was blended with PCL and PLA. PHB has the disadvantage of being stiff and brittle, but the mechanical properties can be significantly enhanced by blending PHB with PCL and PLA.

The combination of thermogravimetric analysis (TGA) and FT-IR spectroscopy[8–10] is a powerful technique for the study of polymer decomposition. Based on such measurements, it is possible to achieve

Department of Physical Chemistry, University of Duisburg-Essen, Schuetzenbahn 70, D-45117 Essen, Germany
E-mail: christian.vogel@stud.uni-due.de

a quantitative and qualitative characterization of the decomposition products.

The results of the thermogravimetric analysis of PHB[11–14], PCL[13,15] and PLA[13,16,17] in nitrogen atmosphere were reported in several publications. PHB decomposes mainly to crotonic acid and at higher temperatures to carbon dioxide and propene. 5-hexenoic acid and ε-caprolactone are the main degradation products of PCL. PLA decomposes to acetaldehyde, carbon dioxide and carbon monoxide and at higher temperature to a dilactide. We have studied the decomposition of PHB in oxygen atmosphere[14] and detected the additional evolution of carbon monoxide. Thus, also the thermal degradation of PLA and PCL in oxygen atmosphere will be of interest because in future applications biodegradable polymers will be decomposed by waste combustion in air.

To identify the overlapped absorption bands in the FT-IR gas-phase spectra of the decomposition products 2D-correlation analysis was used in some cases. Generalized 2D correlation spectroscopy, first proposed by Noda,[18,19] is a versatile tool for the analysis of a set of spectroscopic data collected for a system under some type of external perturbation.

Experimental Part

Materials

Bacterially synthesized poly(3-hydroxybutyrate) (Sigma-Aldrich, $M_w = 437{,}000$ g/mol), poly(L-lactic acid) containing 10% meso-lactic acid (NatureWorks LLC, Minnesota, USA) and poly(ε-caprolactone) (Sigma-Aldrich, $M_w = 65{,}000$ g/mol, $M_n = 42{,}500$ g/mol) were used without further purification for the TGA/FT-IR experiments. For the preparation of blends the polymers were dissolved in chloroform p.A. and films were cast by evaporation of the chloroform at 35 °C.

Instrumentation

TGA experiments were carried out on a Netzsch STA 409 PC *Luxx*® system coupled by a heated transfer line (volume 50 mL; heated at 180 °C) to a Bruker Tensor™ IR spectrometer with a liquid-cooled mercury cadmium telluride (MCT) detector. Samples of 28–42 mg were heated with a heating rate of 10 °C/min. from 40 to 720 °C. The flow rate of the carrier gas (nitrogen or oxygen) was 50 mL/min. FT-IR gas-phase spectra were collected with a spectral resolution of 4 cm^{-1} and 32 scans were coadded per spectrum. The TG response and the Gram-Schmidt curves, which represent the amount of decomposition products at a particular temperature, were recorded simultaneously to the FT-IR spectra.

2D Correlation Spectroscopy

The 2D correlation spectra were calculated with the *2Dshige* software (http://sci-tech.ksc.kwansei.ac.jp/~ozaki/main-eng.htm). 2D Correlation spectroscopy has a characteristic pair of synchronous $\Phi(\nu_1, \nu_2)$ and asynchronous $\Psi(\nu_1, \nu_2)$ 2D correlation spectra in which the two spectral variables ν_1 and ν_2 are wavenumbers. In the case of TGA/FT-IR measurements the spectral data sets are collected under the external perturbation of increasing temperature. The mathematical fundamentals of 2D correlation analysis are described in detail elsewhere.[18,19]

Results and Discussion

Degradation of Poly(ε-caprolactone)

Figure 1 A shows the TGA and Gram-Schmidt curves of PCL measured on-line to the thermal decomposition in nitrogen atmosphere. The TGA curve displays a one-step degradation. There is a slight initial increase of the TGA trace detectable, which is determined by thermal buoyancy of the sample holder. This small artefact is also observable in the other degradation experiments. The weight loss observed in the TGA trace is accompanied by a simultaneous increase in the Gram-Schmidt curve at about 360 °C. After a

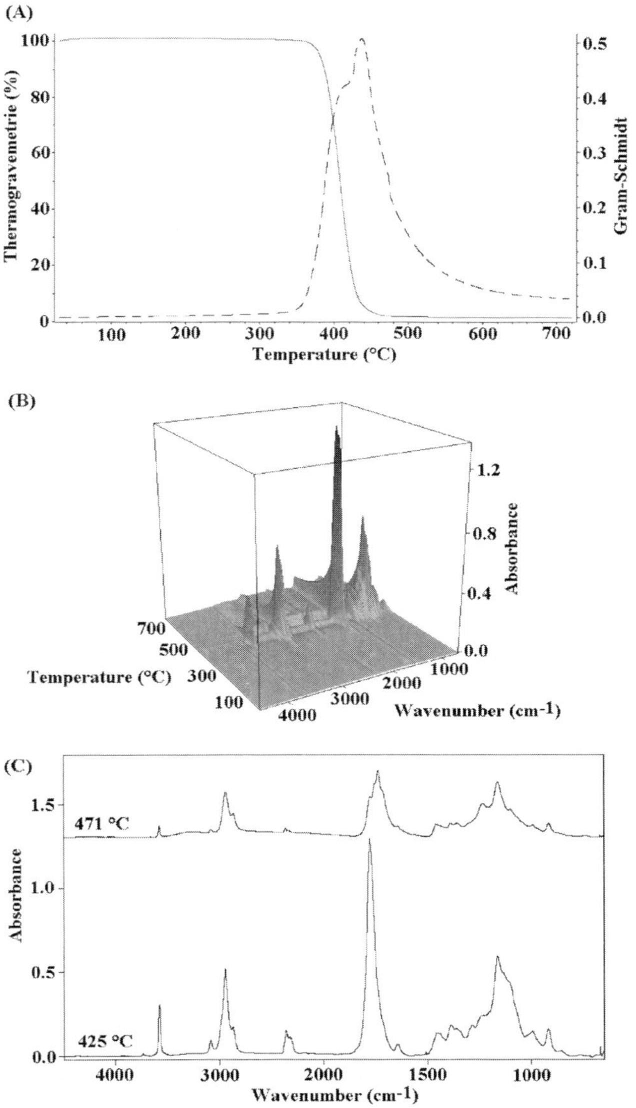

Figure 1.
TGA (—) and Gram-Schmidt (---) curves of the thermal degradation of PCL in nitrogen (A), temperature-dependent 3D FT-IR spectra (B) and characteristic spectra recorded at different temperatures (C).

shoulder at 400 °C the Gram-Schmidt curve reaches a peak maximum at about 450 °C and then decreases until the end of the experiment at 720 °C.

Figure 1 A agrees with the 3D FT-IR spectra of the degradation process shown in Figure 1 B. Similar to the TGA curve, beyond 340 °C the first decomposition products are detectable. Figure 1 C shows the FT-IR spectra recorded at 425 °C and 471 °C. The spectrum recorded at 425 °C can be assigned to the main decomposition product 5-hexenioc acid and its di- and trimers with bands at 3575 cm^{-1} $\nu(OH)$, 3087 cm^{-1} $\nu(H\text{-}C{=}C)$, 2945 cm^{-1} $\nu_{as}(CH_2)$, 2877 cm^{-1} $\nu_s(CH_2)$, 1778 cm^{-1} $\nu(C{=}O)$,

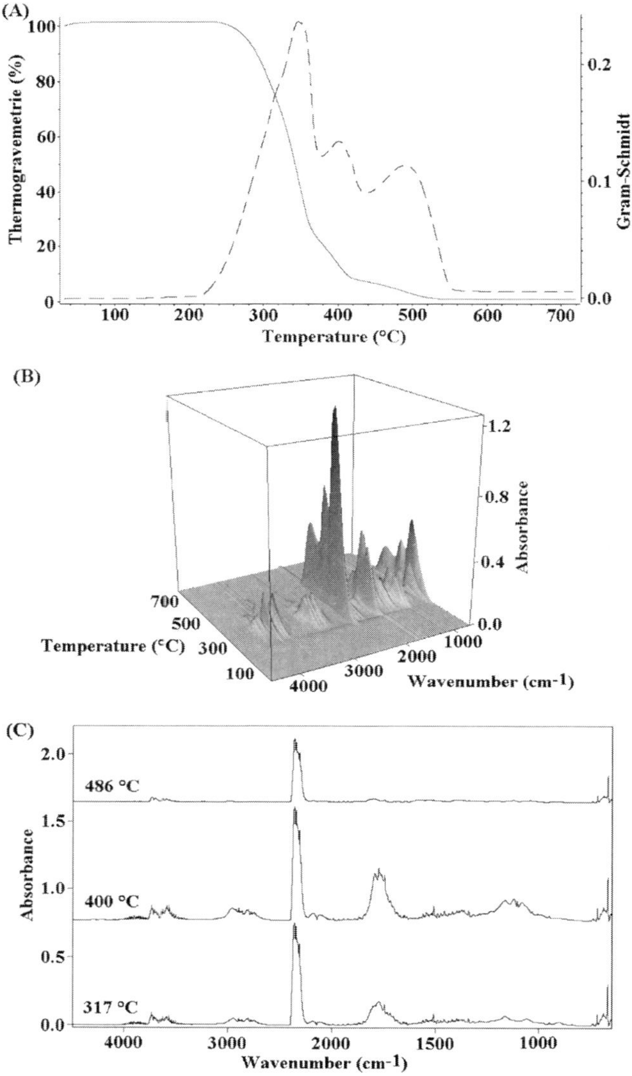

Figure 2.
TGA (—) and Gram-Schmidt (---) curves of the thermal degradation of PCL in oxygen (A), temperature-dependent 3D FT-IR spectra (B) and characteristic spectra recorded at different temperatures (C).

1642 cm^{-1} ν(C=C), 1432 cm^{-1} δ(CH_2), 1164 and 1107 cm^{-1} ν(C–O) and 950 and 915 cm^{-1} γ(R–C=CH_2). Additionally, carbon dioxide (2349 cm^{-1}) and traces of carbon monoxide (2176/2117 cm^{-1}) could be detected. Furthermore, at temperatures >430 °C ε-caprolactone is evolved (Figure 1C, 471 °C), which can be identified by its characteristic band at 1736 cm^{-1} ν(C=O). With a further rise in temperature the evolution of 5-hexenoic acid and ε-caprolactone are decreasing and increasing, respectively.

Figure 3 presents the synchronous (A) and asynchronous (B) 2D correlation spectra (2400-1600 cm^{-1}) of the thermal degradation of PCL in nitrogen atmosphere in the temperature range 143–430 °C. In the synchronous spectrum only the ν(C=O)

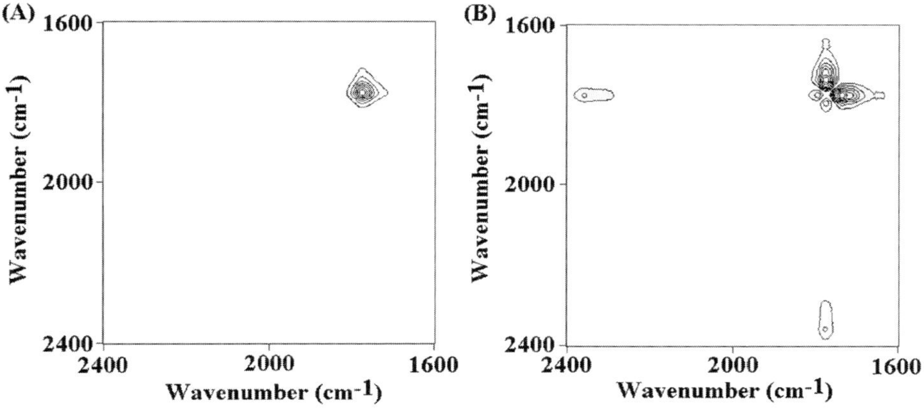

Figure 3.
Synchronous (A) and asynchronous (B) 2D correlation spectra (2400–1600 cm^{-1}) calculated from the temperature-dependent FT-IR spectra of the PCL decomposition in nitrogen atmosphere for the temperature range 143–430 °C (negative cross peaks are grey-shaded).

auto peak of 5-hexenioc acid at 1778 cm^{-1} is displayed. In contrast, several cross peaks are visible in the asynchronous spectrum. The positive cross peak (1778, 2350) indicates that 5-hexenoic acid develops before carbon dioxide. Furthermore, the negative cross peak (1736, 1778) indicates that 5-hexenoic acid is evolved before ε-caprolactone. Figure 2 A shows the TGA and Gram-Schmidt curves of the thermal decomposition of PCL in oxygen atmosphere. In contrast to the one-step decomposition in nitrogen the TGA curve decreases in three steps (220 °C, 360 °C and 420 °C). The corresponding Gram-Schmidt curve also shows three peaks in the range of the three steps of the TGA-curve. In the 3D FT-IR spectra (Figure 2 B) decomposition products are first observable beyond 220 °C. The spectra measured within the temperature range of the three steps of the TGA-curve are displayed in Figure 2 C. The spectrum at 317 °C reflects strong band intensities for carbon dioxide (ν(O=C=O) 2349 cm^{-1}, δ(O=C=O) 667 cm^{-1} and the combination vibration at 3724 cm^{-1}) Additionally, carbon monoxide (2176/2117 cm^{-1}) and water (ν(H-O) at 3900–3400 cm^{-1} and δ(H-O) at 1800–1300 cm^{-1}) could be identified. The rest of the bands belongs to short-chain acids and their di- and trimers: 3589 cm^{-1} ν(O–H), 2952 cm^{-1} $\nu(CH_2)$, 2816 cm^{-1} ν(CH), 1778 cm^{-1} ν(C=O), 1164 and 1107 cm^{-1} ν(C–O). 2D correlation spectroscopy did not provide additional information about these absorption bands.

Degradation of Poly(L-lactic acid)

The thermal decomposition of PLA was also analyzed in nitrogen and oxygen atmosphere. Figure 4 A shows the TGA and Gram-Schmidt curves obtained in nitrogen atmosphere. The degradation takes place in one step beyond 300 °C. In the Gram-Schmidt trace also only one peak is observable, but decomposition products are detectable at higher temperatures. The 3D FT-IR spectra are displayed in Figure 4 B and decomposition products are also detectable beyond 300 °C. Initially mainly acetaldehyde (3475 cm^{-1} ($2 \times \nu$(C=O)), 2968 cm^{-1} ($\nu(CH_3)$), 2740 cm^{-1} (ν(CHO)), 1762 cm^{-1} (ν(C= O)), 1414 + 1371 cm^{-1} ($\delta(CH_3)$), 1127 cm^{-1} (ν(C–O)), carbon dioxide (2349 cm^{-1}) and carbon monoxide (2174/2116 cm^{-1}) develop (Figure 4 C, 352 °C). At higher temperature (Figure 4 C, 425 °C) also lactide (3,6-dimethyl-1,4-dioxan-2,5-dione) is detectable by bands at 3008 cm^{-1} (ν(CH)), 2952 cm^{-1} ($\nu_{as}(CH_3)$), 2893 cm^{-1} ($\nu_s(CH_3)$), 1795 cm^{-1} (ν(C=O)),

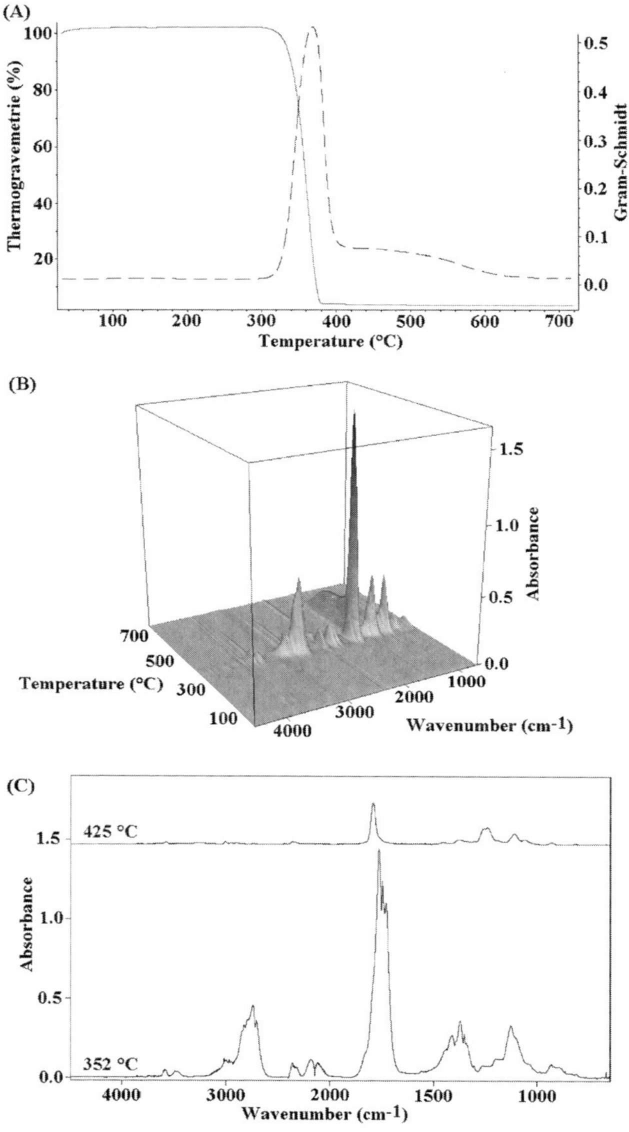

Figure 4.
TGA (—) and Gram-Schmidt (---) curves of the thermal degradation of PLA in nitrogen (A), temperature-dependent 3D FT-IR spectra (B) and characteristic spectra recorded at different temperatures (C).

1365 cm^{-1} ($\delta(CH_3)$), 1248 cm^{-1} and 1108 cm^{-1} (ν(C–O–C)) and 935 cm^{-1}.[20] Water (3900–3400 cm^{-1} and 1800–1300 cm^{-1}) and methane (3200–2900 cm^{-1} and 1400–1200 cm^{-1}) are detectable only in traces. In order to extract more information 2D correlation spectroscopy was applied to these spectral data. In Figure 6 A and B the synchronous and asynchronous spectra in the 2400–1600 cm^{-1} region are shown for the temperature range 143–430 °C. In the synchronous spectrum only an auto peak of the ν(C=O) absorption is shown. The asynchronous spectrum provides more

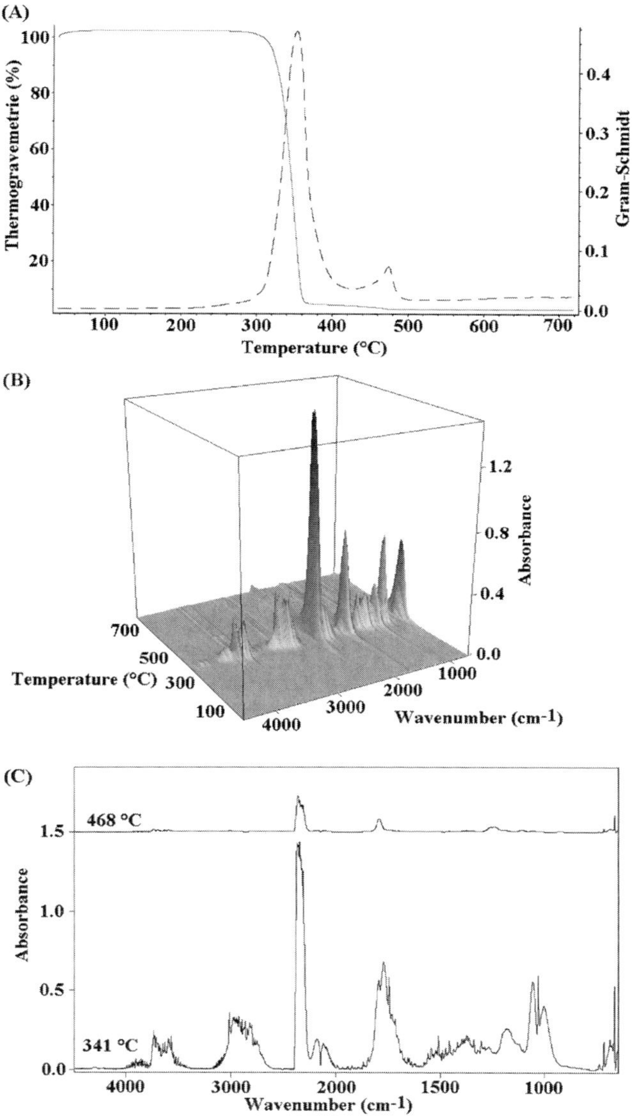

Figure 5.
TGA (—) and Gram-Schmidt (---) curves of the thermal degradation of PLA in oxygen (A), temperature-dependent 3D FT-IR spectra (B) and characteristic spectra recorded at different temperatures (C).

detailed information. The negative cross peak (1795, 2350 cm^{-1}) and the positive cross peak (1743, 2350 cm^{-1}) indicate that carbon dioxide develops before lactide but after acetaldehyde. Also carbon monoxide is evolved before lactide as indicated by the negative cross peaks at (1795, 2185) and (1795, 2111). Cross peaks are also observable for different ν(C=O) absorption bands. The negative cross peak (1795, 1743) belongs to lactide (1795 cm^{-1}) and acetaldehyde (1743 cm^{-1}) and allows the conclusion that lactide develops after acetaldehyde. Furthermore, a negative cross peak at (1795, 1831) is detected. The band at 1831 cm^{-1} develops simultaneously with a band at 1778 cm^{-1} and this doublet can be tentatively assigned to acetic anhydride.[21]

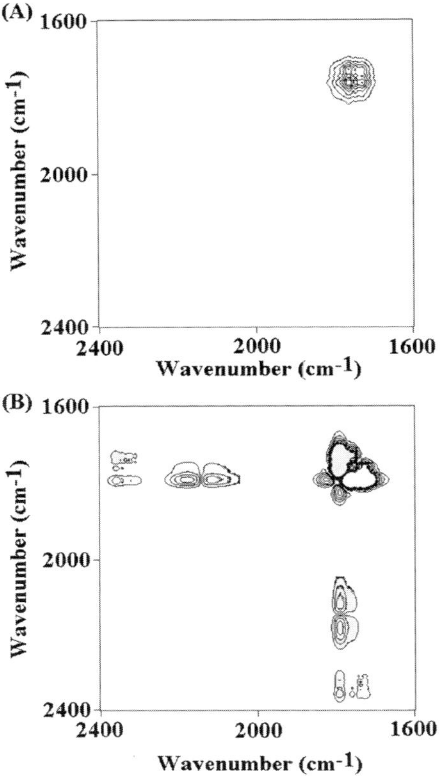

Figure 6.
Synchronous (A) and asynchronous (B) 2D correlation spectra (2400–1600 cm^{-1}) calculated from temperature-dependent FT-IR spectra of the PLA decomposition in nitrogen atmosphere for the temperature region 143–430 °C (negative cross peaks are grey-shaded).

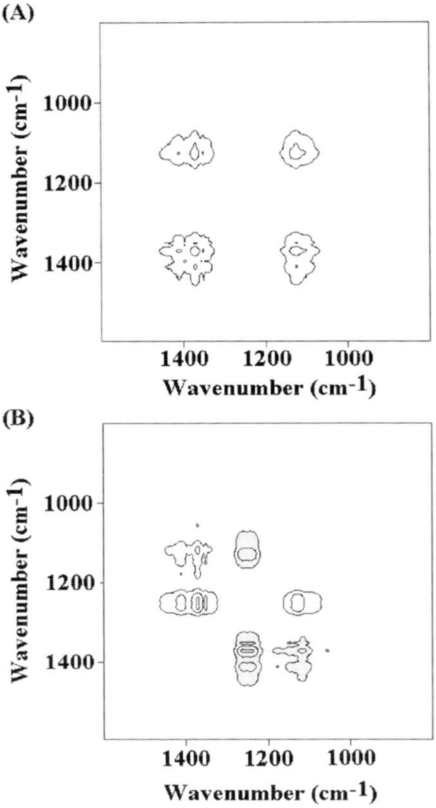

Figure 7.
Synchronous (A) and asynchronous (B) 2D correlation spectra (1600–800 cm^{-1}) calculated from temperature-dependent FT-IR spectra of the PLA decomposition in nitrogen atmosphere for the temperature region 143–430 °C (negative cross peaks are grey-shaded).

Thus, lactide is obviously evolved after acetic anhydride. Acetic acid[22] is also detectable as illustrated by the ν(O–H) band at 3586 cm^{-1}. The ν(C=O) band of acetic acid, however, is overlapped by the band of acetic anhydride at 1778 cm^{-1}. In Figure 7 A and B the synchronous and asynchronous spectra between 1600–800 cm^{-1} are displayed for the temperature range 143–430 °C. The synchronous spectrum shows two auto peaks and corresponding positive cross peaks at 1376 cm^{-1} and 1126 cm^{-1}. Both bands belong to acetaldehyde and therefore develop simultaneously. The asynchronous spectrum shows a positive cross peak at (1126, 1251) and negative cross peaks at (1108, 1376) and (1251, 1376). This also proves that acetaldehyde develops before lactide.

Furthermore, the degradation process of PLA in oxygen atmosphere was analyzed. The TGA and Gram-Schmidt curves of the decomposition of PLA in oxygen atmosphere are shown in Figure 5 A. The TGA curve shows a large and a small weight-loss step at 300 °C and 470 °C, respectively. The Gram-Schmidt curve has a large peak maximum at 355 °C and a much smaller one at 475 °C. Figure 5 B displays the corresponding 3D FT-IR spectra. For a better identification of the characteristic decomposition products at different temperatures the spectra recorded at 341 °C and 468 °C are mapped in Figure 5 C. Similar to the degradation of PCL in oxygen PLA also oxidizes in the first step to carbon dioxide, carbon monoxide, water and short-chain acids (absorption band assignments are described in detail in the section on PCL) and their di- and trimers. The small TGA step at 470 °C is characteristic of the evolution of carbon dioxide and lactide. Thus, also in oxygen atmosphere lactide develops in the decomposition of PLA at higher temperatures.

Degradation of Poly(3-hydroxybutyrate) Blends

Figure 8 A shows the TGA and Gram-Schmidt curves of a PHB/PCL (50:50 wt%) blend in nitrogen atmosphere. The weight

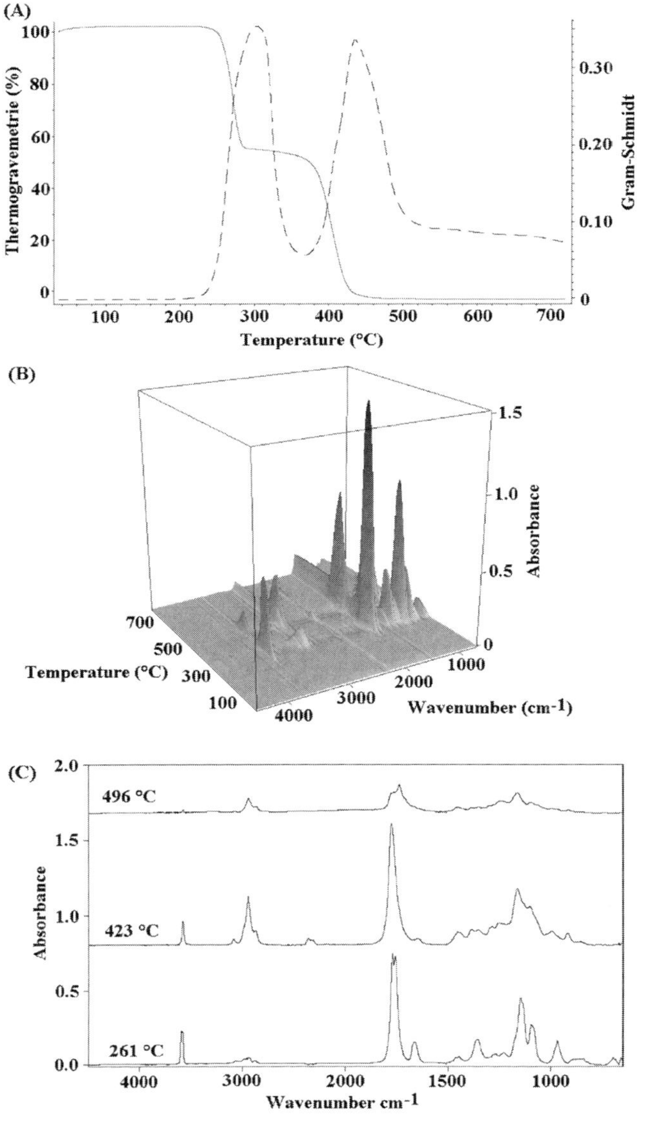

Figure 8.
TGA (—) and Gram-Schmidt (---) curves of the thermal degradation of a PHB/PCL (50:50 wt%) blend in nitrogen (A), temperature-dependent 3D FT-IR spectra (B) and characteristic spectra recorded at different temperatures (C).

loss occurs in two sharp steps with onset temperatures at 260 °C and 380 °C. Accordingly, the Gram-Schmidt curve presents two peaks between 230–370 °C and 370–520 °C. Beyond 520 °C the Gram-Schmidt trace continues at an elevated level. Thus, even at 720 °C decomposition products are detectable.

Figure 8 B displays the corresponding 3D FT-IR spectra of the PHB/PCL (50:50 wt%) blend decomposition in nitrogen. Two temperature regions with high spectral intensities can be distinguished around 300 °C and 450 °C. Individual spectra of these temperature ranges are illustrated in Figure 8 C. The spectrum at 261 °C shows

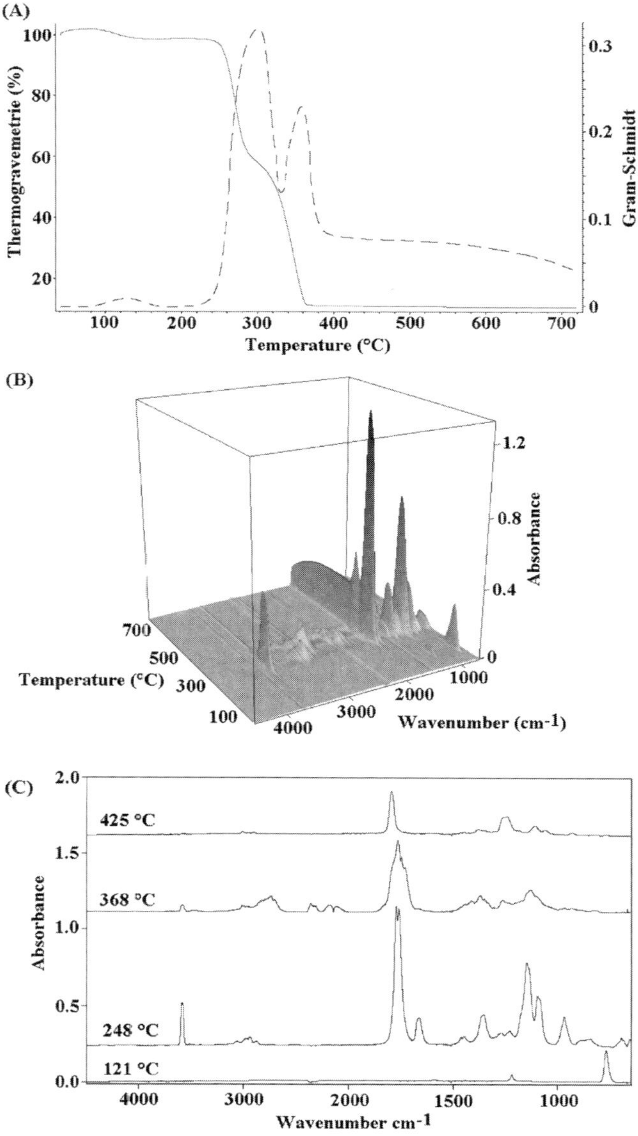

Figure 9.
TGA (—) and Gram-Schmidt (---) curves of the thermal degradation of a PHB/PLA (50:50 wt%) blend in nitrogen (A), temperature-dependent 3D FT-IR spectra (B) and characteristic spectra rcorded at different temperatures (C).

the main decomposition product of PHB, crotonic acid, which can be identified by its typical absorption bands: ν(O–H) 3589 cm^{-1}, ν(C–H) from 3074 to 2865 cm^{-1}, ν(C=O) 1760 cm^{-1}, ν(C=C) 1660 cm^{-1}, δ(C–H) 1450 cm^{-1}, $\delta(CH_3)$ 1357 cm^{-1}, ν(C–O) 1145 + 1097 cm^{-1} and γ(trans-C=C) 967 cm^{-1}. The absorption at 691 cm^{-1} (γ(cis-C=C)) is characteristic for the cis-configuration of crotonic acid. At 423 °C 5-hexenoic acid, the main decomposition product of PCL, can be identified by the absorption bands discussed in the previous section on PCL. Additionally, carbon dioxide (2349 cm^{-1}) and traces of carbon monoxide (2176/2117 cm^{-1}) are detectable. Furthermore, at temperatures > 430 °C ε-caprolactone is evolved (Figure 8 C, 496 °C) and can be identified by its characteristic ν(C=O) absorption at 1736 cm^{-1}. Thus, the PHB/PCL (50:50 wt%) blend decomposes in a step-wise mechanism similar to the individual blend components.

Figure 9 A shows the TGA and Gram-Schmidt curves of a PHB/PLA (50:50 wt%) blend in nitrogen atmosphere. Again, the TGA curve has two sharp weight-loss steps at 260 °C and 330 °C. For both steps corresponding peaks are detectable in the Gram-Schmidt trace. A small weight loss is also detectable at about 100 °C. In Figure 9 B the 3D FT-IR spectra for this decomposition process are mapped.

Two temperature ranges of high absorption intensities around 310 °C and 370 °C can be identified preceded by a region of low-intensity spectra at about 100 °C. Individual spectra of these regions are shown in Figure 9 C. The spectrum at 121 °C with only two bands at 1220 cm^{-1} and 771 cm^{-1} could be identified as chloroform which was not completely removed from the blend during sample preparation. The spectrum at 248 °C reflects the major decomposition product of PHB, crotonic acid. At 386 °C acetaldehyde, carbon dioxide and carbon monoxide - the decomposition products of PLA - are detectable. Finally, at 425 °C the PLA-specific product lactide could be detected. Thus, in analogy to the PHB/PCL blend, the PHB/PLA blend also decomposes in two steps which correspond to the individual blend components.

Conclusion

PCL and PLA undergo completely different degradation processes in nitrogen and oxygen atmosphere. This is in contrast to PHB where the main decomposition product is crotonic acid in both environments and otherwise only small differences could be found for the degradation in nitrogen and oxygen.[14]

As already reported,[13,15] PCL decomposes in nitrogen atmosphere at 340 °C to 5-hexenioc acid and carbon dioxide and at temperatures >430 °C ε-caprolactone is evolved. Additionally, traces of carbon monoxide could be identified. 2D correlation spectroscopy established that 5-hexenoic acid develops before carbon dioxide.

PLA decomposes in nitrogen atmosphere to acetaldehyde, carbon dioxide, carbon monoxide, water and methane. With increasing temperature lactide becomes the major degradation product. This is in agreement with the results reported in other publications.[16,17] By 2D correlation spectroscopy it was found that acetaldehyde develops before carbon dioxide and lactide.

In oxygen atmosphere, on the other hand, PCL and PLA decompose into carbon dioxide, carbon monoxide, water and short-chain acids. At higher temperatures the formation of lactide was observed for PLA.

The polymer blends PHB/PCL (50:50 wt%) and PHB/PLA (50:50 wt%) decompose in nitrogen atmosphere in two-step mechanisms which resemble the individual blend components. In both blends PHB was decomposed first mainly to crotonic acid and after this process PCL and PLA degraded at higher temperatures to their established decomposition products. No shift in the onset temperatures of the decomposition of the individual homopolymers was detectable due to blend formation.

Acknowledgements: The authors thank the Dr. Jost-Henkel-Stiftung (Düsseldorf, Germany) for financial support and Prof. Y. Ozaki and his research group for the supply of polymer samples and helpful discussions. We also acknowledge the instrumental and experimental assistance for the TGA/FT-IR measurements by the Department of Inorganic Chemistry of the University of Duisburg-Essen (Prof. Dr. M. Epple and Dr. R. Weiss).

[1] R. A. Gross, B. Klara, *Science* **2002**, *297*, 803.
[2] G. C. Eastmond, *Adv. Polym. Sci.* **1999**, *149*, 59.
[3] R. Auras, B. Harte, S. Selke, *Macromol. Biosci.* **2004**, 4, 835.
[4] A. Steinbüchel, Y. Doi, Eds., *Biopolymers, Vol. 4: Polyesters III*, Wiley-VCH, Weinheim **2001**, p. 235.
[5] P. J. Hocking, R. H. Marchessault, "Polyhydroxyalkanoates", in: *Biopolymers from Renewable Resources*, D. L. Kaplan, Ed., (Springer-Verlag, Berlin **1998**,) p. 220.
[6] Y. Doi, A. Steinbüchel, Eds., *Biopolymers, Vol. 3b: Polyesters II*, Wiley-VCH, Weinheim **2001**.
[7] G. A. M. van der Walle, G. J. M. de Koning, R. A. Weusthuis, G. Eggink, *Adv. Biochem. Eng. /Biotechnol.* **2001**, *71*, 263.
[8] K. Pielichowski, J. Njuguna, *Thermal Degradation of Polymeric Materials*, Rapra Technology, **2005**, 12.
[9] C. A. Wilkie, *Poly. Deg. and Stab.* **1999**, *66*, 301.
[10] J. Böcker, *Spektroskopie*, Vogel Verlag, Würzburg **1997**, p. 181.
[11] H. Morikawa, R. H. Marchessault, *Can. J. Chem.* **1981**, *59*, 2306.
[12] N. Grassie, E. J. Murray, P. A. Holmes, *Poly. Deg. and Stab.* **1984**, *6*, 127.
[13] Y. Aoyagi, K. Yamashita, Y. Doi, *Poly. Deg. and Stab.* **2002**, *76*, 53.
[14] C. Vogel, S. Morita, H. Sato, I. Noda, Y. Ozaki, H. W. Siesler, *Appl. Spectrosc.* **2007**, *61*, 755.
[15] O. Persenaire, M. Alexandre, P. Degée, P. Dubois, *Biomacromolecules*, **2001**, 2, 288.
[16] I. C. McNeill, H. A. Leiper, *Poly. Deg. and Stab* **1985**, *11*, 309.
[17] F.-D. Kopinke, M. Remmler, K. Mackenzie, M. Möder, O. Wachsen, *Poly. Deg. and Stab.* **1996**, *53*, 329.
[18] I. Noda, *Appl. Spectrosc.* **1993**, *47*, 1329.
[19] I. Noda, Y. Ozaki, *Two-Dimensional Correlation Spectroscopy* (John Wiley and Sons, Chichester,West Sussex, **2004**).
[20] C. N. Tam, P. Bour, T. A. Keiderling, *J. Am. Chem. Soc.* **1996**, *118*, 10285.
[21] G. Wu, C. van Alsenoy, H. J. Geise, E. Sluyts, B. J. van der Veken, I. F. Shiskov, L. V. Khristenko, *J. Phys. Chem. A* **2000**, *104*, 1576.
[22] R. C. Herman, R. Hofstadter, *J. Chem. Phys.* **1938**, *6*, 534.

In situ ATR-FTIR Spectroscopy of Poly(ethylene terephthalate) Subjected to High-Temperature Methanol

*Jean-Michel Andanson, Sergei G. Kazarian**

Summary: In situ ATR-FTIR (attenuated total reflection Fourier transform infrared) spectroscopy combined with a high-pressure cell has been applied to measure IR spectra of polymers subjected to superheated or near-critical methanol (100–200 °C). Spectra of poly(ethylene terephthalate) (PET) have been measured under exposure to high temperature methanol using this ATR-FTIR approach to understand at a molecular level the recycling process of PET. The evolution of the structure and the morphology of the polymer has been studied during the methanolysis. The quantity of trans PET conformer is generally used as a tracer of the crystallinity of PET. During depolymerisation of PET the evolution of crystallinity and of trans conformer appears to be different. The dependence of the rate constant on reaction temperature was correlated by Arrhenius plot, which shows activation energy of 5.8 kJ/mol.

Keywords: alcoholysis; FT-IR; infrared spectroscopy; methanol; poly (ethylene terephthalate); polymer recycling; supercritical fluids

Introduction

Poly(ethylene terephthalate) PET is one of the most common consumer plastics used and 796 000 tonnes was collected in Europe for recycling in 2005. The growing interest in PET recycling is due to the huge number of commercial applications such as textile fibres, films, bottles, food packaging materials. PET wastes can be melted and then reformed but products have lower properties. The degradation of the polymer by methanol at high temperature and pressure conditions (methanolysis) appears to be more promising for this task.[1–3] Methanolysis has been developed and patented by Lotz et al.[4] and Gruschke et al.[5]; all recycling processes and patents of PET have been reviewed recently.[6,7] Chemical recycling can be used to turn PET wastes into monomers, from which PET can be resynthesised using three methods (hydrolysis, glycolysis and methanolysis).[6,8] In hydrolysis, the polymer is decomposed into terephthalic acid and ethylene glycol.[9] Monomer, bis-2-hydroxyethyl terephthalate and oligomer are obtained if the glycolysis (with ethylene glycol) is applied to depolymerise the PET.[10] With the last chemical process, the methanolysis of PET generates mainly two products: dimethyl terephthalate and ethylene glycol which provides the advantage of easier purification than hydrolysis or glycolysis. Another advantage of the last process for PET recycling is the possibility of build an installation in the polymer production line. Waste PET coming up in the production cycle is employed and the monomers won back by methanolysis are restored back to the production process.

Vibrational spectroscopy can be used in situ to analyse at molecular level the structure of the polymer during the degradation. Thermal degradation of PET[11] and hydrolytic degradation of PET at 90 °C[12] have

Department of Chemical Engineering, Imperial College London, London, SW7 2AZ, UK
E-mail: s.kazarian@imperial.ac.uk

been studied by Fourier transform infrared spectroscopy to observe the crystallinity and the morphology of the polymer. Vibrational spectroscopy, Raman and IR (infrared) appear to be a good choice for studying the morphology and structure of PET.[12–18] Recently, ATR-FTIR (attenuated total reflection Fourier transform infrared) spectroscopy combined with a high-pressure cell has been utilized to study simultaneous sorption and swelling of polymers subjected to supercritical CO_2[19,20] and to study the decomposition of PET in superheated water.[18] The same apparatus can also be employed to measure the in situ evolutions of crystallinity and morphology of the polymer during methanolysis of a thin film of PET between 100 and 200 °C. Several studies have been published recently on the morphology and the structure of PET. Cole et al. have measured and analysed several samples thermally crystallized over a wide range of conditions, to study the crystallization and structure of PET.[21] X-ray diffraction has also been used recently on PET fibres to elucidate the effect of the microstructures of oriented chains on the thermal deformations.[22]

The overall aim of our studies is to understand how material is transformed during the recycling process. Infrared spectroscopy is able to measure the evolution of structure and the morphology of the polymer during this process. Our spectral analysis has been mainly concentrated on the evolution of the ratio between the gauche and trans conformation (Figure 1) and the change in the crystallinity of the PET film.

Experimental Part

A high-pressure cell was used to study in-situ interactions between liquid methanol and the PET film. The cell is fixed on the top-plate of a modified Golden Gate™ accessory (Specac, Ltd.) which utilises a diamond as the ATR crystal. The schema of the experimental setup has been given in Figure 2. Under total internal reflection the light is reflected and an evanescent field is developed at the interface. The depth of penetration is of the order of a micrometer and is dependent on the refractive index of the crystal and the sample, the incident angle of light, and the wavelength of the light. This phenomenon gives us the possibility to study the surface of a material at micrometric depth.

An Equinox 55 spectrometer (Bruker Optics) with a MCT detector was used to measure ATR-FTIR spectra with resolution of 2 cm^{-1}. All other technical details were fully presented in the previous paper.[18] To obtain a good contact between the PET film and ATR crystal, a thin film of PET was melted at 250 °C during half hour, the temperature was slowly decreased until ambient temperature was reached. Under these conditions, the PET film cast on the crystal is mainly in amorphous form. After that, the high pressure cell is attached to the top-plate of the accessory and the cell is completely filled by liquid methanol at

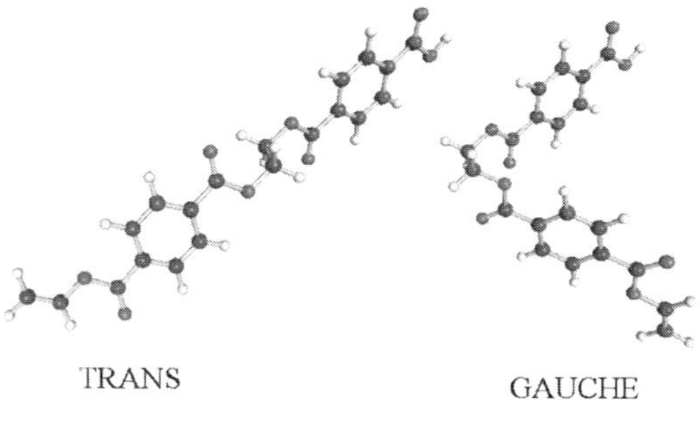

Figure 1.
Dimer of ethylene terephthalate in gauche and trans conformation.

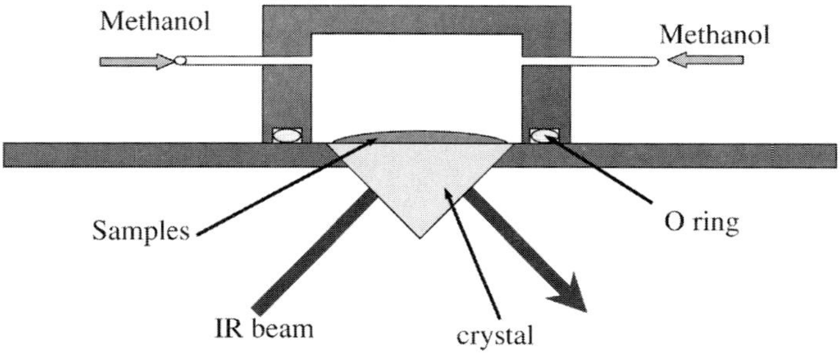

Figure 2.
Schematic presentation of a diamond ATR-IR accessory for studying polymeric materials subjected to high-pressure methanol.

ambient temperature. Finally, the system is heated up to the desired temperature, the pressure rises as expected with the increasing of the temperature. Experiments were done at four different temperatures: 125, 150, 175 or 200 °C and the pressure in the cell were between 1 and 10 MPa. All experiments have been done repeated twice.

Results and Discussion

The ATR-FTIR spectra of the PET film at 20 °C (more amorphous structure) and at 150 °C (more crystalline structure), spectra of methanol and seven different spectra of PET during methanolysis at 150 °C are displayed on Figure 3. Infrared spectra of a range of PET have been examined with minutiae by Cole et al.[23,24] They particularly drew attention to the structure of the polymer and compared several PET spectra corresponding to various states of order: highly amorphous, drawn at 80 °C to different draw ratios, and thermally crystallized under different conditions. Table 1 presents the proposed assignments of some of the main bands of PET, this indexation has also been previously examined in more detail by Holland et al.[25]

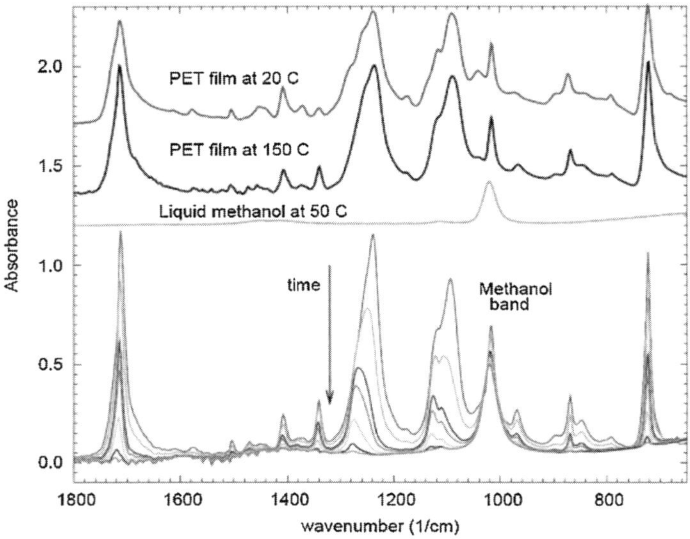

Figure 3.
ATR-FTIR spectra of PET, methanol and PET during methanolysis at 150 °C. Seven spectra of PET film have been measured at following time intervals: 0, 30, 60, 120, 240, 480 and 960 minutes.

Table 1.
Assignment of main vibrations of PET.

Wavenumber (cm^{-1})	Assignment
722	Out of plane of benzene group
845	CH_2 rocking of glycol
870	Out of plane of benzene group
897	CH_2 rocking of glycol
970	C-O stretching of glycol
1017	In-plane vibration of benzene
Broad band ca. 1090	Mainly due to ester C=O stretching
Broad band ca. 1230	Mainly due to ester C=O stretching
1340	CH_2 wagging of glycol
1370	CH_2 wagging of glycol

Several simple observations can be achieved using Figure 3. In this wavenumber range, the C-O stretching of methanol at circa 1015 cm^{-1} is the only intense band of methanol and all others bands disappear in less than ten hours. The methanol band appears in the spectra from the beginning of the experiments. In this case, the diffusion of methanol inside the PET film is much faster than the depolymerisation process. These experiments have been repeated several times and the presence of methanol from the beginning of the experiment is not due to presence of methanol in the interfacial area between the diamond and the PET film, but rather to methanol inside the PET film. At the end of the process, after 15 hours, the only band observable is the methanol band. Under these conditions it is not possible to comment on the product of the methanolysis but only on reactants.

Most visible differences between the spectra of the crystalline and amorphous structures are the shapes of large bands around 1100 and 1250 cm^{-1} which are too complex to be simply decomposed by a single Gaussian or Lorentzian function. The intensities of some of the smaller bands are also changing substantially between the crystalline and amorphous form. More comment will be added on this point afterward. The last observation using Figure 3 is on drastic change in the shape of the broad band at 1100 cm^{-1} between 0 and 30 minutes. To see the progression of the shape of both broad bands at 1250 and 1100 cm^{-1} the evolution of spectra have been shown in more detail in Figure 4. Vibrations involved in the broad band at 1100 cm^{-1} are mainly attributed to the deformation of the benzyl group. A strong change in the spectra could be due to an evolution in the structure and/or morphology of the polymer at molecular level.

The two bands of CH_2 wagging at 1340 and 1370 cm^{-1} correspond respectively to the trans and gauche conformations. Atkinson et al. have investigated the glass form of PET with both wagging bands, a structural alteration occurs on cooling and the change from trans to gauche conformations of the ethylene units in PET.[26] Another interesting band to examine is the aromatic skeletal stretching band at 1410 cm^{-1} which is not influenced by the crystallinity rate or the morphology of PET.[23] In fact, Cole et al. have shown a small sensitivity of this band to the environment but they have explained why it was the best choice to have as an internal reference anyway.[21] Ideally, a reference band should not show any sensitivity and in the case of PET the band at 1410 cm^{-1}, which is associated with the ring in-plane deformation, is generally used to normalize spectra of the polymer.[27] As often for quantitative analysis using ATR-FTIR spectroscopy, the contact between the sample and the crystal is a crucial because the results depend on the absorbance of bands which are directly correlated to the quality of the contact between the sample and the crystal. As presented in the Figure 4, spectra have been recorded regularly with the time and the absorbance shows a homogenous and continuous evolution. This result and the

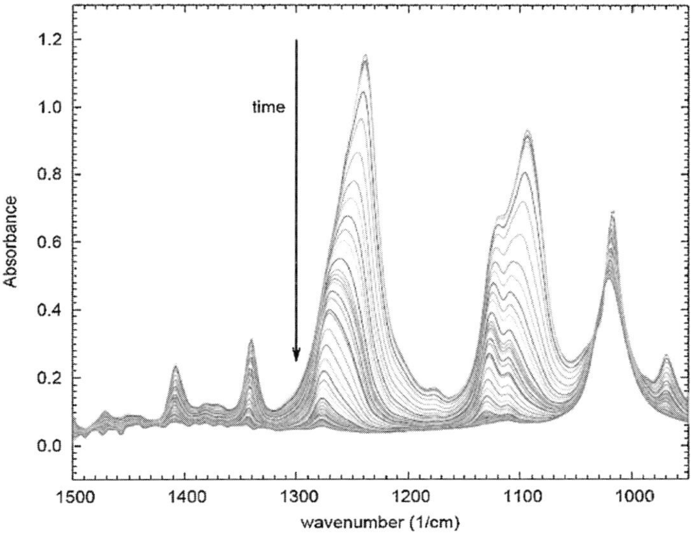

Figure 4.
Evolution with time (15 hours) of PET/methanol ATR-FTIR spectra at 150 °C.

fact that experiments have been reproducible show that the kinetic measure is not spurious and is not due to the penetration of methanol between the film and the crystal which should be a random process. Moreover, a first attempt was done by compression of a PET film on the diamond and not by casting it, in this case, the methanol can diffuse more easily between the PET film and the crystal, and the results obtained were random and the evolution of the absorbance of the spectral bands was not continuous.

To observe more clearly the shape evolution of the spectral region, subtraction spectra are presented in Figure 5. Spectra

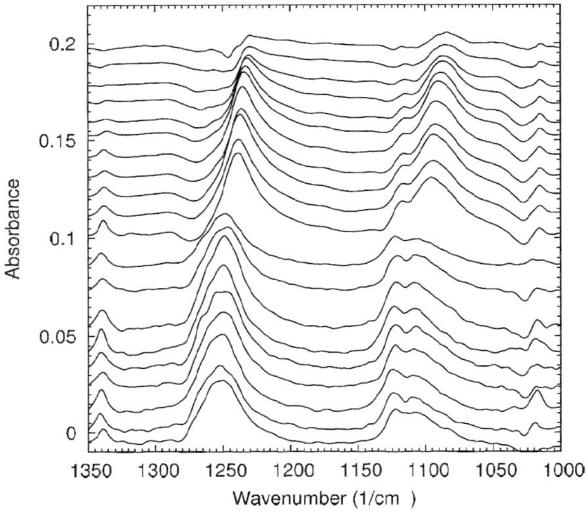

Figure 5.
Difference ATR-FTIR spectra of PET/methanol at 150 °C obtained during the first 2 hours of the depolymerisation (from top to bottom).

presented in the figure have been calculated by the systematic subtraction of the earlier one. The 10 spectra on the top correspond to the subtraction obtained in the first 100 minutes of depolymerisation at 150 °C and the other 10 spectra represent the evolution in next 2 hours. The beginning of the depolymerisation process shows a decrease of the band at 1180 cm^{-1} and 1235 cm^{-1}, both bands have been attributed to a ring mode or CH_2 twist and ring-ester in plane mode respectively. After that, the bands at 1260, 1240, 1125 and 1110 cm^{-1} are decreasing during the methanolysis. All these bands also result from some ring mode.

Decomposition of spectra by mathematical function can give additional information about the evolution of the PET during the process. An integration of each band has been calculated for each measured spectra to achieve a quantitative analysis of average morphology and of the structure of PET during methanolysis. Using the same approach, Pardal et al. have published very recently the evolution of trans and gauche conformations during glycolys of PET by ATR-FTIR spectroscopy.[28] They has measured an increase of the trans form in proportion with the comparison of band at 1340 cm^{-1} (trans) and 1370 cm^{-1} (gauche). In our experiments the signal is too low (sample with high rate of crystallinity) to extract any quantitative information from the band at 1370 cm^{-1}. But as discussed previously the deformation in plan of benzyl ring at 1410 cm^{-1} can be used as an internal reference. Under this condition the evolution of absorbance of CH_2 wagging in trans conformation at 1340 cm^{-1} and of the internal reference (band at 1410 cm^{-1}) has been reported on Figure 6. The evolution of the absorbance of the C=O stretching band, circa 1720 cm^{-1} and the absorbance of Gaussian at 1245 cm^{-1} has been also illustrate in the same figure. The broad band at 1250 cm^{-1} has been decomposed by two Gaussian functions centred around 1245 and 1265 cm^{-1}.

Pardal et al.[28] have shown the same evolution of trans/gauche proportion: an increase in the ratio of trans/gauche during the beginning the alcoholysis by diethylene glycol. In our case, the PET film has been melted before the alcoholysis, and the film has been largely crystallized in trans conformation. As Schmidt et al. have

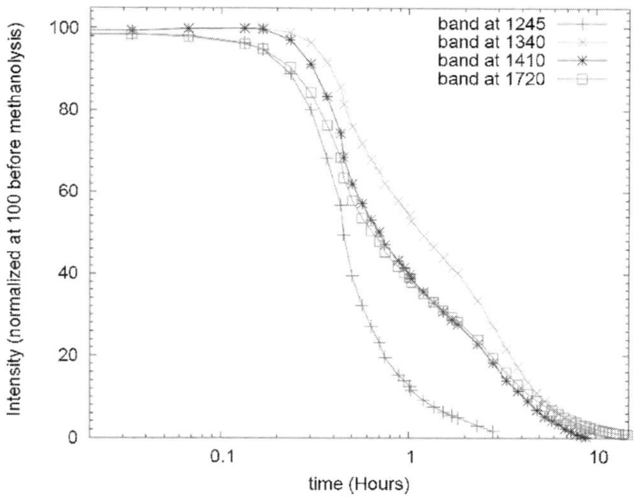

Figure 6.
Normalized absorbance of PET bands during degradation of PET by methanol at 150 °C: band at 1245 cm^{-1} shows the evolution of crystallinity of PET, band at 1340 cm^{-1} represents the evolution of trans conformation of PET, band at 1410 cm^{-1} is used as an internal reference for the quantity of PET; band at 1710 cm^{-1} is the C=O stretching mode.

demonstrated by two dimensional NMR, in crystalline PET the trans content is 100 percent and in the amorphous PET it is 14 percent (±5 percent).[29] The PET film used in the depolymerisation is mainly in crystalline form. To quantify the evolution of the gauche conformer, the CH_2 rocking can also be used. In fact, Cole et al. have shown for this vibration the frequency is around 900 cm^{-1} for the gauche conformer and 850 cm^{-1} for trans conformer. With Figure 6, the comparison between the absorbance of band at 1245 cm^{-1} and the reference band (at 1410 cm^{-1}) shows a heterogeneous evolution of the structure during the decomposition of the polymer by methanol. For clarity reason, the evolution have not been added to the Figure 6, but the evolution of the absorbance of this band is similar to the band at 1245 cm^{-1}. As the band at 1410 cm^{-1} is considered as the internal reference, we can observe that the percentage of the trans conformation is increasing during the process and at the same time the amount of gauche conformation is decreasing.

We have carried out calculations of quantum chemistry using density functional theory for the two conformations of di-ethylene terephthalate at the B3-LYP/6-31G(d) level using the PC GAMESS program.[30,31] Full structure optimizations were performed and used to evaluate harmonic frequencies and infrared intensities of the both conformation. The optimized structures of the two conformer trans and gauche are shown on Figure 1. As it has been explained previously,[32] the absorbance calculated by these methods are altered by large uncertainties. However, these methods are very useful for comparison of positions and/or absorbance the bands of molecular vibrations of different conformers or isomers.

On Figure 7, corrected frequencies and intensities are represented for both trans and gauche conformer of di-ethylene terephthale. The most intense band in infrared for both conformer is around 1250 cm^{-1} and it corresponds to an in plane benzyl vibration. This vibration is more important in trans than in gauche conformation. We have already measured the evolution of the percentage of trans conformation in PET film with the band at 1340 cm^{-1}, and evolutions of intensities bands at 1340 cm^{-1}

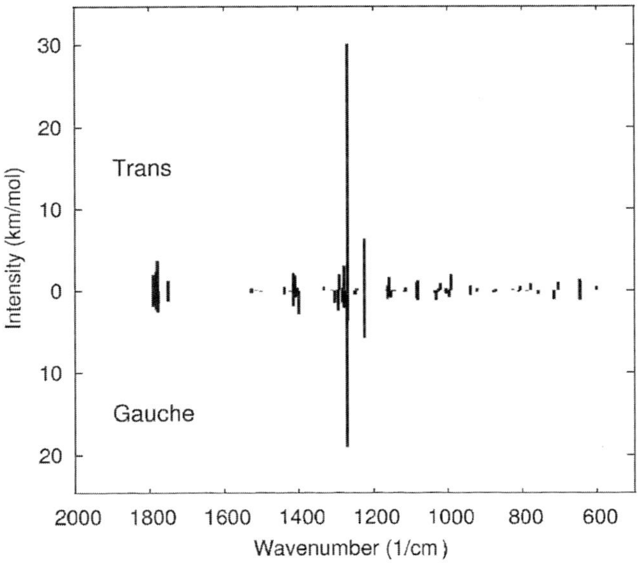

Figure 7.
Scaled harmonic frequencies and infrared intensities of trans and gauche di-ethylene terephthate. Each frequency has been corrected using the scaling factor calculated by Scott and Radom (0.9614 for B3-LYP/6-31G(d)).[33]

(sensitive to trans conformation) and 1250 cm^{-1} are obviously dissimilar. The evolution of the absorbance of the band around 1250 cm^{-1} cannot be explained by the difference of intensities between trans and gauche conformers. Nevertheless, the intensities of the crystalline and amorphous forms are probably also relatively different.

Not only can the absorbance of one band be used to observe the evolution of molecular interactions but the wavenumber of the band can also be very informative. The shift of the C=O stretching band has been employed to detect the variation of crystallinity in the system. Jang et al. exploited this shift to observe the progression of the crystallinity in the polymer blend.[34] For PET, the wavenumber of the ν(C=O) band can be used to distinguish between the crystal (lower frequency) and amorphous (higher frequency) form. As we have shown previously in the paper, the absorbance of the C=O stretching band has the same evolution as the reference band, it's possible to observe the evolution of the shift of the band with the change of absorbance. The position of the band shifts to the higher wavenumber region much slower than the evolution of the absorbance of the same band. It should indicate that during the process the PET film is transformed to a less crystalline form and probably this evolution of the crystallinity ratio is slower than the evolution of trans/gauche conformation. This analysis is in good agreement with the observation that can be achieved if it is assumed that the band at 1245 cm^{-1} is reflecting the crystallinity of the system.

As Huang et al.[35] have shown by solid-state nuclear magnetic resonance (NMR) and differential scanning calorimetry techniques there exists three-domain models containing crystalline (all *trans*), constrained noncrystalline (*trans* rich), and amorphous phase (*gauche* rich), they clearly showed that it is not possible to connect directly an evolution of trans conformation with an increasing of crystallinity. It is not possible to link directly the evolution of the ratio of trans/gauche conformation with the amount of crystallinity. In previous studies on hydrolytic degradation of PET,[12,36–39] authors have explained a 'chemicrystallisation' process to describe the variation of the density due to variation of trans conforma-

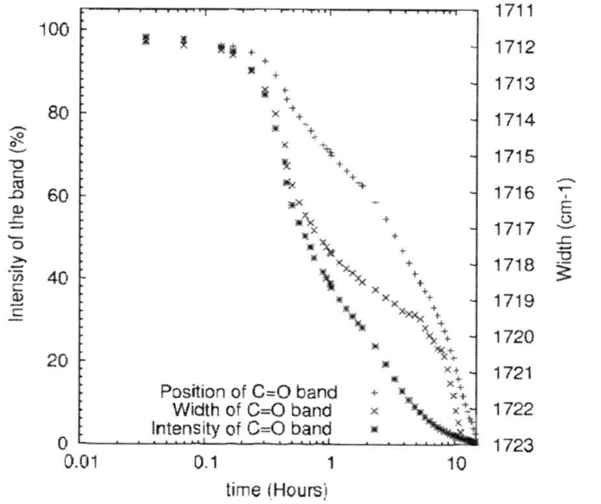

Figure 8.

Evolution of the 3 parameters position, width and intensity (absorbance) of the Gaussian function used to analyse the band of the carboxyl group's stretching vibration during the decomposition process of the PET film. (The width is decreasing from 10.3 to 3.5 cm^{-1}).

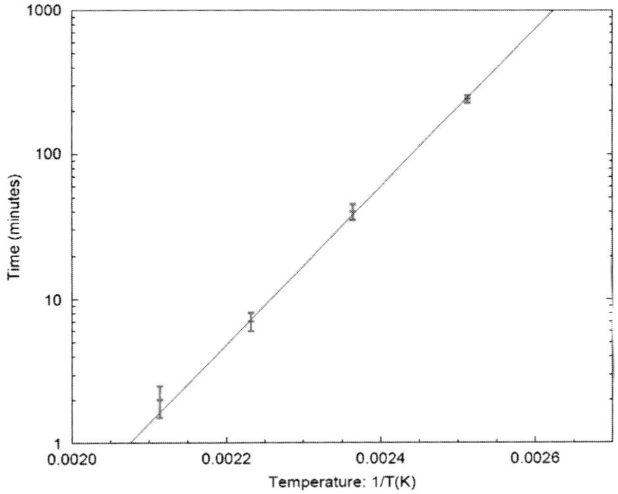

Figure 9.
Degradation rate of PET film under methanol between 100 and 200 °C.

tion in the polymer during the degradation. It appears in fact with a third "phase" that hydrolysis and methanolysis used the same way for the depolymerisation of PET. All efforts to understand both the evolutions of structure and morphology have clearly demonstrated a non correlation between the evolution of crystalline/amorphous phase and trans/gauche conformation. The quantity of the trans form is decreasing rapidly during the process opposing the amount of amorphous form which is increasing. Nevertheless, it is important to notice also that the evolution of the molecular weight during the methanolysis has not been taken into account in this analysis. As the evolution of the crystallinity and the conformation of the PET, this phenomenon might also affect the evolution of the spectra during the depolymerisation process.

Methanolitic degradations have been also studied and analysed at three other temperatures (125, 175 and 200 °C) and similar evolutions have been found. The half life time of PET (using an internal reference band at 1410 cm^{-1}) under methanol pressure are presented in Figure 9. The time of degradation depends exponentially on the inverse of temperature as expected by using a simple Arrhenius law. The process is 100 time faster at 200 °C than at 125 °C.

$$K = A\ \exp(-E_a/RT(K))$$

Arrhenius equation

Where K is the rate constant of the chemical process, T the temperature (in Kelvin) and E_a the activation energy.

As discussed recently by Mishra[40] to the methanolysis of depolymerization of PET waste using zinc acetate in the presence of lead acetate as the catalyst at 120–140 °C in a closed batch reactor, the methanolysis can be modelled by an Arrhenius equation as a simple reaction. In our case, we have found low activation energy of 5.8 $kJ \cdot mol^{-1}$.

Conclusions

ATR-FTIR spectroscopy has been applied in situ during the degradation of PET by methanol at 125, 150, 175 and 200 °C. Infrared spectroscopy is able to observe interactions at molecular level. These experiments gave us the possibility to observe in situ the evolution of morphology of the polymer. The opportunity to observe in situ a complex phenomenon like the degradation of polymer (which is directly related to the polymer recycling process) has been demonstrated with the possibility to observe at same time the proportion of trans/gauche conformer using an internal reference. The evolution of the degree of crystallinity has not been quantified. During the degradation the trans conformations have been destroyed more slowly than the average PET and at the same time the proportion of crystalline form was decreasing. The evolution of the degradation time depends exponentially on the inverse of temperature.

In summary, the in situ spectroscopic approach developed here shows good promise in evaluating of degradation processes of polymers subjected to high-temperature methanol. We hope that this approach can contribute to the optimisation of existing polymer recycling processes under these conditions and could also be used for the design of new processes using high-pressure gases and supercritical fluids.

Acknowledgements: This work was supported by the European Commission (SurfaceT project NMP2-CT-2005-013524) (http://www.icmab.es/surfacet/). We thank Dr. G. G. Martirosyan for his help in the initial stage of this investigation and Dr. N. Everall for the samples of PET.

[1] M. Goto, *Journal of physics. Condensed matter* **2002**, *14*(44), 11427.
[2] Y. Yang, Y. Lu, H. Xiang, Y. Xu, Y. Li, *Polymer Degradation and Stability* **2002**, *75*(1), 185.
[3] H. Kurokawa, M-a. Ohshima, K. Sugiyama, H. Miura, *Polymer Degradation and Stability* **2003**, *79*(3), 529.
[4] R. W. G. Lotz, N. DA, US *Patent* 3,321 510 **1967**.
[5] H. Gruschke, W. Hammerschick, H. Medem, US *Patent* 3,403,115 **1968**.
[6] D. Paszun, T. Spychaj, *Ind. Eng. Chem. Res.* **1997**, *36*(4), 1373.
[7] D. E. Nikles, *Macromolecular materials and engineering* **2005**, *290*(1), 13.
[8] C. Y. Kao, B. Z. Wan, W. H. Cheng, *Ind. Eng. Chem. Res.* **1998**, *37*(4), 1228.
[9] J. R. Campanelli, *J. Appl. Polym. Sci.* **1993**, *48*(3), 443.
[10] J. Y. Chen, *Journal of applied polymer science* **1991**, *42*(6), 1501.
[11] E. Bormashenko, R. Pogreb, A. Sheshnev, E. Shulzinger, Y. Bormashenko, S. Sutovski, Z. Pogreb, A. Katzir, *Polymer Degradation and Stability* **2001**, *72*(1), 125.
[12] C. Sammon, J. Yarwood, N. Everall, *Polymer Degradation and Stability* **2000**, *67*(1), 149.
[13] N. Brantley, S. Kazarian, C. Eckert, *J. Appl. Polym. Sci.* **2000**, *77*, 764.
[14] O. S. Fleming, K. L. A. Chan, S. G. Kazarian, *Vib. Spectrosc.* **2004**, *35*, 3.
[15] O. S. Fleming, S. G. Kazarian, *Appl. Spectrosc.* **2004**, *58*(4), 391.
[16] O. S. Fleming, S. G. Kazarian, E. Bach, E. Schollmeyer, *Polymer* **2004**, *46*(9), 2943.
[17] S. Kazarian, N. Brantley, C. Eckert, *Vib. Spectrosc.* **1999**, *19*, 277.
[18] S. G. Kazarian, G. G. Martirosyan, *Phys. Chem. Chem. Phys.* **2002**, *4*, 3759.
[19] N. M. B. Flichy, S. G. Kazarian, C. J. Lawrence, B. J. Briscoe, *J. Phys. Chem. B.* **2002**, *106*(4), 754.
[20] A. R. C. Duarte, L. E. Anderson, C. M. M. Duarte, S. G. Kazarian, *J. Supercrit. Fluids* **2005**, *36*, 160.
[21] K. C. Cole, A. Ajji, E. Pellerin, *Macromol. Symp* **2002**, *184*, 1.
[22] J. K. Keum, H. H. Song, *Polymer* **2005**, *46*(3), 939.
[23] K. C. Cole, A. Ajji, É. Pellerin, *Macromolecules* **2002**, *35*(3), 770.
[24] K. C. Cole, A. Ajji, É. Pellerin, *Macromolecular Symposia* **2002**, *184*, 1.
[25] B. J. Holland, *Polymer* **2002**, *43*(6), 1835.
[26] J. R. Atkinson, F. Biddlestone, J. N. Hay, *Polymer* **2000**, *41*(18), 6965.
[27] D. J. Walls, *Appl. Spectrosc.* **1991**, *45*(7), 1193.
[28] F. Pardal, G. Tersac, *Polymer Degradation and Stability*, **2007**, in press.
[29] K. Schmidt-Rohr, W. Hu, N. Zumbulyadis, *Science* **1998**, *280*(5364), 714.
[30] A. V. Nemukhin, B. L. G. A. A. Granovsky, *Moscow University Chemistry Bulletin* **2004**, *45*(2), 75.
[31] A. A. Granovsky, *http://classic.chem.msu.su/gran/gamess/index.html* PC GAMESS version 7.1.
[32] J. M. Andanson, J. C. Soetens, T. T. Tassaing, M. Besnard, *J. Chem. Phys.* **2005**, *122*(17), 174512.
[33] A. P. Scott, L. Radom, *J. Phys. Chem.* **1996**, *100*(41), 16502.
[34] J. Jang, *Polymer* **1997**, *38*(16), 4043.
[35] J. M. Huang, P. P. Chu, F. C. Chang, *Polymer* **2000**, *41*(5), 1741.
[36] M. Edge, M. Hayes, M. Mohammadian, N. S. Allen, T. S. Jewitt, K. Brems, K. Jones, *Polymer Degradation and Stability* **1991**, *32*(2), 131.
[37] N. S. Allen, M. Edge, M. Mohammadian, K. Jones, *European Polymer Journal* **1991**, *27*(12), 1373.
[38] N. S. Allen, M. Edge, M. Mohammadian, K. Jones, *Polymer Degradation and Stability* **1993**, *41*(2), 191.
[39] N. S. Allen, M. Edge, M. Mohammadian, K. Jones, *Polymer Degradation and Stability* **1994**, *43*(2), 229.
[40] S. Mishra, *Polymer international* **2003**, *52*(3), 337.

Orientation Dependent FT Raman Microspectroscopy on Hemp Fibers

S. K. Kovur, K. Schenzel, W. Diepenbrock*

Summary: FT Raman microspectroscopy was used for polarization experiments on strained hemp fibre cells. The cellulosic plant fibers were macerated with alkaline and enzymatic solutions. Those cleaned and refined single fiber cells were subjected to micro tensile tests as well as to polarization measurements under the FT Raman microscope. Mechanical parameters of the fiber cells (e.g. E-modulus) were determined and changes in orientation of the –(C–O–C)– structure units of the cellulose were considered with respect to fiber stress and molecular fiber structures. Intensity ratios R_1 and R_2 calculated on the polarized micro FT Raman spectra of the strained fibers describe the order parameter $\langle P_2 \rangle$ and $\langle P_4 \rangle$ allowing the quantitative determination of the orientation of the structure units –(C–O–C)– of fiber cellulose with respect to the fiber cell axis.

Keywords: fiber microstraining experiments; hemp fibers; orientation of cellulose structure units; polarized FT Raman microspectroscopy

Introduction

Natural cellulose fibers of annual plants - like hemp and flax - are increasingly being used for various industrial applications. They are recognised as potential replacement for glass fibers because of their biodegradability and their high strength-to-weight ratio.

The so-called bast fibers consist of different hierarchical microstructures, whereby the microfibrils of cellulose serve as basic units. They are embedded in a matrix of hemicelluloses and form the different cell wall layers of an elementary single fiber cell, which is 10–30 μm in diameter in case of hemp fibres. The single fiber cells are bonded together with pectins and small amounts of lignin framing the next level of microstructure, the technical fibers, with diameters of 50–100 μm. These filaments are fixed together with a pectin-lignin matrix that forms the fiber bundles in the cortex of the plant stems. Thus, bast fibers are bundles of individual strands of fibers held together by a pectin-lignin interface.[1]

The removal of these binding material results in the refining of the fiber bundles, yielding filaments or single fiber cells respectively. An important characteristic of the fiber cells is their composition of highly structured cell wall layers, e. g. primary (P) and secondary (S) ones. The S layer contains the most cellulose mass and with regard to the mechanical support of the plant fibers it is the most important one. In its microstructure one has to differentiate between three additional layers S1, S2, and S3. They are distinguished by their microfibril textures which are represented by different orientation angles of the cellulose microfibrils with respect to fiber cell axis. These microstructural characteristics together with molecular substructures of the fiber cellulose, which may exist in terms of the polymorphic forms, cellulose I and/or cellulose II respectively,[2] strongly influence mechanical parameters of the plant fibers like tensile strength and E-modulus.[3–6]

Martin Luther University Halle-Wittenberg, Department of Natural Science III, Institute of Agricultural and Nutritional Science, D-06099 Halle, Germany
E-mail: karla.schenzel@landw.uni-halle.de

Since the molecular orientation affects the physical properties of macromolecular systems it is very important to understand the mechanisms that govern their orientation and relaxation of orientation. Polarized Raman microspectroscopy is a powerful technique suitable to characterize the orientation of macromolecules since it can provide information about the degree of orientation of chemical groups and structure units in crystalline or semicrystalline polymers. One of its main advantages is the feasibility to record high quality spectra of high spatial resolution. Low diameters of the focused laser beam allow the investigation of single fiber cells. In addition, the second and fourth terms, $\langle P_2 \rangle$ and $\langle P_4 \rangle$, of the Legendre polynomial expansion of the orientation distribution function can be determined from polarized Raman spectra. Thus, for systems of uniaxial or fiber type symmetry a general procedure to determine $\langle P_2 \rangle$ and $\langle P_4 \rangle$ coefficients, the so-called order parameters, has been developed in detail by Bower[7] and Jen et al.[8]

Thereupon, it was the aim of our FT Raman microscopic investigations to follow changes in the orientation of structure units of cellulose microfibrils of hemp fibers due to fiber stress experiments and with respect to changes of the molecular fiber composition and substructures of the fiber cellulose caused by enzymatic and alkaline fiber treatments.

Experimental Part

Plant Material

Hemp (*Cannabis sativa*, L.; *cv*. USO 31) was cultivated at experimental station of the Martin-Luther-University Halle-Wittenberg. After dew-retting the fibers were extracted mechanically from the stems on industrial premises. The extracted fiber bundles, 50–100 μm in diameter, were refined by hand. Following the mechanically isolated fibers were cleaned and refined chemically by alkaline, C_{NaOH} (w/w%) = 6% and 25%, as well as by enzymatic treatments with hemicellulase (*Aspergillus niger*, 1.5 Umg^{-1}) and cellulase (*Aspergillus niger*, 1.3 Umg^{-1}) always after an alkaline treatment using 6% NaOH solution. Finally, the fibers were washed with distilled water and dried at room temperature.

Environmental Scanning Electron Microscopy (ESEM)

ESEM micrographs of the isolated and refined fibers were recorded using the electron microscope XL30, ESEM-FEG (Philips) with different magnifications.Typically it was equipped with environmental SE-detectors. The fiber micrographs were recorded at a pressure of 0.13 kPa and an accelerating voltage of 15.0 keV. Fibers were measured in the wet mode preserving their original characteristics without making the samples conductible.

NIR FT Raman Microspectroscopy

Near Infrared Fourier Transform (NIR FT) Raman spectroscopic measurements were carried out using the BRUKER RFS/100S spectrometer coupled directly to a NIKON microscope, Eclipse E 400. The Nd:YAG laser ($\nu_0 = 1064$ nm) was used for Raman excitation. Raman back scattering was detected by a nitrogen cooled Ge-diode. A rotable $\lambda/2$ plate serving as polarizer was included into the laser pathway between spectrometer and microscope. On the terms indicated the laser was oriented parallel (x) or perpendicular (y). An analyzer – placed into the pathway of the Raman back scattering towards the interferometer – allowed the detection of the polarized scattered radiation along the x and y directions.

Details of the fibre stress experiments under the Raman microscope were described elsewhere.[6] For fiber tensile tests in combination with polarization experiments fiber cells of around 15–30 μm were fixed to a fiber card at a micro straining rig. The axis of the fiber cell was oriented parallel to the electric field vector of the incident laser, as shown in Figure 1.

The fiber straining experiments in combination with polarization investigations require a strong compromise between the

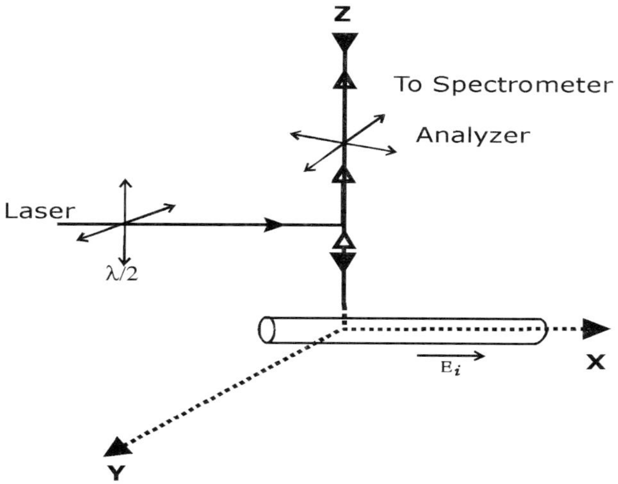

Figure 1.
Coordinate system used for the recording of the micro FT Raman spectra of the hemp fiber cells. E_i = vector of the incident electric field.

need for short spectra accumulation times during the straining process as well as the need for spectra with good S/N ratio and sharp Raman signals to determine real peak positions and peak intensities. By view of this phenomenon each spectrum was accumulated over 120 scans at each level of strain following the wavenumber shift of the Raman signal at 1096 cm^{-1}.

For each polarization measurement a strain increment of 100 μm was chosen. Real time spectra were recorded always before starting the measurement of a new fiber cell and when several spectral changes were observed. Then, the sample position under the microscope were monitored and the fibre position as well as the focal plane were readjusted to get similar spectra.

All micro FT Raman spectra were treated with the Bruker software, OPUS 4.2. The spectra were vector-normalized, baseline corrected and smoothed (9-pt). Raman peak positions and band intensities were determined by curve-fitting (Levenberg-Maquardt algorithm) considering a linear baseline in the frequency range 1050–1150 cm^{-1}.

Results and Discussion

As shown by the ESEM images in Figure 2, typical differences in the microstructure of the fiber bundles were detected due to different alkaline and enzymatical fiber

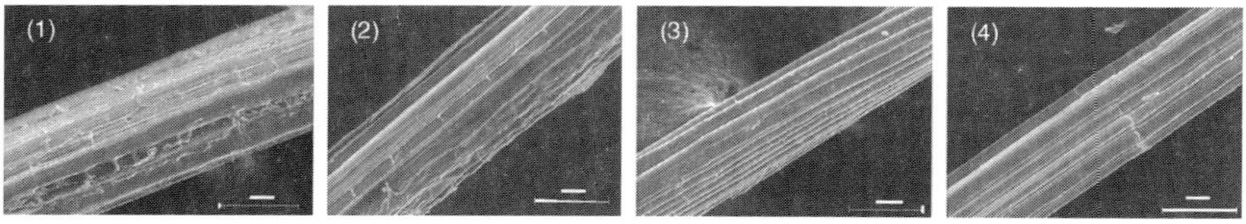

Figure 2.
ESEM images of untreated (1), 6% NaOH (2) and 25% NaOH treated (3) as well as of cellulase treated (4) hemp fibers.

treatments. A surface cleaning and refining process of the fibers bundles was observed and can be followed in the images (1)–(4).

The alkaline and enzymatical fiber treatments caused changes in molecular fiber composition, particularly by removing of small amounts of waxes and remaining parts of lignin from the fiber surfaces. Also the degradation of hemicelluloses and pectines, which serve as matrix materials for the cellulose microfibrils and tie the single fiber cells to a bundle structure, was yielded.

The strongest effect of surface cleaning and fiber refining was achieved if the fibers were treated enzymatically with cellulase afterwards an alkaline treatment with 6% NaOH solution, compare ESEM image (4). A fiber bundle characterized by another substructure of the fiber cellulose is represented in image (3). This bundle was cleanded from waxes and lignin. However the substructure of the cellulose was transformed from cellulose I into the polymorphic form -cellulose II- due to an alkaline fiber treatment using a 25% NaOH solution. The polymorphic transformation followed by FT Raman spectroscopy and microspectroscopy was described earlier.[9] As a result of fiber cleaning and refining the uniaxial morphology of the hemp fiber cells became obvious by the ESEM investigations.

Assuming an uniaxial symmetry structure for the microfibrils of the fiber cellulose the orientation of molecular structure units of cellulose with respect to the fiber axis can be determined quantitatively utilizing polarized micro Raman spectra of the fiber cells.[7,8]

In this work, changes in orientation of –(C–O–C)– structural units of fiber cellulose of hemp fibers were examined with respect to fiber straining and fiber composition by means of FT Raman microscopy. The orientation dependent behaviour of –(C–O–C)– units of cellulose was indicated by intensity changes of the Raman signal of the ν(C–O–C) stretching vibration, as shown in Figure 3 (left). The vibrational mode ν(C–O–C) is mainly assigned to the 1,4-β (C–O–C) glycosidic linkages in the cellulose chains [11] and has already proved its high stress sensitivity for cellulose fibers.[6,12] The highest intensity of the corresponding Raman signal at 1096 cm^{-1} was detected if the electric field vector (E_i) of the incident laser beam was oriented parallel to the fiber axis. By turning the E_i vector to the perpendicular direction, the signal intensity of the ν(C–O–C) mode was decreasing, compare Figure 3 (left). Thereupon a mainly parallel orientation with respect to the fiber cell axis has to be assumed for the glycosidic linkages and the glucopyranose units (GPUs) which form

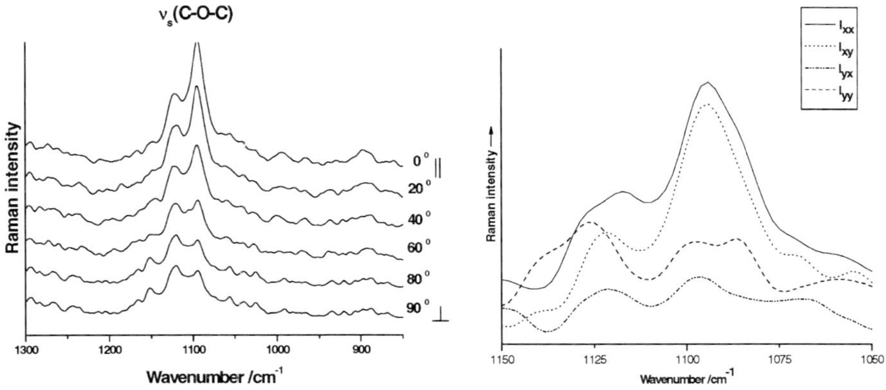

Figure 3.
Orientation dependent FT Raman spectra of the ν(C–O–C) mode of a hemp fiber cell (left) and polarized spectra of the ν(C–O–C) mode (right).

the molecular skeletons of the cellulose microfibrils.

Here not only the orientation sensitivity was surveyed on the strained fiber cells. Additionally polarization experiments in the FT Raman microscope were carried out to determine changes in orientation of the –(C–O–C)– structure units quantitaively. Thus the order parameters $\langle P_2 \rangle$ and $\langle P_4 \rangle$ were determined for these molecular units, always with respect to the level of fiber strain and molecular fiber composition.

Following the works of Pezolet and co-workers on spider silk fibers[10] a quantitative determination of $\langle P_2 \rangle$ and $\langle P_4 \rangle$ for systems of uniaxial symmetry can be yielded from the intensity ratios R_1 and R_2 which are obtained directly from the polarized micro Raman spectra:

$$R_1 = \frac{I_{z(xy)\overline{z}}}{I_{z(xx)\overline{z}}} \qquad (1)$$

$$R_2 = \frac{I_{z(yx)\overline{z}}}{I_{z(yy)\overline{z}}} \qquad (2)$$

Four polarized micro FT Raman spectra were recorded for the ν(C–O–C) mode of a hemp fiber cell on each level of increasing fiber strain up to fracture. In Figure 3 (right) the spectra are represented illustrating one level of fiber strain. They were identified by their polarization configurations: $Z(XX)\overline{Z}$, $Z(XY)\overline{Z}$, $Z(YY)\overline{Z}$ and $Z(YX)\overline{Z}$, illustrated in Figure 1 and described in accordance with the Porto[13] notation. The intensity ratios R_1 and R_2 were calculated from peak-height intensities of the signal of the ν(C–O–C) mode in the polarized spectra. Average values of the ratios were obtained over 5 samples.

In this manner, changes in orientation of the cellulose structural units –(C–O–C)– of the hemp fibers were followed up with increasing fiber strain and with respect to different chemical compositions of the plant fibers. In Figure 4 the changes are descibed by means of the order parameter $\langle P_2 \rangle$ for the alkaline and enzymatically treated hemp fiber cells. The values of this parameter for the unstrained fibers are in the range between 0.18 ± 0.03 and 0.29 ± 0.03. Knowing that the limiting values of $\langle P_2 \rangle$ are 1 and -0.5 for perfect orientation at 0° and 90° from the fiber axis, the obtained values indicate that the –(C–O–C)– units of the unstrained fibers are arranged

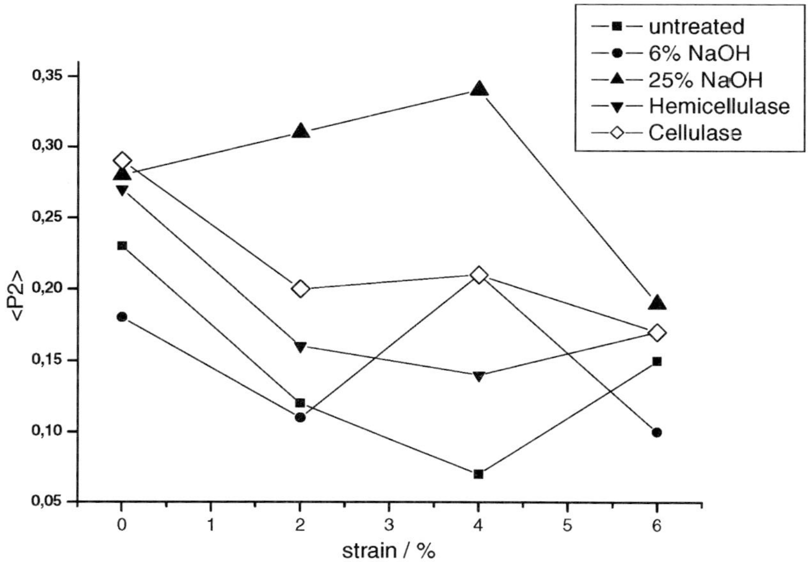

Figure 4.
Changes of the orientation parameter $\langle P_2 \rangle$ of the –(C–O–C)– units of alkaline and enzymatical treated hemp fiber cells with respect to fiber strain.

neither parallel nor perpendicular to the fiber cell axis.

The orientation angle between the fiber axis and the –(C–O–C)– units of unstrained fibers are bigger for the enzymatical and 25% NaOH treated fibers than for total untreated ones. Only unstrained fibers treated with 6% NaOH solution showed a smaller angle and therewith a more parallel orientation of the –(C–O–C)– units to the fiber axis.

With increasing fiber strain the orientation angles of –(C–O–C)– units increase except for fibers treated with 6% and 25% NaOH solutions. For these fibers an increase of the $\langle P_2 \rangle$ values was obtained indicating smaller angles between –(C–O–C)– units and the fiber axes, coincident with a more parallel orientation of the cellulose microfibrils. It was concluded, that the cellulose structures become more isolated and get a higher capability for incorporating mechanical stress in terms of the stretching of the 1,4-β –(C–O–C)– glycosidic linkages owing to the removal of lignin by the alkaline fiber cleaning processes. Thus, a decrease of the orientation angle between –(C–O–C)– units and fiber cell axis was recorded at higher strain levels.

Conclusions

Orientation dependent FT Raman micro spectroscopic experiments were carried out on strained fiber cells of alkaline and enzymatical treated hemp fibers beside morphological investigations by means of ESEM. With respect to chemical and enzymatical fiber treatments a fiber cleaning and refining process was observed in the micro structure of the fibers.

Order parameters of the molecular structure units –(C–O–C)– of cellulose of the alkaline and enzymatical treated hemp fibers were determined with respect to the levels of fiber strain by means of polarization investigations in the FT Raman microscope.

Exemplarily, changes of the $\langle P_2 \rangle$ values were discussed with respect to changes in fiber morphology and chemical fiber composition. It was shown, that the –(C–O–C)– units of the cellulose microfibrils of unstrained fibers lie inbetween a parallel and a perpendicular orientation to the fibre cell axis.

Only the fibers treated with 6% and 25% NaOH solutions showed a decrease in the orientation angle of the –(C–O–C)– units with respect to the fiber axis with increasing fiber strain, indicating the more parallel orientation of the microfibrils at higher strain levels. Here, the cleaned and isolated cellulose structures represent their capability for higher stress incorporation. Thereby, the decrease of the orientation angle was caused.

In contrast, the increase of the orientation angle was observed for enzymatical treated fibers at higher strain levels reflecting the loss of order within the cellulose structures owing to the removal of matrix materials and shortening of cellulose chains by cellulase treatments.

Acknowledgements: Financial support of Deutsche the Forschungsgemeinschaft (DFG), project no. SCHE 415/11-1, 2 is gratefully acknowledged.

[1] H. J. Purz, H.-P. Fink, H. Graf, *Das Papier* **1998**, *6*, 315.
[2] R. D. Gilbert, J. F. Kadla, in: "*Biopolymers from Renewable Resources*", D. L. Kaplan, Ed., Springer Berlin, Heidelberg, New York **1998**, 47–90.
[3] S. Pleasants, W. J. Batchelor, I. H. Parker, *Appita J.* **1998**, *51*, 373.
[4] B. M. Prasad, M. M. Sain, *Mater. Res. Innovat.* **2003**, *7*, 231.
[5] A. Bhatnagar, M. Sain, *J. Reinforced Plastics Composites* **2005**, *24*, 1259.
[6] P. Peetla, K. C. Schenzel, W. Diepenbrock, *Appl. Spectrosc.* **2006**, *6*, 682.
[7] D. I. Bower, *J. Polym. Sci. Phys.* **1972**, *10*, 2135.
[8] S. Jen, N. A. Clark, P. S. Pershan, E. B. Priestly, *J. Chem. Phys.* **1977**, *66*, 4635.
[9] K. Schenzel, S. Fischer, *Cellulose* **2001**, *8*, 49.
[10] M.-E. Rousseau, T. Lefevre, L. Beaulieu, T. Asakura, M. Pezolet, *Biomacromol.* **2004**, *5*, 2247.
[11] H. G. M. Edwards, D. W. Farwell, D. Webster, *Spectrochim. Acta* **1997**, *53A*, 2383.
[12] S. J. Eichhorn, J. Sirichaisit, R. J. Young, *J. Mat. Sci.* **2001**, *36*, 3129.
[13] T. C. Damen, S. P. S. Porto, B. Tell, *Phys. Rev.* **1966**, *142*, 570.

The Structure of Aminopropylsiloxane Polymerized in DC Electric Field

Lahorija Bistričić,[*1] Vesna Volovšek*[2]

Summary: The Raman spectra of aminopropylsiloxane (APS) polymerized on PVC substrate in DC electric field were measured to study the structure of deposited films. The modelling of low wavenumber Raman spectra was achieved using contribution of acoustic phonons, intramolecular vibrations as well as optical phonons related to librational motions of ladders. Results suggest that electric field influences the orientations of ladders increasing thus the degree of order in microdomains.

Keywords: acoustical phonons; aminopropylsiloxane; bose peak; DC electric field; libration

Introduction

Aminopropylsilanetriol (APST) is frequently used as a coupling agent for functionalization of surfaces like semiconductors, nanopariticles, optical fibres etc. The molecules of APST are metastable units which undergo self-condensation to thermodynamically more stable units with Si–O–Si siloxane bonds by elimination of water. The bonding of APST with the surface of substrate depends on the conformation of polymerized structure. Controlled hydrolysis of aminopropylsiloxane polymer (APS) provides polysiloxanes with different structures.

Recently, we have investigated the vibrational dynamics and structure of APS polymer using DFT calculations.[1] The analysis of low wavenumber polarized Raman spectra of APS polymerized on PVC substrate suggests medium range ordered ladder structure, stacked in layers with different orientation of ladders.[2] Ladder polymers are of particular interest in the field of advanced materials due its incontestable superiority in thermo- or radiation-resistant and mechanical properties.[3] The synthesis of structure-controlled ladder polymers has been a challenge in the polymer research field.

According to DFT calculations APST molecule has permanent electric dipole moment.[4] When the DC electric field is applied to polar molecules, the dipoles tend to orient themselves along the direction of the field. Therefore, the polymerization in electric field is considered a useful method to control the molecular orientation in polymers.[5–8]

In this work we investigated the structure of APS polymerized on PVC substrate in DC electric field. The study of Raman scattering from intramolecular and intermolecular modes in APS films was thus performed to obtain the information regarding polymer structure.

Experimental Part

Aminopropylsilanetriol (APST) in 25% aqueous solution was purchased from ABCR Company. APS polymer films around 1 mm thick were prepared by deposition of APST aqueous solution on PVC substrate. The condensation process was going on very slow. The condensation was carried out at room temperature in dark for approximately four days. The polymerization in electric field was carried out on PVC

[1] Faculty of Electrical Engineering and Computing, University of Zagreb, Unska 3, Zagreb, 10000, Croatia
Fax: 385 1 6129 605;
E-mail: lahorija.bistricic@fer.hr
[2] Faculty of Chemical Engineering and Technology, University of Zagreb, Marulićev trg 19, Zagreb, 10000, Croatia

substrate that is laid between two metal plates of condenser linked up with high voltage DC power. The electric field intensity in range 10^3–$3.5 \cdot 10^5$ V/m was obtained by adjusting the output voltage in high voltage DC power as well as the distance between the plates.

The Raman spectra of APS from 3500–10 cm^{-1} were recorded on a Dilor Model Z24 triple monochromator coupled with an IBM AT computer. The 514.5 nm line of a Coherent INNOVA-400 argon ion laser was used for excitation. Laser power at the sample was below 200 mW. The spectral resolution was 2 cm^{-1}. 90° scattering geometry has been used. The incoming light was polarized vertically with respect to the scattering plane. The polarization of the scattered light was analyzed in vertical (VV) and horizontal (VH) direction.

Results and Discussion

The Raman spectra of APS polymerized in various electric field intensities are compared in Figure 1. The intensity of each spectrum is normalised to the peak intensity of the 1452 cm^{-1} band assigned to the CH_2 scissoring vibrations. It can be noted that there are no significant difference among those spectra in the frequency region above 300 cm^{-1}. All spectra comprise the characteristic vibrations of ladder APS polymer observed at 1144, 520 and 451 cm^{-1}.[1] According to our normal mode analysis based on DFT calculations the band at 1144 cm^{-1} is assigned as in phase Si-O-Si stretching vibrations in the rungs of the ladder structures.[1] The fundamental at 520 cm^{-1} can be described as the ring breathing mode of silicon atoms, while those observed at 451 cm^{-1} corresponds to the ring breathing mode of oxygen atoms.[1]

The low frequency Raman spectra of all samples are dominated by very strong Bose band.

Our previous analysis has shown that the low wavenumber Raman spectra of APS polymerized on PVC substrate without electric field consist of two contributions: a broad vibrational band centred around 200 cm^{-1} and very strong Bose band with

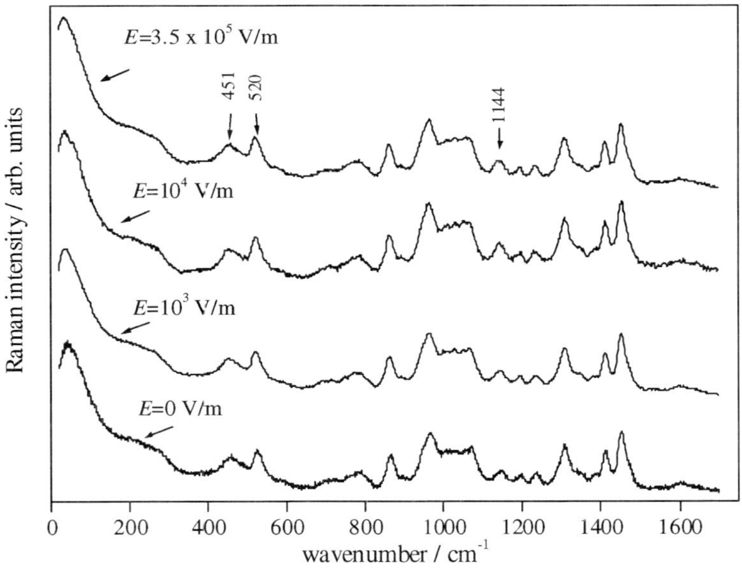

Figure 1.
Raman spectra of APS polymerized in different electric field intensities. The intensity of each spectrum is normalised by the peak height of the 1452 cm^{-1} intramolecular band.

maximum measured at 42 cm^{-1}.[2] The broad vibrational band consists from contribution of normal modes related to the quadripolar mode of silicooxygen rings as well as to deformations of propyl groups.[1] The contribution of intramolecular vibrations in the frequency region below 100 cm^{-1} can be neglected because DFT calculations predict very low Raman activity for those modes.[1]

We have shown that the nature of Bose band in APS polymerized on PVC substrate without electric field can be explained by model elaborated by Martin and Brenig[9] and then modified by Malinovsky and Sokolov[10] (MB model). According to this model, Bose band has been closely related to the acoustic transversal (TA) and longitudinal (LA) phonons in the medium range ordered structure. The medium range order implies that the arrangements of structural units are not completely random but have some correlation.[11]

It is well established that the observed low frequency Raman spectrum of disordered materials consists from quasi-elastic scattering $I_{QE}(\nu)$, Bose peak $I_B(\nu)$ and low-frequency vibrational modes $I_V(\nu)$:[2]

$$I_{\exp}(\nu) = I_{QE}(\nu) + I_B(\nu) + I_V(\nu). \tag{1}$$

It is usual to define the reduced Raman intensity:

$$\begin{aligned} I_R(\nu) &= I_{\exp}(\nu)\cdot\nu\cdot[n(\nu,T)+1]^{-1} \\ &= I_{RQE}(\nu) + I_{RB}(\nu) + I_{RV}(\nu), \end{aligned} \tag{2}$$

where $n(\nu, T) = [\exp(hc\nu/k_BT) - 1]^{-1}$ is a temperature Bose factor.

For easier comparison with various theoretical models we used another expression for the reduced Raman intensities:

$$I_{RED\exp}(\nu) = \frac{I_R(\nu)}{\nu^2} \tag{3}$$

$I_{RED\exp}(\nu)$ was calculated according relations (2) and (3) using experimental Raman spectrum $I_{VH}(\nu)$ normalised by the peak height of the 1452 cm^{-1} band. Our aim was to fit particular model to an observed spectrum $I_{RED\exp}(\nu)$.

In MB model, in the region $\nu \geq 10$ cm^{-1} where the quasi-elastic light scattering is negligible, the redefined intensity of Bose band $I_{RB}(\nu)$ can be factorized:

$$I_{RB}(\nu) = C(\nu)\cdot g(\nu). \tag{4}$$

The well known Debye expression $g(\nu) \propto \nu^2$ is used for the density of states of acoustic phonons. $C(\nu)$ describes the coupling between the light and vibrational modes:

$$C(\nu) = \nu^2\cdot[g_{TA}(\nu)\cdot E_{TA} + g_{LA}(\nu)\cdot E_{LA}], \tag{5}$$

where $g_i(\nu)$ is space Fourier transforms of the correlation function of disorder and E_i is light-vibration coupling parameters for transversal and longitudinal acoustic phonons. The best results in modelling the Bose band was achieved with exponential correlation function of disorder $G_{dis}(\nu) = \exp(-r/R_c)$, ($R_c$ is so-called structure correlation radius), which leads to:

$$\begin{aligned} g_{TA}(\nu) &= \left(\nu^2 + \nu_{TA}^2\right)^{-2} \text{ and } g_{LA}(\nu) \\ &= (\nu_{TA}/\nu_{LA})^5\cdot\left(\nu^2 + \nu_{LA}^2\right)^{-2}, \end{aligned} \tag{6}$$

where $\nu_i = c_i/\pi\cdot R_c$ and c_i is respective (TA) or (LA) sound velocity. Such function of disorder corresponds to the structure of continuous network, which consists of the ordered microregions of the size ~2 R_c.[10]

The model function $I_{REDcalc}(\nu)_{E=0}$ for fitting the reduced Raman spectra has been defined as:

$$\begin{aligned} I_{REDcalc}(\nu)_{E=0} &= I_0 + C(\nu) \\ &+ I_V\frac{(\Gamma_V/2)^2}{(\nu-\nu_V)^2+(\Gamma_V/2)^2} \\ &\cdot\frac{1-\exp(hc\nu/k_BT)}{\nu} \end{aligned} \tag{7}$$

The background noise in Raman spectra I_0 has been taken into account in modeling as a free parameter. The molecular low-frequency vibrational band in Raman spectra is presented with Lorentzian $I_V(\Gamma_V/2)^2/\left((\nu-\nu_V)^2+(\Gamma_V/2)^2\right)$.

In the Raman spectra of all samples polymerized in electric field the maximum

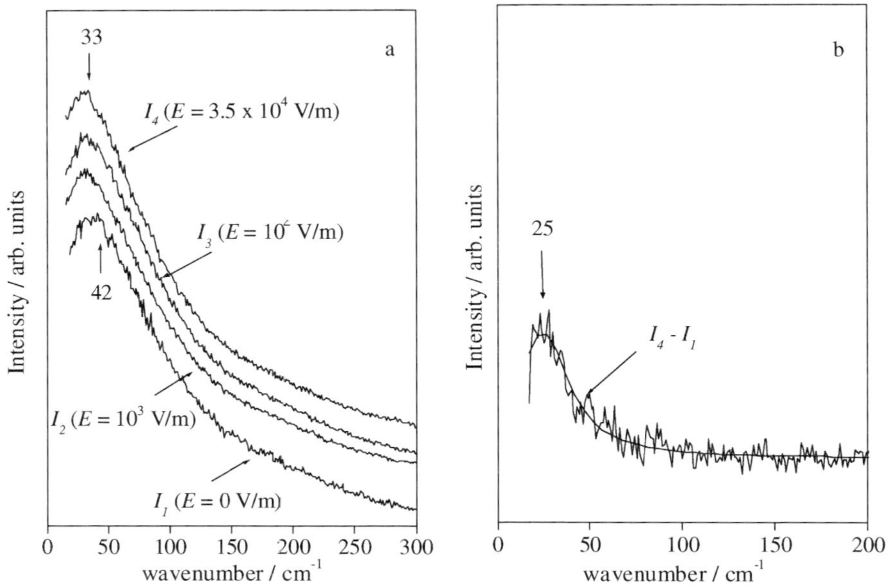

Figure 2.
a) The reduced I_{VH} Raman spectra for different intensities of electric field; b) The difference Raman spectrum fitted to the Lorentzian band.

of Bose band is shifted to 33 cm^{-1}. The model function represented with relation (7) could not reproduce the observed reduced low frequency I_{VH} Raman spectra (Figure 2a). In an attempt to solve this problem we took the difference spectra between the reduced Raman intensities of the samples polymerized in various intensities of electric field and without electric field. All these spectra show the Lorentzian band centred at around 25 cm^{-1} (Figure 2b). Therefore, we included the additional vibrational band to the model function. The following expression for reduced Raman intensity $I_{REDcalc}(\nu)_{E\neq0}$ was used for data modeling:

$$I_{REDcalc}(\nu)_{E\neq0} = I_{REDcalc}(\nu)_{E=0} + I_L \frac{(\Gamma_L/2)^2}{(\nu-\nu_L)^2+(\Gamma_L/2)^2} \cdot \frac{1-\exp(hc\nu/k_BT)}{\nu} \tag{8}$$

The model parameters were estimated through least-squares fit to the observed spectral amplitudes with uncertainties calculated according to the actual photon count. The model depends on eleven parameters. In six of them it is non-linear. The best fit parameter values are derived by iterative optimization routine, followed by the Jackknife procedure to estimate the parameter uncertainties.[12] The model is tested by calculated goodness of fit G. It is common to accept the model with $G >$ 0.001.[12] Best fit parameter values, corresponding uncertainties and goodness of fit G are given in Table 1. The results of the fitting of reduced Raman intensity for APS polymerized in electric field 10^3 V/m are shown in Figure 3a.

As can be seen from the results in Table 1 and Figure 3a the main contribution to the Bose band arises from transversal acoustic phonons TA. The frequency ν_{TA} of transversal acoustic phonons in disordered media lies between 10–80 cm^{-1} and depends on chemical composition, pressure as well as thermal history of the sample.[10] In the reduced Raman spectra of APS the position of $\nu_{TA} \sim 50$ cm^{-1} and the value of coupling

Table 1.
Bose, vibrational and librational contribution to the reduced low-wavenumber Raman spectra in dependence on electric field intensity.

E[a)]	0	10^3	10^4	$3.5 \cdot 10^5$
$I_o \pm \sigma_{Io}$	111.50 ± 7.58	212.54 ± 3.04	238.94 ± 4.37	146.14 ± 3.78
$E_{TA} \pm \sigma_{ETA}$	144.14 ± 1.97	143.78 ± 1.64	150.06 ± 2.16	144.46 ± 2.80
$\nu_{TA} \pm \sigma_{\nu TA}$	51.30 ± 1.24	50.54 ± 0.57	49.96 ± 0.62	48.55 ± 0.62
$E_{LA} \pm \sigma_{ELA}$	71.28 ± 36.62	0.86 ± 0.68	1.95 ± 0.17	6.00 ± 0.52
$\nu_{LA} \pm \sigma_{\nu LA}$	17.83 ± 1.46	9.21 ± 1.15	10.21 ± 1.32	11.65 ± 1.27
$I_V \pm \sigma_{IV}$	321.36 ± 30.72	213.20 ± 9.48	277.95 ± 15.00	240.80 ± 13.74
$\nu_V \pm \sigma_{\nu V}$	200.10 ± 2.65	197.23 ± 1.38	198.44 ± 1.56	195.45 ± 1.79
$\Gamma_V \pm \sigma_{\Gamma V}$	143.85 ± 18.66	98.67 ± 7.74	120.94 ± 9.49	119.45 ± 10.98
$I_L \pm \sigma_{LV}$	–	929.92 ± 200.00	658.40 ± 247.50	550.20 ± 292.00
$\nu_L \pm \sigma_{\nu L}$	–	26. 77 ± 1.42	28.02 ± 2.21	28.18 ± 8.00
$\Gamma_L \pm \sigma_{\Gamma L}$	–	32.76 ± 3.80	29.03 ± 5.37	27.18 ± 2.60
G	0.140	0.176	0.305	0.010

[a)](E in V/m; $I_o \cdot 10^{-5}$, $\sigma_{Io} \cdot 10^{-5}$, E_{TA}, σ_{ETA}, $E_{LA} \cdot 10^{-3}$, $\sigma_{ELA} \cdot 10^{-3}$, $I_V \cdot 10^{-3}$, $\sigma_{IV} \cdot 10^{-3}$, $I_L \cdot 10^{-3}$, $\sigma_{L'V} \cdot 10^{-3}$ in arb. units; ν_{TA}, $\sigma_{\nu TA}$, ν_{LA}, $\sigma_{\nu LA}$, ν_V, $\sigma_{\nu V}$, Γ_V, $\sigma_{\Gamma V}$, ν_L, $\sigma_{\nu L}$, Γ_L, $\sigma_{\Gamma L}$ in cm^{-1}).

coefficient E_{TA} are independent of electric field. The frequency of longitudinal acoustic mode in sample polymerized without electric field is calculated at 17.83 cm^{-1}.

The frequencies of longitudinal modes in samples polymerized in electric field moves towards lower values and are in the range 9.21–11.65 cm^{-1}. The coupling coefficient E_{LA} has the greatest value without electric field, and significantly decreases in electric field. It is important to emphasise that the contribution of longitudinal acoustic phonon to the Bose band is dominated with position of ν_{LA} (relations 5. and 6.) and not with E_{LA}. Consequently, in spite of the significant decrease of E_{LA} in the electric field the contribution of LA phonon remains important (Figure 3b).

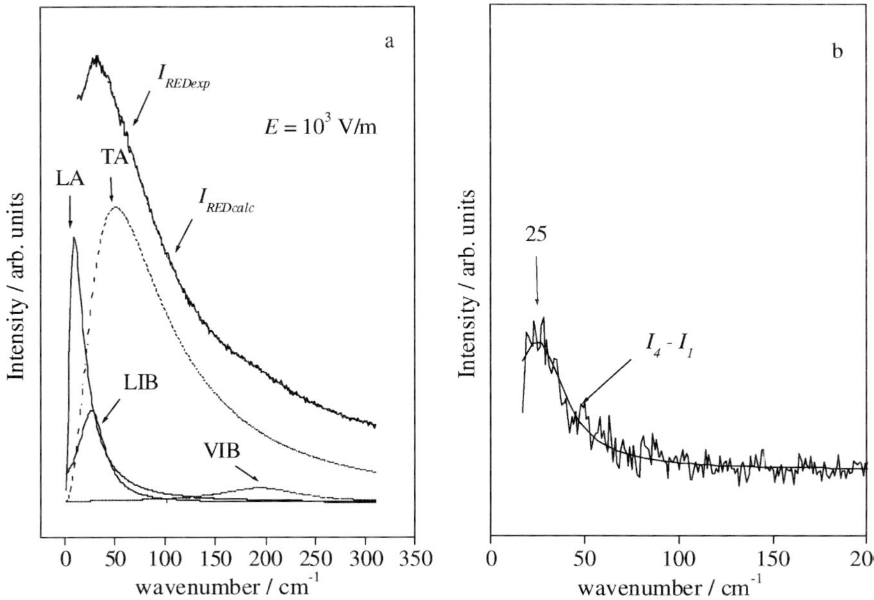

Figure 3.
a) Results of fitting of the reduced Raman intensity at $E = 10^3$ V/m (LA-longitudinal acoustic phonons; TA-transversal acoustic phonons; LIB-librational contribution; VIB-vibrational contribution); b) The dependence of longitudinal acoustic phonons LA on electric field intensity.

The similar results are obtained in studies that analyze the influence of pressure on fast dynamics and elastic properties of different materials including polymers.[13] The variations of TA modes under pressure are significantly smaller than that of LA modes. Detailed studies of disordered materials demonstrated that variation in Bose peak frequency under pressure is stronger than the variation in the sound velocity.[13] The size of ordered microdomains, or the structural correlation range can be estimated from the position of acoustic phonons and the sound velocity. Assuming that the velocity of longitudinal acoustic wave in APS samples changes slower than the frequency ν_{LA}, the structure correlation radius R_c increases in samples polymerized in electric field. In that case the dimensions of ordered domains in longitudinal direction are greater.

The Bose band can also be analysed using the depolarization ratio of the recorded low frequency Raman spectra. According MB model disorder-induced Raman scattering is ascribed to the electrical and mechanical disorder. As shown by Novikov *at al.*,[14] the strain tensor can be divided into diagonal and traceless parts. The traceless tensor is responsible for the depolarized Raman intensity I_{VH} and for part of the polarized Raman intensity I_{VV}; the diagonal tensor is responsible for the polarised intensity I_{VV}. The depolarization ratio $\rho = I_{VH}/I_{VV}$ is equal to 0.75 or 0, according to whether transversal or longitudinal modes, respectively, produce the scattering. Obviously, ρ lies between these limits if both kinds of modes scatter. The depolarization ratio of Bose band in all APS samples in range 0.50–0.56 also indicates the transverse-like character of acoustic vibration contributing to the spectra in the low frequency region.

The conformational and vibrational analysis of APST molecule has shown that there are two stabile conformers *trans* and *gauche*.[4] The *trans* conformer is the predominant form present in APS polymer. The calculated dipole moment of *trans* APST molecule is $6.86 \cdot 10^{-30}$ Cm. Therefore, the electric field-induced orientation of aminopropyl segments during polymerization is expected. The analysis and the modeling of low frequency Raman spectra confirmed the existence of vibrational band at $\sim$28 cm^{-1} (Table 1 and Figure 2b). According to the general principle that in the Raman scattering the intensities of rotatory fundamentals are always much stronger than those of translatory motions, the observed band probably arises due to the librational motions of ladders.[15]

Conclusion

The Raman light scattering study has been undertaken to elucidate the structure of APS polymerized in DC electric field. The analysis shows that applied electric field has no influence on the intramolecular vibrations. The changes are pronounced in the low wavenumber region. The low frequency Raman spectra consist from Bose band, optical phonons and intramolecular vibrations. The modelling of Bose band has shown the shift of longitudinal acoustical phonons toward lower frequencies as well as a possible enhancement of structure correlation radius R_c. The observed optical phonon centred around 28 cm^{-1} due to the libration of ladders suggests the orientational ordering in microdomains.

[1] V. Volovšek, L. Bistričić, V. Dananić, I. Movre Šapić, *J. Mol. Struct* **2007**, *834–836*, 414.

[2] V. Volovšek, L. Bistričić, K. Furić, V. Dananić, I. Movre Šapić, *J Phys. Conference Series* **2006**, *28*, 135.

[3] H. Tang, J. Sun, J. Jiang, X. Zhou, T. Hu, P. Xie, R. Zhang, *J. Am. Chem. Soc.* **2002**, *124*, 10482.

[4] L. Bistričić, V. Volovšek, V. Dananić, I. Movre Šapić, *Spectrochimica Acta Part A* **2006**, *64*, 327.

[5] S. Kurihara, K. Iwamoto, T. Nonaka, *Polymer* **1998**, *39*, 3565.

[6] A. Böker, K. Schmidt, A. Knoll, H. Zetti, H. Hänsel, V. Urban, V. Abetz, G. Kraush, *Polymer* **2006**, *47*, 849.

[7] Y. Tsori, D. Andelman, C. Y. Lin, M. Schick, *Macromolecules* **2006**, *39*, 289.

[8] G. J.-F. Demets, H. E. Toma, *Electrochem. Comun.* **2003**, *5*, 73.

[9] A. J. Martin, W. Brenig, *Phys. Status Solidi* **1974**, *64*, 163.

[10] V. K. Malinovsky, A. P. Sokolov, *Solid State Comm.* **1986**, 57, 757.
[11] J. Schroeder, W. Wu, J. L. Apkarin, M. Lee, L.-G. Hwa, C. T. Moynihan, *J. Non-Cryst. Solids* **2004**, 349, 88.
[12] B. Efron, *The Jackknife, the Bootstrap and other resampling Plans*, SIAM, Capital City Press, **1994**.
[13] B. Begen, A. Kisliuk, V. N. Novikov, A. P. Sokolov, K. Niss, A. Chauty-Cailliaux, C. Alba-Simionesco, B. Frick, *J. Non-Cryst. Solids* **2006**, 352, 4583.
[14] V. N. Novikov, E. Duval, A. Kisliuk, A. P. Sokolov, *J. Chem. Phys.* **1995**, 102, 4691.
[15] G. Turrell, *Infrared and Raman spectra of Crystals*, Academic Press, **1972**.

 DOI: 10.1002/masy.200850524

Infrared Intensity Studies in Fluorinated Macromolecules

S. Radice,[1] G. Canil,[1] P. Toniolo,[1] P.A. Guarda,[1] S. Petricci,[1] A. Milani,[2] M. Tommasini,[2] C. Castiglioni,[2] G. Zerbi[2]*

Summary: Functionalized perfluoropolyethers are investigated by infrared spectroscopy with the aim to give a quantitative estimate of the carboxylic acid groups species; since hydrogen bonded moieties have been observed, also the relative distribution of the associated species has been determined. The study is based both on measured and on theoretically predicted infrared intensities, as obtained by density functional theory (DFT) calculations on model systems.

Keywords: density functional theory simulations; hydrogen bonds; infrared spectroscopy; perfluoropolyethers

Introduction

Fluorinated polymers are high performance materials used in many technological fields. Functional groups are sometimes introduced in fluoropolymers in order to improve their adhesion on high energy surfaces or their compatibility with hydrogenated polymers. Perfluoropolyethers (PFPEs) are a peculiar family of fluid polymers, with outstanding properties such as high thermal and chemical stability and very low glass transition temperature (down to −120 °C), resulting in a wide working temperature range. These polymers are used in high-tech applications like high vacuum technology, special lubrication, aerospace and electronic industries, etc. Their synthesis and their chemical/physical properties have been described in previous papers.[1]

In the last decades PFPEs bearing functional groups were developed for specific applications such as coating or treatment of surfaces, film lubrication, and as additives and building blocks for the synthesis of new copolymers.[2] Moreover, functional groups (in particular carbonyl groups) are present in the intermediates of the PFPEs manufacture process. They also result from degradation of the PFPE chains in severe application conditions.

Vibrational spectroscopy is one of the most powerful analytical techniques used to identify and quantify functional groups in fluorinated polymers and especially in tetrafluoroethylene (TFE) based polymers.[3,4] Even if IR spectroscopy was frequently employed in case of PFPEs for the detection of carbonyl groups or carboxylic moieties[5,6] and references therein, the data regarding the absolute IR intensities of their characteristic bands and their evolution while changing the chemical environment are still to be investigated.

In this work we present experimental spectroscopic data obtained from samples of linear perfluoropolyether macromolecules containing different concentrations of –COOH end groups. These data are rationalized with the help of DFT simulations carried out on molecular models.

[1] Solvay Solexis Research & Technology, Viale Lombardia 20, 20021 Bollate (MI) (Italy)
Fax: (+39)0238356355;
E-mail: stefano.radice@solvay.com
[2] Dipartimento di Chimica, Materiali e Ingegneria Chimica "G. Natta", Politecnico di Milano, piazza Leonardo da Vinci, 32, 20133 Milano, (Italy)

On this basis a method is proposed for measuring the relative populations of free and associated macromolecules. The measure of absolute infrared intensities also provides a direct check of the reliability of the calculated intensity values.

Experimental Part

A polydispersed perfluoropolyether with backbone structure $-(CF_2O)_n(CF_2CF_2O)_m-$, average molecular weight of 8000 amu and having, on average, about 0.7 $-CF_2COOH$ end groups per chain was used as the source of carboxylic groups (corresponding to an absolute concentration of carboxylic end groups of 88.9 meq/Kg evaluated by acidimetric analysis and also checked by ^{19}F NMR spectroscopy).

This product was diluted with a commercial Fomblin® Z, which has the same backbone structure (and comparable molecular weight) but only perfluorinated chain ends (mainly $-CF_3$ groups). In this way several samples with $-CF_2COOH$ concentration in the range 0.05–10 meq/kg have been obtained. Some of the solutions (obtained by weighting) have been checked also by acidimetric analysis.

Spectra of the solutions and of neat Fomblin® Z have been recorded by FT-IR Nicolet Nexus instrument, with a spectral resolution of 2 cm^{-1}, coadding 256 scans for sample. A sealed cell for liquids with CaF_2 windows and optical path-length of 1 mm has been used.

Theoretical calculations have been carried out by means of Gaussian03 code[7] in the framework of Density Functional Theory (DFT). B3LYP functional[8] and the 6-311G** basis set have been used for all the calculations here reported. Moreover, in the investigation of hydrogen bonded dimers, we have also checked the influence of diffuse functions (6-311++G**) and basis set superposition error on the calculated IR spectra obtaining very small corrections with respect to the previous calculations. These results have not been introduced in the following discussion.

Results and Discussion

Experimental Observations

Optical density data and frequencies characteristic of the functional groups have been obtained subtracting from the IR spectra of the solutions the IR spectrum of neat Fomblin® Z. The characteristic vibration of the C=O stretching of perfluorinated carboxylic acids, near 1800 cm^{-1}, is about 100 cm^{-1} higher than in the case of hydrogenated homologues. Moreover, two main bands whose intensity ratio depends on the concentration of the functional groups (see Figure 1) can be easily observed in this spectral region. The component at higher wave-number (1820 cm^{-1}) is assigned to carboxylic acid group of "free" molecules (FM) while a second stronger band (1780 cm^{-1}) can be ascribed to CO groups involved in hydrogen bonds. Due to the high stabilization energy arising from the formation of a cyclic complex bonded through two Hydrogen Bonds (see a sketch of the structure in Figure 1, b (II)) we propose to assign this band to cyclic dimers (2HB). Other fluorinated carboxylic acids show a similar behavior in the infrared and a marked red-shift of the C=O stretching is observed upon formation of dimers.[9]

For increasing concentrations, the relative intensity of the 2HB band with respect to the FM band grows monotonically. This observation suggests that as soon as we increase the amount of the solute, the number of 2HB dimers increases, as it could be expected from simple arguments based on the likelihood that two –COOH groups approach within the interaction range.

A third component can be also identified in the CO stretching region as a shoulder on the higher frequency side of the strong (2HB) band. Its intensity seems to be only weakly dependent on the degree of concentration of the solution. Based on results from theoretical modeling (see below), this shoulder can be attributed to partially bonded dimers i.e. to dimers where only one hydrogen bond is linking the two molecules (see Figure 1, III). The occurrence of at least three components in the

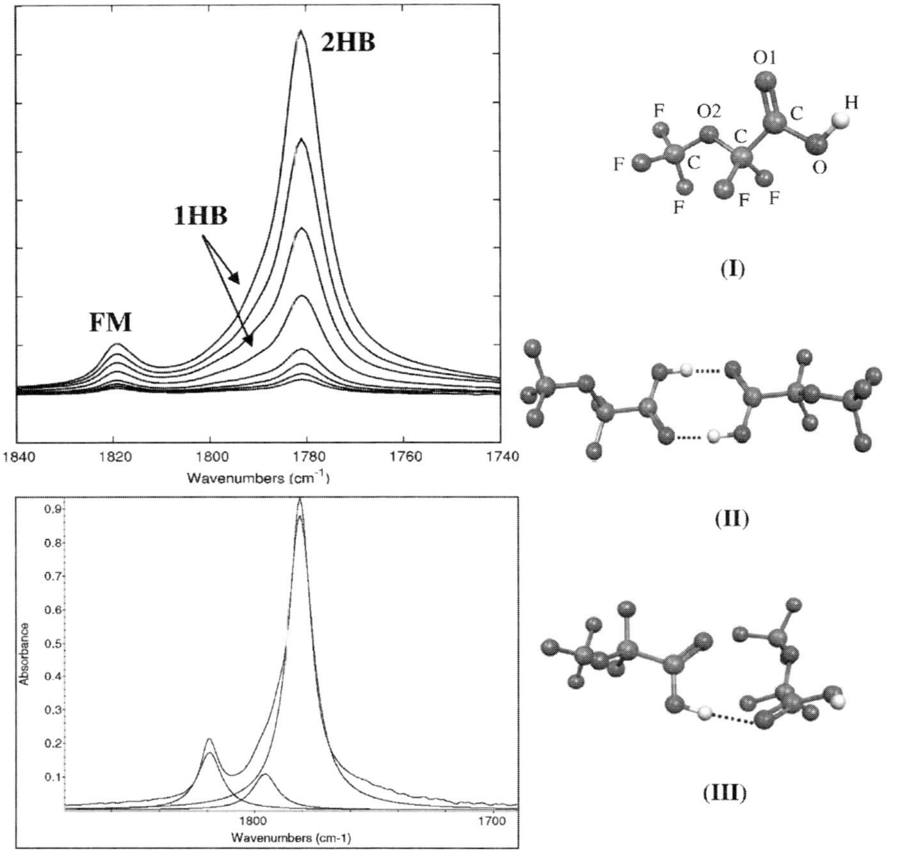

Figure 1.
Left: (upper) Experimental IR Spectra at increasing concentrations (after subtraction of the spectrum of the solvent) (lower): example of a curve fitting. Right: structures of the model molecule (I) and of the two hydrogen bonded dimers considered in this work (II) and (III).

CO stretching region is supported by a curve fitting (see Figure 1) which reproduces very well the experimental shape of IR spectra. Notice that the curve fitting procedure (frequency range 1900–1700 cm^{-1}, Lorentzian band-shapes, Lab Calc® software, Galactic Industries) has been necessary in order to measure both the optical densities and integrated areas of the three components observed.

The presence of free and associated species in the samples under study is also confirmed by the observation of characteristic features in the OH stretching region.

Theoretical Modeling

First principles calculations have been carried out with the following aims:

- to confirm the band assignments proposed on the basis of the experimental findings.
- to obtain a reliable prediction of the relative CO stretching transition dipole strength of free and associated molecules (both 2HB and 1HB).

This second point opens the way for building a model (based on measured IR intensities) that allows to estimate the fraction of hydrogen bonded molecules as a function of the total concentration of perfluorinated carboxylic acid in the solution.

The first step requires the choice of a (possibly small) molecular model. In the sketch (I) of Figure 1 a short chain is illustrated (CF_3–O–CF_2–COOH), that is supposed to mimic one functionalized end of the macromolecule under study (perfluoropolyether functionalized with carboxylic acid end group). The chemical groups that are supposed to be mainly affected by HB association are indeed present in our molecular model. Before analyzing the effect of intermolecular hydrogen bonding on the frequency and intensity of the C=O stretching band it is necessary to check whether conformational effects could introduce frequency shifts and/or intensity changes on the IR spectrum of the "free" molecule. Even if the sterical hindrance of fluorine atoms lowers the number of possible stable conformations (e.g. with respect to the case of hydrogenated species), we have recently demonstrated in the case of partially halogenated ethers that only an accurate consideration of the whole molecular structures corresponding to the different potential energy minima in the conformational space allows for a careful interpretation of the IR spectra.[10] For this reasons, effects ascribed to the presence of different conformers cannot be *a priori* overlooked. The conformational analysis presented here is developed as follows:

- determination of the potential energy surface (PES) as a function of the torsional angle (θ) defined by the sequence of atoms O1-C-C-O2 (see Figure 1.I).
- calculation and comparison of the IR spectra for the structures corresponding to the minima of PES.

As it can be seen from Figure 2.a, the more extended conformation ($\theta = 10°$) is only a relative minimum of the curve while two other (almost energetically equivalent minima) are found for $\theta = 115°$ and $\theta = -115°$. It should be noted that the energy barrier between these minima is quite small (about 0.7 kcal/mol), thus implying the occurrence of the three conformations already for relatively low temperatures. In Figure 2.b the calculated IR spectra of the three conformers are shown in the region of the C=O stretching band.

The effect of the molecular conformation on both frequency and intensity is found to be small; it is indeed completely negligible with respect to the effect expected in the case of intermolecular hydrogen bonding. This effect can be discussed with the help of DFT calculations that have been carried out for the three

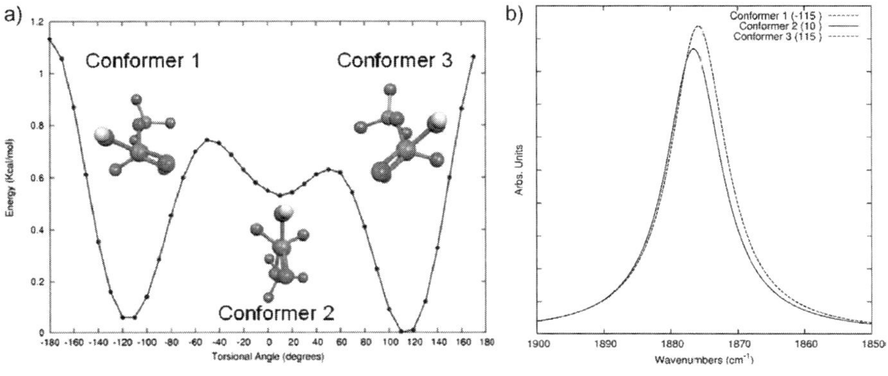

Figure 2.
(a) Torsional potential energy from DFT simulations. (b) Predicted infrared spectra (CO stretching region) for the three equilibrium structures of the molecule.

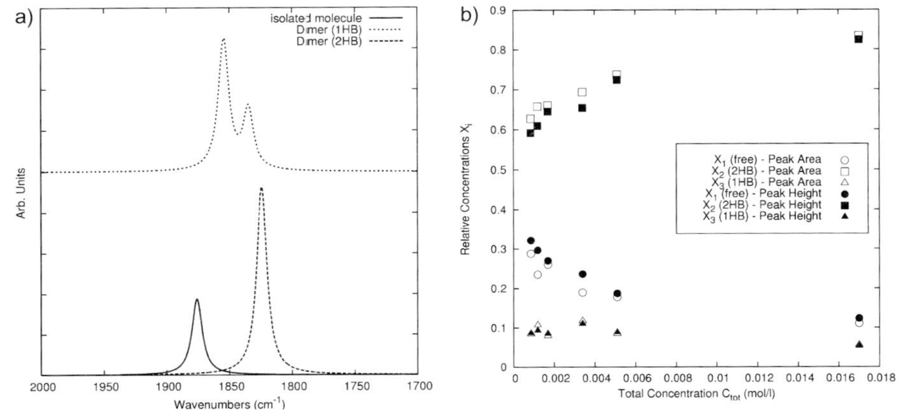

Figure 3.
(a) Predicted infrared spectra (CO stretching region for the isolated model molecule and two dimers (1HB and 2HB, see text). (b) Relative concentration of free and associated CO bonds as a function of the total concentration of functional groups (see text).

model systems represented in Figure 1, were the starting conformation of the model molecule corresponds to the conformer number 2 in Figure 2.a. IR spectra of the free molecule, the 2HB dimer and the 1HB dimer have been calculated and they are compared in Figure 3.

As expected, the frequency trend from DFT is consistent with experimental findings and confirms the assignment proposed above: a downshift in frequency is found for the 2HB dimer with respect to FM while the 1HB dimer shows intermediate frequencies, between FM and 2HB. As for the intensities, it should be noted that the total intensities of the dimers have to be divided by two in order to make a comparison with the CO stretching intensity of the free molecule (i.e. one should refer to the intensity of *one* associated CO bond). The so-normalized CO stretching intensities obtained from DFT calculations are:

$$a_{FM} = 273 \text{ km/mol};$$

$$a_{2HB} = 363 \text{ km/mol};$$

$$a_{1HB} = 361 \text{ km/mol}.$$

These results show that the occurrence of intermolecular hydrogen bonds causes a slight increase of the intensity of the C=O stretching band.

Finally, we note that even if a stable geometry is found for 1HB dimers, a slight displacement from this equilibrium geometry, followed by a structure re-optimization often relaxes into the most stable 2HB dimer. Accordingly, it seems that the occurrence of 1HB dimers should be explained as a consequence of the interactions with the several solvent molecules present in the real system, that sometimes can hinder the relaxation into the more stable 2HB dimers. To this respect it must be also considered that the real molecules are much longer (and than less mobile) than the model molecule chosen in this work.

On the basis of the results obtained so far, a simple model has been built to estimate the evolution of the relative concentration X_i of the different CO species (FM, i = 1; 2HB, i = 2; 1HB, i = 3) as a function of the total concentration C_{tot} of functional groups in a given sample. The following definitions have been applied:

$$C_{tot} = C_1 + C_2 + C_3;$$

$$X_1 + X_2 + X_3 = C_1/C_{tot} + C_2/C_{tot} + C_3/C_{tot} \equiv 1$$

The model makes use of the measured CO stretching intensity ratios Y_i, ($Y_i = I_i/I_{tot}$, $I_{tot} = I_1 + I_2 + I_3$, were I_i is the experimentally determined intensity of the band or component assigned to the i-th species). Relative intensities $R_2 = a_2/a_1$ and $R_3 = a_3/a_1$ of the associated CO oscillators with respect to the free one are introduced in the model taking the results from DFT calculations.

Following the definitions above and considering that $I_i \propto C_i\ a_i$, one obtains a system of three equations in the unknown X_1, X_2, X_3, namely:

$$\frac{I_1}{I_{tot}} = \frac{X_1}{X_1 + X_2R_2 + X_3R_3}$$

$$\frac{I_2}{I_{tot}} = \frac{X_2R_2}{X_1 + X_2R_2 + X_3R_3}$$

$$\frac{I_3}{I_{tot}} = \frac{X_3R_3}{X_1 + X_2R_2 + X_3R_3}$$

This system can be put in the form of an homogeneous linear system of three (linearly dependent) equations. Introducing in the system the theoretical values of the intensity ratio R_i (which are fixed by DFT predictions) and of the experimental ratios Y_i (which are deduced from the spectra of Figure 1.a) the system can be solved for every value of the total concentration. In this way, the evolution of the concentrations of each species can be investigated as a function of the total concentration of the functional groups in the sample.

In Figure 3.b the results of the model are shown both for Y_i values obtained as ratios of experimental peak height (optical densities) and as ratios of integrated band areas. As expected, the relative concentration of the 2HB cyclic dimers increases with total concentration while that of FM decreases; a slowly varying trend is obtained for 1HB dimers.

Moreover, the above obtained concentrations were used to evaluate the equilibrium constant K for the association of carboxylic functional groups, defined as follows:

$$K = ([\text{Complex}])/[\text{FM}]^2$$

Where Complex is defined as [Complex] = ([1HB] + [2HB])/2, since in the complex two functional groups are present. Using the data reported in Figure 3.b, we obtained a value for K of the order of 2000 L/mol, in agreement with literature for vapor phase association of neat trifluoroacetic acid (5000 L/mol)[9] and very different from trifluoroacetic acid in supercritical carbon dioxide (SC–CO_2) (50 L/mol).[9] We explain this behavior considering the low polarity of Fomblin® Z dilutant which is not expected to interact significantly with the functional groups, so stabilizing the complexes.

As a final check, absolute IR intensities have been obtained from the experimental spectra making use of the relative concentrations determined with the model. The values obtained after averaging over the whole set of experimental data at different total concentration are:

281 km/mol (FM); 371 km/mol (1HB);

373 km/mol (2HB).

These values are in very good agreement with DFT results, thus justifying the soundness of this approach.

Conclusions

In this study we presented experimental spectroscopic data on linear perfluoropolyether macromolecules bearing –COOH chain ends at different concentrations. First principles DFT calculations have been carried out on molecular models in order to evaluate the absolute CO stretching IR intensity for free and associated molecules. On this basis, a description of the system in terms of relative populations (associated/non associated molecules) is obtained from experimental IR intensity data. The knowledge of the relative concentration of free and bonded species allows to obtain an experimental determination of the absolute intensity of the CO stretching bands for the different species which favorably compares with the theoretically predicted values. Moreover, a third structure where the two interacting molecules are linked by just one hydrogen bond is observed and investigated.

[1] G. Marchionni, G. Ajroldi, G. Pezzin, "*Comprehensive Polymer Science*", S. L. Aggarwal, S. Russo, Eds., Second Supplement, Pergamon, London **1996**, 347–388.

[2] C. Tonelli, P. Gavezotti, E. Strepparola, "Linear perfluoropolyether difunctional oligomers: chemistry, properties and applications", *Journal of Fluorine Chemistry* **1999**, *95*, 51–70.

[3] M. Pianca, E. Barchiesi, G. Esposto, S. Radice, "End groups in fluoropolymers", *Journal of Fluorine Chemistry* **1999**, *95*, 71–84.

[4] Klaus Lunkwitz, Uwe Lappan, Ulrich Sheler, "Modification of perfluorinated polymers by high-energy irradiation", *Journal of Fluorine Chemistry* **2004**, *125*, 863–873.

[5] Jim Liang, Larry S. Helmick, "Tribochemistry of PFPAE Fluid on M-50 Surfaces by FTIR Spectrsocopy", *Tribology Transactions* **1996**, *39*, 705–709.

[6] David J. Carrè, Jonathan A. Markowitz, "The reaction of Perfluoropolyalkyether Oil with FeF3, AlF3 and ALCL3 at Elevated Temperatures", *ASLE Transactions* **1985**, *28*, 40–46.

[7] Gaussian 03, Revision C.02, M. J. Frisch, G. W. Trucks, H. B. Schlegel, G. E. Scuseria, M. A. Robb, J. R. Cheeseman, J. A. Montgomery, Jr., T. Vreven, K. N. Kudin, J. C. Burant, J. M. Millam, S. S. Iyengar, J. Tomasi, V. Barone, B. Mennucci, M. Cossi, G. Scalmani, N. Rega, G. A. Petersson, H. Nakatsuji, M. Hada, M. Ehara, K. Toyota, R. Fukuda, J. Hasegawa, M. Ishida, T. Nakajima, Y. Honda, O. Kitao, H. Nakai, M. Klene, X. Li, J. E. Knox, H. P. Hratchian, J. B. Cross, V. Bakken, C. Adamo, J. Jaramillo, R. Gomperts, R. E. Stratmann, O. Yazyev, A. J. Austin, R. Cammi, C. Pomelli, J. W. Ochterski, P. Y. Ayala, K. Morokuma, G. A. Voth, P. Salvador, J. J. Dannenberg, V. G. Zakrzewski, S. Dapprich, A. D. Daniels, M. C. Strain, O. Farkas, D. K. Malick, A. D. Rabuck, K. Raghavachari, J. B. Foresman, J. V. Ortiz, Q. Cui, A. G. Baboul, S. Clifford, J. Cioslowski, B. B. Stefanov, G. Liu, A. Liashenko, P. Piskorz, I. Komaromi, R. L. Martin, D. J. Fox, T. Keith, M. A. Al-Laham, C. Y. Peng, A. Nanayakkara, M. Challacombe, P. M. W. Gill, B. Johnson, W. Chen, M. W. Wong, C. Gonzalez, J. A. Pople, Gaussian, Inc., Wallingford CT, **2004**.

[8] A. D. Becke, *J. Chem. Phys.* **1993**, *98*, 5648–5652.

[9] P. W. Bell, et al., *Ind. Eng. Chem. Res.* **2003**, *42*, 6280–6289.

[10] M. Tommasini, C. Castiglioni, A. Milani, G. Zerbi, S. Radice, P. Toniolo, C. Grossi, R. Picozzi, A. Di Meo, C. Tonelli, *J. Fluor. Chem.* **2006**, *127*, 320–329.

Indirect Measurement of the Cooperative Hydrogen Bonding of Polymers Using NMR Quadrupole Relaxation and PFG Methods

Jaroslav Kříž, Jiří Dybal*

Summary: A general method of measurement of polymer hydrogen bond (HB) cooperativity using a low-molecular weight model ligand named marker and two independent methods of ^{2}H NMR is presented. As marker, a deutero-compound chemically similar to the functional groups of one of the polymers is used, e.g. pyridine-d_5 in the investigated interaction of *poly*(4-vinylpyridine) with *poly*(4-vinylphenol) or acetic acid-d_4 in the interaction of polyacrylic acid with *poly*(4-vinylphenol) reported here. The method is based on the fact that a substantial fraction of the marker, originally bound to the groups of one of the polymers, is liberated by the cooperative interaction between the two polymers. For the establishment of the fraction of the bound marker before and after mixing the polymers, ^{2}H NMR quadrupolar relaxation or ^{2}H PFG NMR diffusion measurement can be used with comparable precision. In both these methods, the results must be normalized to a standard viscosity using the relaxation or diffusion of an added inert compound such as $CDCl_3$.

Keywords: cooperative bonding; hydrogen bond; NMR; *poly*(4-vinylphenol); *poly*(4-vinylpyridine)

Introduction

The cooperativity of non-covalent electrostatic, hydrogen bond (HB) or hydrophobic interactions is crucial in the formation of natural or synthetic macromolecular complexes.[1–4] Many individual examples of such cooperative binding were recognized and studied, in particular in the case of HB,[5–10] but systematic work has just started in this field. Several principles governing cooperativity were discovered in the study of electrostatic interactions between oppositely charged macromolecules.[10–17] One of them, dubbed the proximity effect, was found to be operative in the HB cooperativity as well.[18–20] However, systematic verification of this principle as well as the discovery of others needs quantitative studies of the binding degree between the macromolecules in question. This meets serious obstacles in the case of HB.

It is certainly true that a sufficiently strong HB usually changes both frequency and intensity of some infrared bands corresponding to the stretching vibration of the bond bearing the hydrogen. It also changes the position of NMR signals of at least the protons involved in the bond. However, the quantification of these changes is not as easy as it could seem to be. The main difficulty lies in the fact that the polymers involved in HB interactions usually are soluble only in solvents forming HB with them so that no (or almost no) spectroscopic change is brought about by the binding between the polymers.

This difficulty could be seemingly avoided by turning from individual atomic groups of the polymers to their molecular shape or bulkiness, e.g. by measuring their

Institute of Macromolecular Chemistry, Academy of Sciences of the Czech Republic, Heyrovský Sq. 2, Prague 6, Czech Republic
Fax: +420-296809410;
E-mail: kriz@imc.cas.cz

hydrodynamic radius using DLS or PFG NMR. Ironically, the effective size of the macromolecular complexes often decreases with increased degree of their inner binding, obscuring thus the quantitative information we are looking for.

There is an indirect way of circumventing these difficulties, which we already suggested for a special case[18,19] but found to be rather generally applicable in our newer studies. The principle, explained in detail in this article, is the following. The solution of one of the macromolecules in an appropriate solvent is mixed with a low-molecular-weight analogue of the other macromolecule (let us call it marker), in a quantity sufficient to block most (say 98%) of its binding sites. It is preferable to have the marker in a deuterated form. The marker molecules bound to the polymer have a diffusion coefficient (measured by PFG NMR) very different from those in the excess free molecules and, usually, there is also a large difference in the quadrupolar relaxation rate (measured by ^{2}H NMR) of the corresponding deuterons so that, even under conditions of a very fast exchange, the binding degree of the marker can be established by two independent methods. After adding the complementing macromolecule, some of the bound marker molecules are liberated due to the cooperative interaction between the macromolecules. The degree of the freed molecules can be measured by the same method as above and the binding degree between the macromolecules can be calculated.

Materials and Methods

Materials

Poly(4-vinylpyridine) ($M_n = 5100$) and polyacrylic acid ($M_n = 432, 720, 1070, 1300, 1500$ and 1800) were purchased from Polymer Source, Canada. Acetic acid-d$_4$ (AcD), and chloroform-d ($CDCl_3$) were products of Eurorad. These solvents were dried with molecular sieve before use. The measured solutions were transferred into a NMR tube, degassed and sealed.

NMR Relaxation and PFG Measurements

All NMR measurements were done with an upgraded Bruker Avance DPX 300 spectrometer (^{1}H frequency at 300.13 MHz) equipped with a Bruker PFG unit BGU2 and inverse-detection water-cooled PFG probe with a gradient range up to 1500 G/cm. ^{2}H longitudinal relaxation measurements were done at 46.071 MHz with a broadband probe using an inverse-recovery pulse sequence d_1-π_x-d_2-$\pi/2_{x,-x}$-FID, with $d_1 = 8$ s and d_2 incremented in 16 steps from 0.1 to 1.6 s. 2048 points were collected in each of the 48 scans. The measurements were done without lock, usually at night to minimize outer magnetic influences. Each measurement was repeated at least five times and the arithmetic mean of the resulting T_1 values (scattered less than +/− 2%rel.) was taken. PFG diffusion experiments of the deutero-compounds were measured at the ^{2}H resonance, using the decoupler channel of a z-gradient inverse-detection probe. The pulse sequences were both those of pulsed gradient spin-echo (PGSE), i.e. d_1-$\pi/2_x$-gr-d_2-π_y-d_2-gr-FID and pulsed gradient stimulated echo (PGSTE),[21,22] i.e. d_1-$\pi/2_x$-gr-$\pi/2_x$ -d_2-d_2-$\pi/2_x$-gr-FID, with $d_1 = 8$ s and $d_2 = 0.2$ s. The gradient of a 2 ms pulse was incremented from 10 to 50 G/cm in 16 steps. 2048 points were collected in each of the 64 scans. Again, the measurements were done without lock and at least 5 times repeated (scatter +/− 2%rel.). Diffusion experiments with the polymer PVP were done on ^{1}H resonance with a deuterium lock using PGSTE sequence on a special water-cooled inverse-detection z-gradient probe, $d_2 = 0.02$ s and the 1 ms gradient pulse incremented from 5 to 500 G/cm in 16 steps. The measurements were repeated 3 times with the scatter $< 3\%$.

Results and Discussion

As indicated in the Introduction, the indirect method of measuring the degree of binding α of polymeric HB-donor (HBD) groups to complementary HB-acceptor (HBA) groups or vice versa is based on measuring the fraction of a low-molecular-weight ligand (called marker in this paper) liberated from its binding to one of the polymers due to its substitution by the groups of the other polymer. The choice of the marker depends on its availability but, for physical clarity of the results, it should be chemically as similar as possible to the polymer groups by which it is substituted. In such case, the quantity of its liberation by the analogous polymer substantially exceeding that given by statistics is a measure of *net cooperation* of bonds between the polymers.

Theoretically, the competition between the marker in large excess and the analogous groups of the polymer could disrupt the bonding pattern between the polymers to such degree that a weak cooperativity could not be observed in this way. Such case is not very probable, however, because a substantial part of the driving force of cooperative binding is the entropy gain due to the liberation of low-molecular-weight ligands. Such ligands are always present, at least as solvent molecules (it is virtually impossible to dissolve hydrogen-bonding polymers in a solvent, which would not interact with them in some way). Hence, if cooperative binding does occur in usual solution of the polymers, it will very probably occur in the presence of the marker, too.

It is preferable to have it in a fully deuterated form for two main reasons: (1) its signals do not interfere with the ^{1}H NMR spectra of the interacting polymers and (2) longitudinal ^{2}H NMR relaxation is mostly quadrupolar, its rate being dependent exclusively on the local dynamics and usually sharply differing between the free and bound states. In a recent study[18,19] of cooperative binding of *poly*(4-vinylpyridine) (PVP) to *poly*(4-vinylphenol) (PVF), pyridine-d_5 (PD) was added as marker to the solution of PVF. Here, we present a complementary example: in the study of cooperative binding of polyacrylic acid (PAA) to PVP, acetic acid-d_4 (AC) is added as marker to the solution of PVP.

After having explored various media containing THF-d_8, DMSO-d_6, dioxane-d_8, and chloroform-d (CD) in various proportions, in which either PVP was not properly soluble or some of the components interfered as a too strong HB acceptor, we found the solution of 2.6 mol/L of AC in CD as a viable medium for our study (so AC serves both as marker and as co-solvent; at the same time, CD serves as co-solvent and inert for viscosity correction, see below). In the following, we first report on the extraction of necessary diffusion and relaxation parameters of the marker AC under its interaction with PVP and, subsequently, show an example of the study of cooperative binding between PVP and PAA.

PFG Study of AC Diffusion in the System with PVP

We applied the PGSE pulse sequence[21] in ^{2}H NMR for the measure of the diffusion coefficient of AC. The first prerequisite of the applicability of our method is that the exchange between the bound and free AC is much faster than the signal decay due to diffusion.[18] The sure criterion of it is the mono-exponential dependence of the signal intensity on the square of the field gradient applied. Figure 1 shows in a semi-logarithmic graph that this requirement is fulfilled in various systems. Being it so, the molar fraction φ of AC bound by HB to PVP can be obtained from the formula

$$\varphi = \frac{D_{\rm F}^N - D^N}{D_{\rm F}^N - D_{\rm B}^N} \tag{1}$$

where the individual quantities are the diffusion coefficients of AC in the current system (D^N), that of the free AC ($D_{\rm F}^N$) and that of AC bound to PVP ($D_{\rm B}^N$). The subscript N means normalization, i.e. correction[31] for the influence of viscosity of the system. In our case, $D_{\rm F}^N$ is by definition the diffusion coefficient of AC in its 2.6 mol/L solution in CD at 295 K; its actual value is 2.001×10^{-9} m^{2}s^{-1}. Let $D_{\rm C}^S$ be the diffusion coefficient of CD in the same solution. Then

$$D^N = DD_{\rm C}^S/D_{\rm C} \tag{2}$$

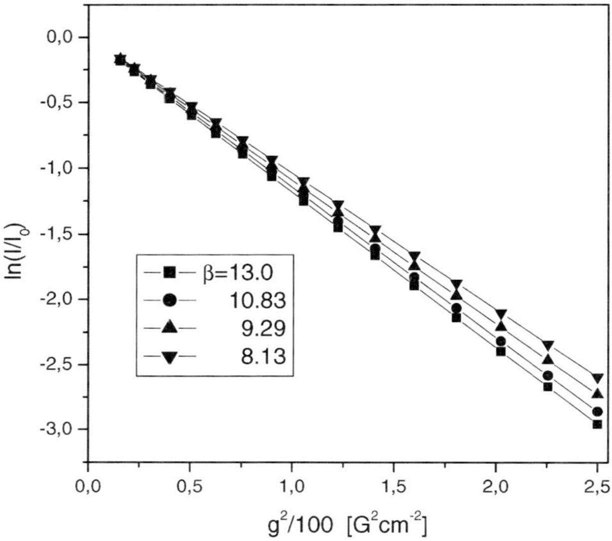

Figure 1.
Logarithmic dependences of relative signal intensity of ^{2}H NMR signals of acetic acid-d_4 (AC) on the square of field gradient in PGSE of PVP-AC systems with the indicated AC/VP molar ratios at 295 K.

where D and D_C are the diffusion coefficients of the given species (AC or PVP) and CD, respectively, measured in the actual system. If diffusion of PVP is measured by ^{1}H PFG NMR, we get D_B of the given system (thanks to quite narrow M_w distribution, we get a single quantity rather than a distribution).

The results of PFG measurements in AC-PVP systems are given in Table 1. For further use but also as a check of the regularity of measurement, the calculated effective equilibrium constant K is added. It is defined as

$$K = \frac{[\mathrm{AC \cdot VP}]}{[\mathrm{AC}][\mathrm{VP}]} = \frac{\varphi}{[\mathrm{VP}]_0(1-\varphi\beta)(1-\varphi)} \quad (3)$$

where $[\mathrm{AC \cdot VP}]$, $[\mathrm{AC}]$, and $[\mathrm{VP}]$ are the actual concentrations of hydrogen-bound groups, free acetic acid and vinylpyridine groups, respectively. As in other papers, $\beta = [\mathrm{AC}]_0/[\mathrm{VP}]_0$.

As it can be seen, the value of K varies only 1.6%rel. around 18.9 L·mol^{-1}, which gives assurance of the reliability of measurement. Calculated from this value (or from φ), the fraction α of the VP groups blocked by the marker is between 0.978 and 0.977 for all examined ratios β, i.e. high enough for our purposes.

Study of ^{2}H NMR Quadrupolar Relaxation of AC in the System with PVP

Like in the previous case, the independent relaxation method of measuring φ presupposes that the exchange between free and bound AC is much faster than the actual relaxation rate of the deuterons in the system. Figure 2 shows in a semi-logarithmic that

Table 1.
Diffusion coefficients (m^{2}s^{-1} × 10^9) of AC (D) and PVP (D_B), the calculated fraction φ of bound AC and the effective equilibrium constant K in various solutions of AC and PVP in CD at 295 K (concentration of PVP given in equivalents of VP groups $[\mathrm{VP}]_0$).

$[\mathrm{VP}]_0$	β	D_B	D_B^N	D	D^N	φ	K
0.20	13.00	0.1833	0.2098	1.630	1.866	0.0753	19.2
0.24	10.83	0.1802	0.2102	1.576	1.839	0.0903	18.8
0.28	9.29	0.1742	0.2099	1.504	1.812	0.1053	19.1
0.32	8.13	0.1682	0.2103	1.429	1.786	0.1202	18.7
0.36	7.22	0.1614	0.2097	1.354	1.759	0.1353	18.8
0.40	6.50	0.1553	0.2098	1.282	1.732	0.1502	18.6

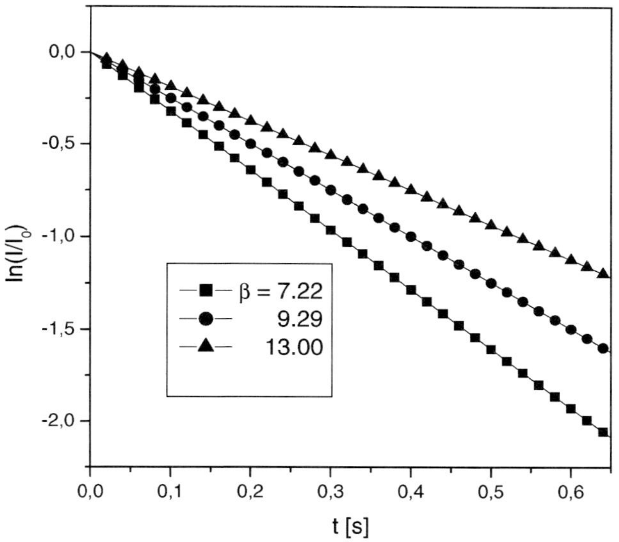

Figure 2.
Logarithmic dependences of relative signal intensity of ^{2}H NMR signals of acetic acid-d_4 (AC) on time in their quadrupolar relaxation in PVP-AC systems with the indicated AC/VP molar ratios at 295 K.

the intensity decay of the NMR signals is strictly mono-exponential, i.e. this requirement is fulfilled, too. Under these conditions, the bound fraction φ of AC can be independently obtained from

$$\varphi = \frac{R_1^N - R_{1F}^N}{R_{1B}^N - R_{1F}^N} \quad (4)$$

where R_1, R_{1B} and R_{1F} are the longitudinal relaxation rates actually measured and in the bound and free state, respectively. Again, the superscript N means normalization to the same viscosity of the standard solution of 2.6 mol/L AC in CD, using in all cases the formula[18]

$$R_1^N = R_1 R_{1C}^S / R_{1C} \quad (5)$$

where R_{1C} and R_{1C}^S are the relaxation rates of CD in the actual and the standard solution, respectively.

In Equation (4), R_1 is actually measured in the given system, R_{1F} corresponds to the standard solution and was established to be $R_{1F}^N = 0.661$ s^{-1}. However, R_{1B}^N is not accessible so easily. The extrapolation method devised earlier[18] does not give reliable results in the present case as the range of β is limited by PVP solubility. Relinquishing partially the independence of both methods, we can utilize the values of φ obtained by diffusion measurements to get R_{1B}^N. Table 2 shows the results for the same systems as above. The values of R_{1B}^N vary less than 1% around 14.309 s^{-1} giving us thus the assurance that this is the right value for the present systems. The use of φ from PFG measurements makes the relaxation method somewhat less independent than desirable. Nonetheless, it is still valuable as it is somewhat more sensitive; it can serve as a check of the PFG method as well.

Table 2.
Longitudinal relaxation rates R_1 (s^{-1}) of AC deuterons in AC-PVP solutions in CD and the calculated values of R_{1B}^N.

$[VP]_0$	β	φ	R_1	R_1^N	R_{1B}^N
0.20	13.00	0.075	1.872	1.685	14.314
0.24	10.83	0.090	2.172	1.890	14.297
0.28	9.29	0.105	2.494	2.095	14.288
0.32	8.13	0.120	2.838	2.299	14.311
0.36	7.22	0.135	3.210	2.504	14.323
0.40	6.50	0.150	3.612	2.709	14.314

Table 3.
Marker relaxation rates R_1, R_1^N (s^{-1}) and diffusion coefficients D, D^N ($m^2s^{-1} \times 10^9$), calculated binding fractions φ of the marker and α of the PAA groups and the resulting cooperativity coefficient ζ for the given polymerization degrees P_n of PAA.

P_n	R_1	R_1^N	φ	D	D^N	φ	α	ζ
5.99	2.632	2.289	0.120	1.483	1.779	0.124	0.183	2.62
10.01	2.791	2.326	0.122	1.488	1.786	0.120	0.190	2.73
14.85	2.856	2.285	0.119	1.496	1.795	0.115	0.214	3.11
18.04	2.754	2.203	0.113	1.439	1.800	0.112	0.244	3.64
20.82	2.669	2.135	0.108	1.443	1.804	0.110	0.268	4.12
24.98	2.635	2.108	0.106	1.447	1.809	0.107	0.281	4.40

The Study of HB Cooperativity in the PVP-PAA System

In the following, we present one series of measurements as an illustration of the method. In this series, $[VP]_0 = 0.40$ eq/L and $[AC]_0 = 2.6$ mol/L as above, i.e. $\beta = 6.5$. The concentration of PAA was always $[AA]_0 = 0.40$ eq/L, i.e. the ratio of AA and VP groups was 1.0. The binding fraction φ of the marker was measured in each case both by the diffusion and relaxation methods. Naming φ_0 the fraction of the marker bound in the system without added PAA (0.1502 in the present case), the fraction α of binding of the AA to VP groups can be obtained from the simple relation

$$\alpha = (\varphi_0 - \varphi)[AC]_0/[AA]_0. \quad (6)$$

Defining the cooperativity coefficient as the ratio of the effective equilibrium constants of binding of PAA groups (K_{AA}) and marker groups (K_{AC}), it easy to show that

$$\zeta = \frac{K_{AA}}{K_{AC}} = \frac{\alpha(1-\varphi)}{\varphi(1-\alpha)}. \quad (7)$$

Table 3 shows the results obtained in this series.

Although the obtained cooperativity coefficients ζ are lower than those, which

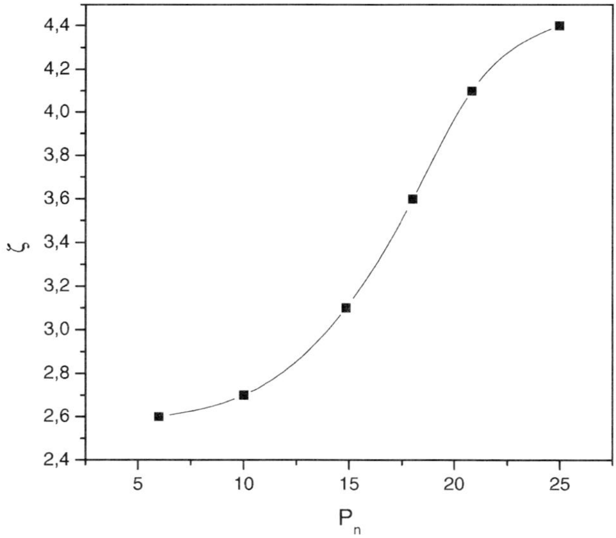

Figure 3.
Dependence of the cooperativity coefficient ζ of PAA groups in their hydrogen bond interaction with PVP on P_n (acetic acid-d_4/chloroform-d, 295 K).

could be calculated for the interaction between PVP and PVF[32], they are still quite substantial. This could be expected from the already studied interaction of PVP with bivalent HB donors[20] crudely modeling the dimer of acrylic acid. Moreover, as shown in Figure 3, the dependence of ζ on P_n appears to have slightly sigmoidal shape indicating[32] that the cooperativity is of a higher order, i.e. does not result from a mere sum of ΔG of individual hydrogen bonds. The reason for this is clearly the proximity effect[20] of neighboring hydrogen bonds.

Conclusions

We have presented two complementary examples of the method quantitatively evaluating the cooperativity of hydrogen bond interaction between two complementary polymers, one of them a HB polymer donor (*poly*(4-vinylphenol) or polyacrylic acid), the other a HB acceptor (*poly*(4-vinylpyridine)). We demonstrated that meaningful results can be obtained with chemically quite different markers as pyridine-d_5 (as crude model of PVP groups) and acetic acid-d_4 (as crude analogue of PAA groups). The basic principle of the method is the measurement of the fraction of the marker bound to one of the polymers before and after adding the second polymer. It evidently can be fulfilled by the two demonstrated ^{2}H NMR methods, namely quadrupolar relaxation of the marker's deuterons and PFG measurement of self-diffusion of the marker molecules, provided that the precision in both cases is high (the scatter of results being 2% rel. or less) and the results are normalized to a standard viscosity using relaxation or diffusion of an added inert compound such as $CDCl_3$.

If the marker is a close low-molecular-weight model of the HB donor or acceptor groups of one of the polymers then the quantity of its liberation by the analogous polymer substantially exceeding that given by statistics is a measure of *net cooperation* of bonds between the polymers.

Although further studies have to be done for its full establishment in other, more difficult fields of use, the presented method can already be considered to be general and, at the same time, probably unique for quantitative evaluation of the type of hydrogen bond cooperativity.

Acknowledgements: The authors thank the Grant Agency of the Academy of Sciences of the Czech Republic for financial support given under the Grant IAA 400500604.

[1] J.-M. Lehn, *Supramolecular Chemistry*, VCH Publishers, Weinheim, Germany, **1995**.
[2] E. Tsuchida, *Macromolecular Complexes, Dynamic Interactions and Electronic Processes*, VCH Publishers, New York **1991**.
[3] J. Mich, Ed., *Modular Chemistry*, Proceedings, NATO ASI Series, Series C, Vol. 499; Kluwer Academic Publishers, **1997**.
[4] A. Weber, Ed., *Structure and Dynamics of Weakly Bound Molecular Complexes*, Kluwer Academic Publishers, **1987**.
[5] G. C. Pimentel, A. L. McClellan, *The Hydrogen Bond*, W.H. Freeman and Co., San Francisco and New York **1960**.
[6] G. R. Desiraju, T. Steiner, *The Weak Hydrogen Bond*, Oxford University Press, Oxford **1999**.
[7] R. Otero, M. Schock, L. M. Molina, E. Laegsgaard, I. Stensgaard, B. Hammer, F. Besenbacher, Angewand. Chemie. Int. Ed. **2005**, *44*, 2270.
[8] P. M. Tolstoy, P. Schah-Mohammedi, S. N. Smirnov, N. S. Golubev, G. S. Denisov, H. H. Limbach, *J. Am. Chem. Soc.* **2004**, *126*, 5621.
[9] M. Arno, L. R. Domingo, *Org. & Biomol. Chem.* **2003**, *1*, 637.
[10] A. G. Fraile, D. G. Morris, A. G. Martinez, S. D. Cerero, K. W. Muir, K. S. Ryder, E. T. Vilar, *Org. & Biomol. Chem.* **2003** *1*, 700.
[11] J. Kříž D. Kurková, J. Dybal, D. Oupický, *J. Phys. Chem. A* **2000** *104*, 10972.
[12] J. Kříž, H. Dautzenberg, *J. Phys. Chem. A* **2001** *105*, 3846.
[13] J. Kříž, J. Dybal, H. Dautzenberg, *J. Phys. Chem. A* **2001** *105*, 7486.
[14] J. Kříž, J. Dybal, D. Kurková, *J. Phys. Chem. B* **2002** *106*, 2175.
[15] J. Kříž, J. Dybal, D. Kurková, *J. Phys. Chem. A* **2002** *106*, 7971–7981.
[16] J. Kříž, H. Dautzenberg, J. Dybal, D. Kurková, *Langmuir* **2002** *18*, 9594.

[17] J. Kříž, J. Dybal, D. Kurková, *J. Phys. Chem. B* **2003** *107*, 12165.
[18] J. Kříž, J. Dybal, *J. Phys. Chem. B* **2005** *109*, 13436.
[19] J. Kříž, J. Dybal, J. Brus, *J. Phys. Chem. B* **2006** *110*, 18338.
[20] J. Kříž, J. Dybal, *J. Phys. Chem. B* **2007**, *111*, 6118–6126.
[21] E. O. Stejskal, J. E. Tanner, *J. Chem. Phys.* **1965**, 42, 288.
[22] J. E. Tanner, *J. Chem. Phys.* **1970**, 52, 2523.

Fourier-Filtering Methods of Interference-Patterned Spectra in Multivariate Calibration and Prediction for Sample Identification and Thickness Determination

Éva Jeszenszky, László Kocsányi, Attila Barócsi, Péter Richter*

Summary: Determining the thickness or identification of polymer materials with building a multivariate calibration model is based on the near infrared spectral information of the material. The spectrum of a thin plastic sheet is modulated by the interference of multiply reflected beams from the boundary surfaces and causes a disturbing signal component. On one hand, this yields unidentifiable samples or introduces large errors in the sample prediction set. On the other hand, interference-patterned spectra have to be excluded from the calibration set. Fourier-transformation of an interference-patterned spectrum vs. wave number leads to a Fourier-spectrum as a function of the optical path length (OPL) containing a well recognizable interference peak. After replacing these interference-components and performing an inverse Fourier-transformation, the spectra can be used for calibration or prediction. Two types of replacing were studied: the spline-interpolation on Fourier-spectrum vs. OPL and a novel method based on linear approximation between Fourier-spectra and thickness values. The effectiveness of each filtering method was tested on low-density polyethylene and polypropylene sheets.

Keywords: Fourier-filtering methods; infrared spectroscopy; interference-patterned spectrum; multivariate calibration; plastics

Introduction

In the recent years, near infrared (NIR) analysis has been steadily growing in popularity because of its ability to provide quick, non-destructive, qualitative and quantitative analytical information on a wide variety of materials from single pure, raw materials to complex products, like foods, textiles, cosmetics, medicals or polymers (see a wide array of applications described in the literature[1–4]). Light absorption in the NIR region (750–2500 nm) is primarily due to overtones and combinations of fundamental bands occurring in the middle infrared region. Lines are typically overlapped and very broad, leading to complex spectra. So univariate calibration methods are not suitable for this type of data. The use of suitable multivariate calibration (MvC) techniques, such as principal component analysis or partial least squares regression requires a relatively large number of calibration samples. The NIR technique with the potential of MvC methods suits excellently the identification or thickness determination of polymer materials.[5–9]

The spectrum of a thin plastic sheet is modulated by the interference of multiply reflected beams from the boundary surfaces and causes a disturbing signal component (cf. Figure 1). As also mentioned by Davies and Grant,[8] this interference pattern on the spectrum could make their analyses impossible and, on one hand, yields unidentifiable samples or introduces large errors in the

Department of Atomic Physics, Budapest University of Technology and Economics, Budafoki 8, H-1111 Budapest, Hungary
E-mail: ejeszy@eik.bme.hu

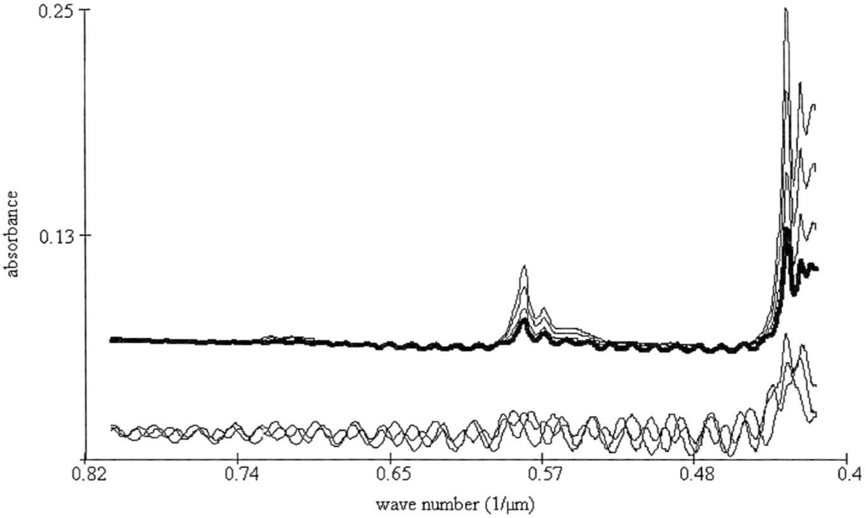

Figure 1.
Absorbance spectra of polypropylene (PP) and low density polyethylene (LDPE) polymer sheets (offsets for better viewing); — 14–38 μm PP (below), ▬ 25.5 μm LDPE, — 42–85.4 μm LDPE (above).

sample prediction set. On the other hand, interference-patterned spectra have to be excluded from the calibration set. In the thickness range where both interference-patterned (IP) and non-interference-patterned (NIP) sample spectra are available, it can be problematic to collect enough useable calibration spectra to build a robust MvC model and predict the thickness of samples with NIP spectra.

The goal of this work study is to overview the filtering methods,[10,11] that results in more accurate identification and thickness determination. Subjecting the IP prediction spectra to filtering fits to the MvC methods as a pre-processing step and allows the reuse of the interference-patterned spectra in the calibration process instead of their exclusion.

Theoretical Background

Considering a plane-parallel absorbing film situated between two non-absorbing media and normal incidence of light, the transmittance is the function of the wave number and thickness and is given by:[12]

$$T(k,h) = \frac{\frac{n_3}{n_1}\tau_{12}^2\tau_{23}^2 x}{1+\rho_{12}^2\rho_{23}^2x^2+2\rho_{12}\rho_{23}x\cos(\delta_{12}+\delta_{23}+\varphi)} \quad (1)$$

where $x(k,h) = \exp(-\alpha'(k)h)$, $\varphi(k,h) = 4\pi\eta(k)hk$, $t_j(k) = \tau_j(k)e^{i\chi_j(k)}$ and $r_j(k) = \rho_j(k)e^{i\delta_j(k)}$ ($j = 12,\ 23$) are the complex transmission and reflection coefficients at the first and second interfaces. This formula is valid for both TE and TM waves if the factor n_3/n_1 is replaced by n_1/n_3 for a TM wave.

Transparent Region

Substituting $x = 1$ and $\delta_{12} = \delta_{23} = 0$ for Eq. 1 yields:

$$T(k,h) = \frac{\frac{n_3}{n_1}\tau_{12}^2\tau_{23}^2}{1+\rho_{12}^2\rho_{23}^2+2\rho_{12}\rho_{23}\cos(4\pi\eta hk)} \quad (2)$$

Then, the transmittance vs. the wave number k becomes a periodic function with

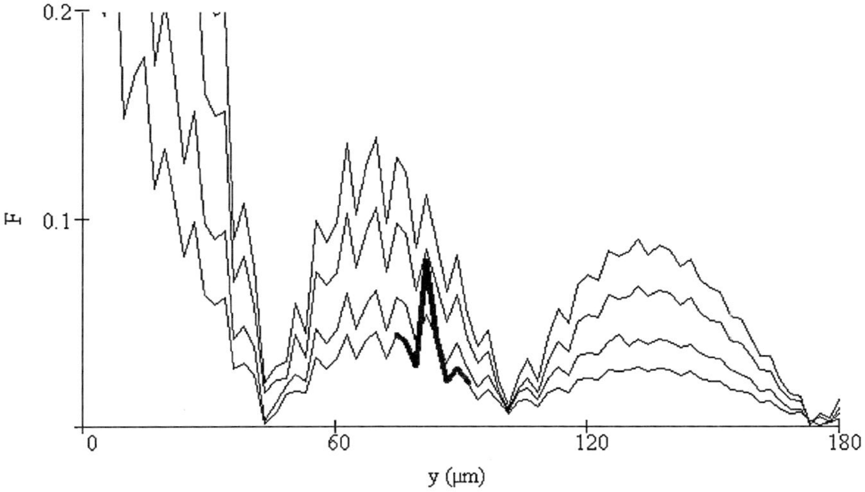

Figure 2.
Fourier-spectra of 25.5–85.4 μm LDPE sheets with ▬▬ interference peak.

a period of $1/(2\eta h)$. This finding is also valid for the absorbance.

Region of Strong Absorption

Multivariate calibration techniques rely on the presence of absorbance in the spectra, which contain the material information of the film. Using the approximation of $x \ll 1$ in the regions of strong absorption, Eq. 1 becomes much simpler:

$$T(k,h) = \frac{n_3}{n_1}\tau_{12}(k)^2\tau_{23}(k)^2\exp(-\alpha'(k)h) \quad (3)$$

Then, the absorbance of the plastic film (with negligible reflectance and scattering) is a linear function of the thickness (according to the Lambert-law):

$$A(k,h) = \log(T(k)) \approx \alpha(k)h - \beta(k) \quad (4)$$

Fourier-Transformed Spectra

Fourier-transformation of a transmittance or an absorbance (A) spectrum vs. wave number (k) leads to a Fourier-transformed spectrum (Fourier-spectrum) as a function of the OPL vertical to the sheet (y). The Fourier-spectrum (F) of an IP spectrum contains a well recognizable interference-peak (cf. Figure 2). Inhomogenities cause a peak-broadening, so a region near to the peak maximum has to be cut off and replaced. After performing an inverse Fourier-transformation, the spectra can be used for calibration or prediction.

Unfortunately, in the Fourier-range of the interference components, there are components related to spectral information of absorbance, on which MvC is based.

Replacing by Interpolation

Since the exact wave-number dependence of the complex refractive index of the sheet is not known, the exact Fourier-spectrum vs. OPL function is not known either. So replacing on the basis of the remaining part of the Fourier-spectrum can be done by interpolation on an own smoothed spectrum (see flow chart on Figure 3). With this method, replaced data become smooth and poor in details (see Figure 4), but can be used for identification and thickness determination too.

Replacing by Linear Regression

In case of thickness determination, another relation can be taken into account: a linear approximation between Fourier-transformed

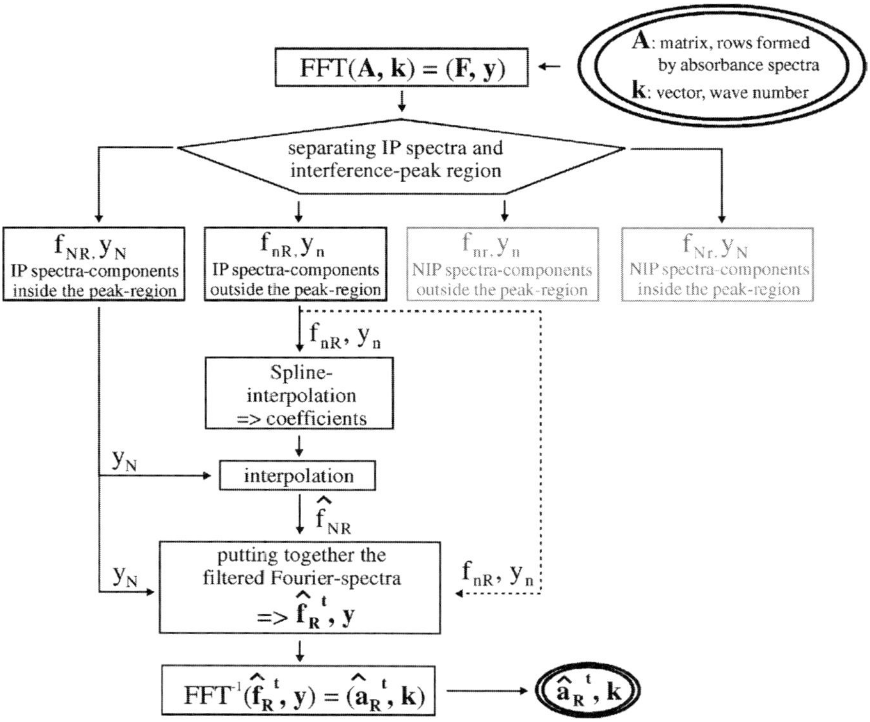

Figure 3.
Flow chart of replacing by interpolation.

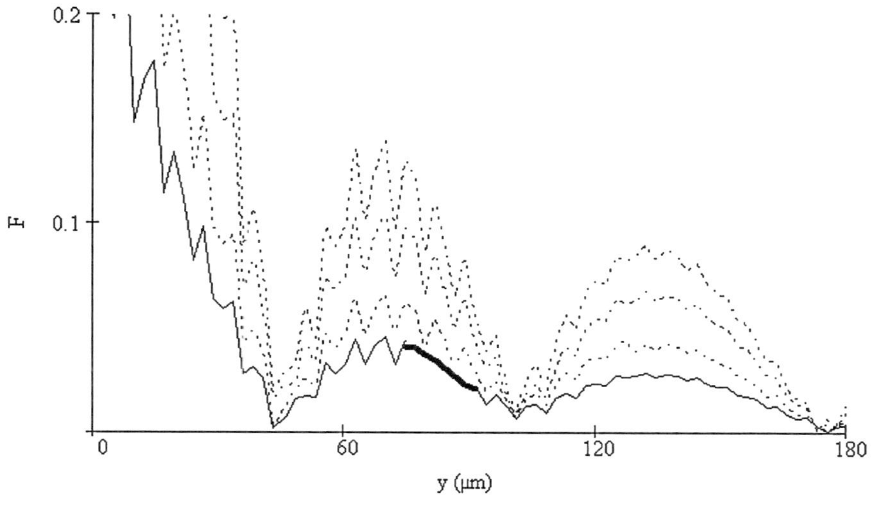

Figure 4.
Replacing by Spline-interpolation; — 25.5 μm LDPE with ▬ replaced components, - - - 42–85.4 μm LDPE.

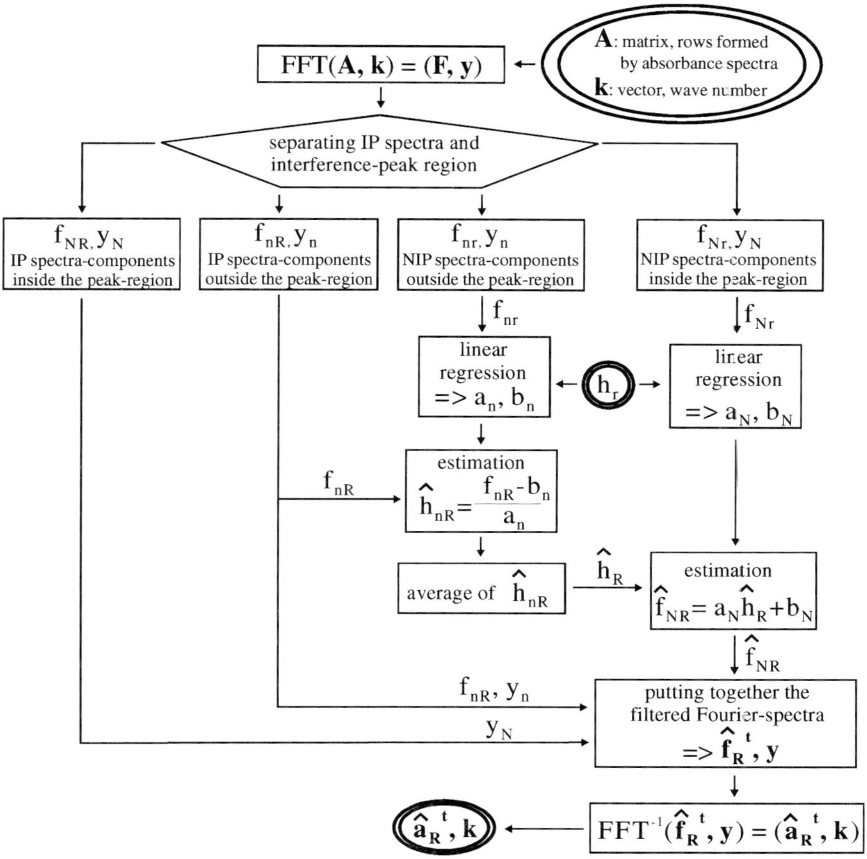

Figure 5.
Flow chart of replacing by linear regression.

absorbance spectra (from now Fourier-spectra, F), and thickness values (h). Its theoretical background is given by the Lambert-law (Eq. 4) and the linearity of the Fourier-transformation. With this method, all but the interference-patterned Fourier-spectra – which contain components at the deleted peak – and the corresponding thicknesses can be used for calculating regression coefficients and, finally, for replacing. If the IP-spectrum is in the prediction set, and its related thickness is not known, an estimation can be made. This means, that we calculate regression coefficients not only at the deleted, but at all other locations. Then we estimate thickness values at these locations and finally average them. Flow chart of this filtering method is shown on Figure 5.

This latter replacing method, compared to the previous one, yields more detailed filtered Fourier-spectra (see Figure 6).

In some cases, baseline corrections (for example a 2nd derivative process) have to be performed before the Fourier-transformation for reducing the strong disturbance of scattering.

Experimental Part

Materials

Low density polyethylene (LDPE) samples produced especially for us by Tisza Chemical Works (TVK, Hungary) were available with thicknesses ranging from about 20 to 80 μm. This means that some of the samples were thin enough to show IP spectra and

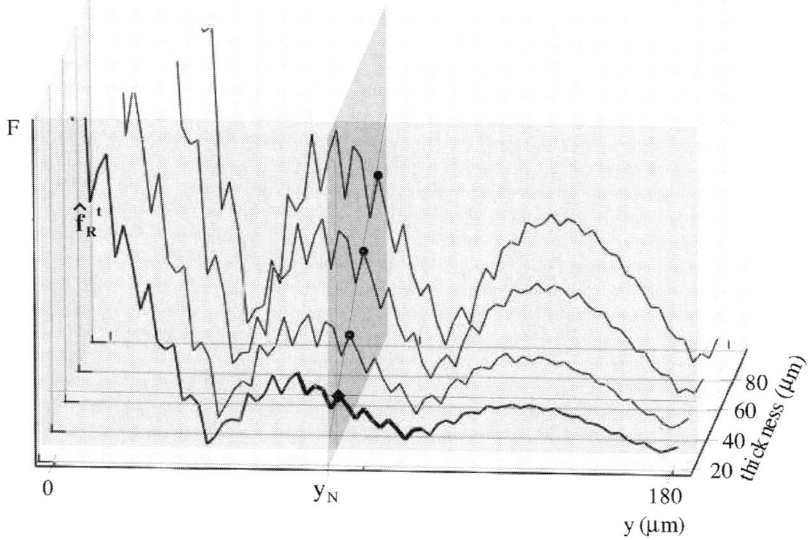

Figure 6.
Replacing by linear regression on Fourier-spectra vs. related thicknesses; ▬▬ 25.5 μm LDPE with ▬▬ replaced components, — 42–85.4 μm LDPE.

some of them were too thick or their quality was not as good as to show this pattern. Polypropylene (PP) samples from Radici Film Hungary under the trade name Biofol-KS were available with thickness ranging from about 15 to 40 μm all of them showing interference-patterned spectra. LDPE and PP samples were used for identification. The spectra of these two polymer materials contain absorbance peaks near to each other. LDPE samples were used for the thickness determination. Both types of samples were split into two sets, one for calibration (40 PP, 40 LDPE) and the other for validation (6 PP IP, 16 LDPE).

Thickness Determination

The thickness of the samples was measured at the same position used for spectral measurement by means of a digital micrometer (Mitutoyo) with accuracy of 1 μm. Thickness of the validation samples was measured 8 times, then averaged.

Spectrophotometer

Transmission spectra were collected by means of an acoustooptic tunable filter (AOTF) based spectrophotometer (type SINIS A, Siemens) equipped with an optical fibre. The spectral range of the measurement was 1200–2400 nm, with a sampling point at every 2 nm.

Software

The data were recorded and exported in txt format by using the application tool (APT) software of SINIS A. Additional calculations were performed in Mathcad,[13] and the MvC was made with GRAMS.[14]

Identification Method

We used PCA method[15,16] with cross validation (out: 1 files) and the data were mean centred and variance scaled. The analysis was performed on the second derivative of spectral data (gap: 16 points). Identification was provided by using the Mahalanobis-distance.[15,16]

Calibration Procedure for Thickness Determination

To build the calibration model, we used the PLS2 method[15,16] with cross validation (out: 1 file). The model was made with different parameters. Data were set or not mean-centered (MC) and variance-scaled (VS), regions of strong absorbance were selected or not, and the analysis was made on second derivative of spectral data with various gaps (4, 8, 16). To make predictions on the validation set, factors to be used were 1 or 2. The calibration model was chosen on the basis of the correlation coefficient between original vs. predicted thickness of the calibration set. The F-test ($\alpha = 0.01$) was applied to determine the statistical significance of outliers.

Filtering Parameters

The window-width parameter of the Fourier-filtering was 36 μm. The second derivative process have to be made before the filtering.

Calibration Model Performance: Comparison of SEPs

Calibration model performance was studied by taking the precision of the thickness determination on validation samples with standard error of prediction (SEP). Since both calibration models were validated on the same sample set, the calculated errors were correlated. Statistical tests were performed on a pair wise significance comparison of SEPs as recommended by Fearn.[17] Based on Fearn's criteria, comparisons of SEPs lead to SEP pairs with significant or non-significant differences.

Results and Discussion

We performed the above described calibration and validation procedure on different pre-processed spectra. Without Fourier-filtering, there is a possibility to make a calibration on all unfiltered spectra in the calibration set and, after that, a calibration on a restricted set from which part of the spectra (probably with interference pattern) could be excluded from the final calibration-model as outliers. In our case, no outliers were found but we built and tested a model in which spectra with interference pattern were excluded by Fourier-analysis. Influence of filtering in prediction set was tested by identification of the sheets and by thickness determination on thin (about 20 μm) samples showed very large amplitude interference-patterns.

We studied the influence of different filtering methods on the calibration model performance. So calibration models were built and tested on unfiltered or filtered spectra with components replaced by interpolation and on filtered spectra with components replaced by linear regression.

Table 1 shows results of these predictions.

In the case of filtered spectra, all of the PP and the LDPE foils could be easily identified. In case of the model on unfiltered PP spectra, separation was impossible, i.e., at least 5 out of the 6 foils were identified incorrectly. In case of the model

Table 1.
Prediction results.

Calibration model built on	Identification (correct/incorrect) for				Thickness determination (SEP values) for			
	Unfiltered PP	Filtered PP	Unfiltered LDPE	Filtered LDPE	Unfiltered predict. spectra thin LDPE	Filtered, replacem. by interp. thin LDPE	Filtered, replacem.by lin. regr. thin LDPE	All LDPE predict. spectra
Unfilt. cal. spect.								
all	1/5	6/0	13/0	13/0	11.8	3.22	0.68	1.26
NIP	1/5	4/2	13/0	13/0	10.74	–	–	1.4
Filtered, replacement by interp.	6/0	6/0	13/0	13/0	–	–	–	1.4
Filtered, replacement by lin. regr.	–	–	–	–	–	–	–	0.63

on unfiltered LDPE spectra, all samples could be identified correctly. The reason of this prediction difference is due to the smaller interference-modulation in the LDPE-spectra (in which case the longer wavelength end can be used).

The filtering by Spline-interpolation on thin prediction spectra results in a significant improvement in thickness determination. Thickness error of thin LDPE sheets is significantly the lowest after subjecting the prediction spectra to a Fourier-filtering based on linear regression.

After studying the influence of different filtering methods on the calibration model performance, the following can be concluded. In general, the calibration on filtered spectra with replacement by interpolation together with calibration on unfiltered spectra turned to perform the worst, whereas calibration on filtered spectra with replacement by regression gave significantly the best results. When unfiltered spectra are used in the calibration model, the calibration set becomes too small to build a well performing model after excluding the IP spectra and it is better to use the entire calibration set. It can be seen that filtering by Spline-interpolation is not a well suited filtering method in this case, but filtering by linear regression is.

Conclusion

Two types of Fourier-filtering were overviewed by testing their performances in identification and thickness determination of thin plastic sheets and in a calibration model. The spectrum of a thin plastic sheet is modulated by the interference of multiply reflected beams from the boundary surfaces and causes a disturbing signal component. This interference pattern on the spectrum could make the identification impossible and introduces large error in thickness determination. On the other hand, interference-patterned spectra have to be excluded from the calibration set. Nevertheless, it is a waste of information in the thickness region where just a few non-interference patterned sample spectra are available. Two types of replacing were studied: the spline-interpolation on Fourier-spectrum vs. OPL and a novel method based on linear approximation between Fourier-spectra and thickness values. The first method can be used for both identification and thickness determination. The latter method can only be used for thickness determination, but yields a more detailed filtered Fourier-spectrum.

It was shown, that the application of Fourier-filtering by Spline-interpolation allows a more precise identification of different sheets (PP and LPDE) as well as their thickness determination in comparison with those methods without filtering. Application of the Fourier-type filtering technique – replacing components of the interference peak using linear regression on non-filtered Fourier-spectra of the calibration set vs. the corresponding thickness values – led to significant improvements in thickness determination of plastic sheets and in calibration performance on the basis of significance comparisons of SEPs.

[1] B. G. Osborne, T. Fearn, *Near Infrared Spectroscopy in Food Analysis*, John Wiley, New York **1986**.

[2] *Near Infrared Spectroscopy: The Future Waves*, A. Davies, P. Williams, Eds., NIR Publications, Chichester **1996**.

[3] D. Burns, E. Ciurczak, *Handbook of Near-Infrared Analysis*, Marcel Dekker, New York **1992**.

[4] International NIR/NIT Conference, Budapest, 1986 Akadémiai Kiadó, Budapest, **1987**.

[5] K. A. Lee Bunding, *Appl. Spectrosc. Rev.* **1993**, *28*, 231.

[6] C. E. Miller, *Appl Spectrosc. Rev.* **1991**, *26*, 277.

[7] R. Feldhoff, T. Huth-Fehre, T. Kantimm, L. Quick, W. van den Broek, D. Wienke, K. Cammann, H. Fuchs, *J. Near Infrared Spectrosc.* **1995**, *3*, 3.

[8] A. M. C. Davies, A. Grant, G. M. Gavrel, R. V. Steeper, *Analyst* **1985**, *110*, 643–647.

[9] C. E. Miller, S. A. Svendsen, T. Naes, *Appl. Spectrosc.* **1993**, *47*, 346–356.

[10] E. Jeszenszky, L. Kocsanyi, P. Richter, *Appl. Spectrosc.* **2004**, *58*, 693–697.

[11] E. Jeszenszky, L. Kocsányi, A. Barócsi, P. Richter, *Appl. Spectrosc.* **2006**, *60*, 162–167.

[12] Z. Knittl, *Optics of Thin Films*, Wiley, London **1976**.

[13] http://www.mathcad.com

[14] http://thermogalactic.com

[15] H. Martens, T. Naes, *Multivariate Calibration*, John Wiley & Sons Ltd., New York **1989**.

[16] B. F. J. Manly, *Multivariate Statistical Methods*, Chapman and Hall, London **1986**.

[17] T. Fearn, *JNIRS News* **1996**, *7*, 5–6.

Interactions and Temperature Transitions of Ethylene Oxide – Propylene Oxide – Ethylene Oxide *tri*-block Copolymers in Water Media

Pavel Schmidt, Jiří Dybal*

Summary: ATR FTIR spectra of two ethylene oxide – propylene oxide – ethylene oxide $(EO)_n$-$(PO)_m$-$(EO)_n$ *tri*-block copolymers (Pluronics) with different lengths of the EO blocks were investigated in water media at various temperatures. The observed wavenumber shifts and intensity changes of the bands of different chemical groups of polymers and of water molecules served as a basis for the estimation of structural changes and interactions of polymers with the surrounding water molecules. Two types of such interactions, i.e. hydrophilic (ether group – water) and a hydrophobic (methyl group – water) are detected. In the copolymer with shorter length of the EO blocks, an interchain H_2O bridge in a liquid crystalline phase was discovered and confirmed by *ab initio* calculations. A model for the structural changes during the temperature transitions is specified.

Keywords: FTIR; polyethers; structure; transition; water-soluble polymers

Introduction

Ethylene oxide – propylene oxide – ethylene oxide $(EO)_n$-$(PO)_m$-$(EO)_n$ *tri*-block copolymers (Pluronics) are widely applied in medicine, biochemistry, pharmacy and have many other industrial, agricultural and technological uses. A good survey of their application can be found in [1]. These copolymers are characterised in water media by typical transition temperatures T_t; below that points they tend to form water solutions, above T_t they aggregate. These effects are phenomenogically well described and there exist also publications trying to use vibration spectroscopy for the explication of the phase changes of copolymers with temperature[2–11]; nevertheless, the structural origin of such behaviour is not yet sufficiently explained.

In this article the structural changes and interactions of two copolymers having different length of EO blocks in water media will be compared and some newly found effects in the spectra will be clarified.

Institute of Macromolecular Chemistry, Academy of Sciences of the Czech Republic, Heyrovský Sq. 2, Prague 6, Czech Republic
E-mail: schmidt@imc.cas.cz

Experimental Part

Materials

Two Pluronic materials, PE6200 (producer BASF, a liquid of nominal average formula $(EO)_5(PO)_{31}(EO)_5$) and F68 (producer FLUKA, a solid of nominal average formula $(EO)_{81}(PO)_{31}(EO)_{82}$) were used. Their EO molar fractions obtained from ^{1}H NMR were 0.373 and 0.837, respectively. Copolymers were measured in water media (their weight fraction of prepared solution $W_p = 0.1$) and as pure compounds. Water of the type LC-MS CHROMASOLV was a product of Riedel-de-Haen.

Measurements and Processing of Spectra

For the measurement of spectra, a FTIR spectrometer Nexus Nicolet 870 purged with dry air and equipped with a cooled MCT detector was used. Samples were measured on a horizontal micro-ATR

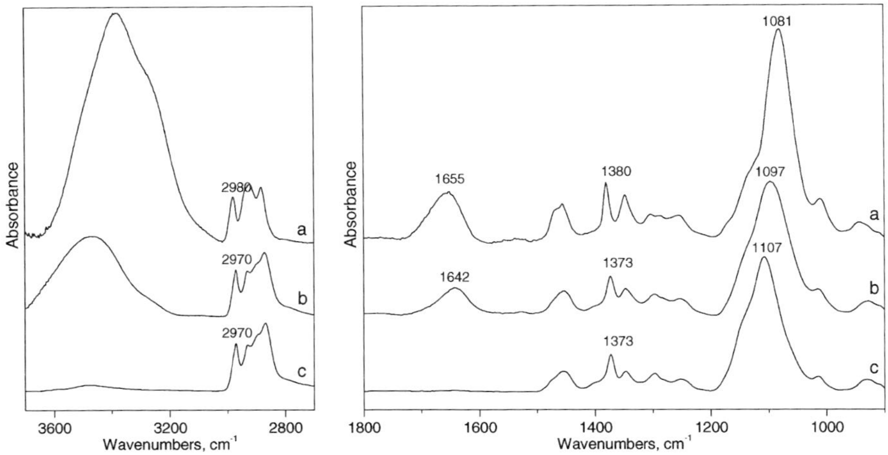

Figure 1.
ATR FTIR spectra of copolymer PE6200. $W_p = 0.1$, at 10 °C (a), at 70 °C (b), spectra of bulk water subtracted. (c) Pure copolymer at 24 °C.

Golden Gate unit (SPECAC), having a control-heated top plate with a diamond window. Spectral resolution was 4 cm^{-1}. The samples were gradually heated and subsequently cooled in the temperature range 5–80 °C in the steps of 2 °C. From the spectra of copolymers in water media, the spectra of bulk water measured at the corresponding temperature were subtracted. For this, the highest reasonable subtraction coefficients k were chosen. Their values were estimated by the criterion that no counter peaks of bulk water intersect the baseline in the region of 3100-3000 cm^{-1} (see Figure 1 and 2, curves a, b). All the spectra were processed by ATR correction and by introducing proper baselines. Integration of intensities of C–H stretching bands was carried out in the range 3020–2800 cm^{-1}, integration of H_2O bending in the range 1760–1560 cm^{-1}. For proper location of the baselines under the bands

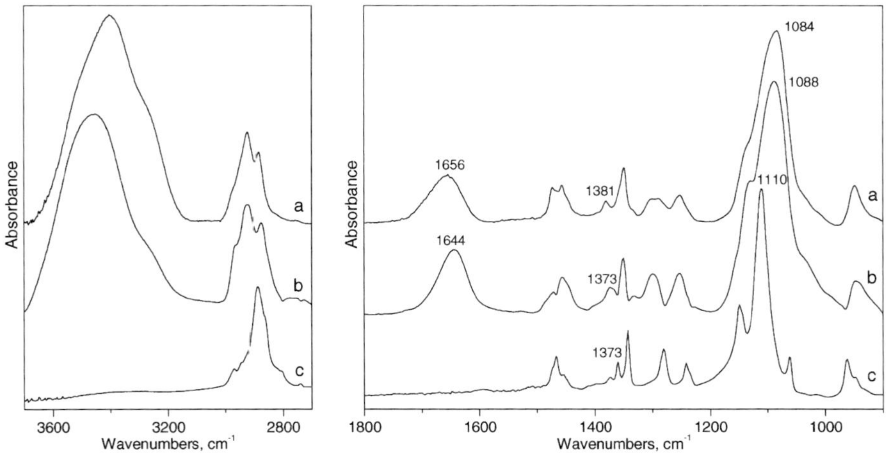

Figure 2.
ATR FTIR spectra of copolymer F68. $W_p = 0.1$, at 10 °C (a), at 70 °C (b), spectra of bulk water subtracted. (c) Pure copolymer at 24 °C.

and for suppressing the rest of water vapour some spectra were smoothed.

Local Concentration Changes during ATR Measurement in Water Media

ATR method appeared to be useful for the measurement of spectra in water media, facilitating infrared measurements of strongly absorbing water solutions and preventing rather unnatural conditions and water evaporation which would appear during a transmission measurement in a thin layer. On the other hand it must have been taken into consideration that by that method spectral information is obtained only from a layer of the material touching closely the surface of the ATR diamond prism. Above the transition temperature and sometimes also below it the copolymer material in the water layer had a tendency to settle down on the prism increasing its local concentration there. Local concentrations of the measured copolymers in water media near the ATR diamond prism are reflected both in values of subtraction coefficients of bulk water and in values of area under the C–H stretching bands (see in Figure 3). They will be discussed later.

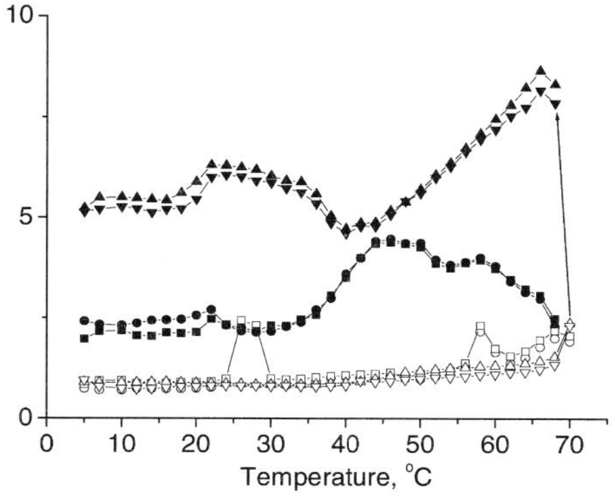

Figure 3.
Local concentration changes of copolymers in the water media near the diamond ATR prism, expressed by the integral intensity of C–H stretching ***A*** under curve 3020–2800 cm^{-1} and by the coefficient ***k*** found for a proper subtraction of bulk water. PE6200: ***A*** heating (○), ***A*** cooling (●); ***(1-k)*** × ***10*** heating (□), ***(1-k)*** × ***10*** cooling (■) F68: ***A*** heating (△), ***A*** cooling (▲); ***(1-k)*** × ***10*** heating (▽), ***(1-k)*** × ***10*** cooling (▼).

Results and Discussion

Typical spectra of $(EO)_n$-$(PO)_m$-$(EO)_n$ *tri*-block copolymers can be found in Figure 1 and 2. Comparing difference spectra of copolymers in water media, it is obvious that wavenumbers of bands and their relative intensities change with temperature. For the polymer solutions below T_t, the C–O stretching vibrations near 1090 cm^{-1} lie at lower, the CH_3 deformation vibration near 1380 cm^{-1} at higher wavenumbers comparing with the situation in water media above T_t, where the wavenumbers of these vibrations approach to those of the pure compounds.

After subtraction of spectra of bulk water, also some residual absorption remains in the spectra with the maxima near 3400 and 1650 cm^{-1} corresponding to water molecules directly interacting with copolymers.

From the Figure 3 it can be concluded that the integral intensity of the C–H stretching bands and the values *(1-k)* derived from the subtraction coefficient of bulk water sufficiently correlate together, both expressing the local concentration of copolymers near the diamond prism. Beginning at the low temperature, the concentrations increase with heating only slightly for PE6200 and F68 copolymers. It appears that both the solutes below T_t and the aggregated materials till the temperature ~60 °C are in the whole liquid layer on the diamond prism mostly homogenously dispersed. At the temperature ~70 °C and/or during the further course of cooling the aggregated materials begin to sediment. The amount of the sediment on the ATR prism is temperature dependent. The maximum local copolymer concentration near prism approaches for PE6200 to a quadruple and for F68 to an octuple compared with the originally prepared homogeneous solutions.

Wavenumber Shifts of CH_3 Bending and C–O Stretching Vibrations

Changes in the structure and interactions of $(EO)_n$-$(PO)_m$-$(EO)_n$ *tri*-block copolymers in water media with temperature can be

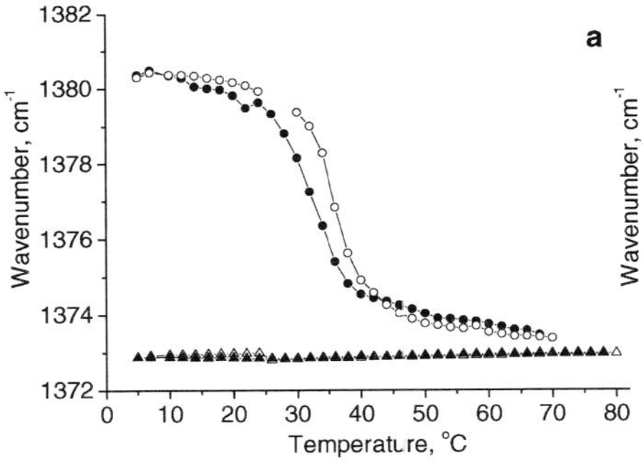

Figure 4.
Wavenumber shifts of methyl deformation (a) and C–O stretching (b) vibration in copolymer PE6200 with temperature. Water medium, heating (○), Pure, heating (△); Water medium, cooling (●), Pure, cooling (▲)

well characterised using the wavenumber shifts of maxima of the appropriate infrared bands. The important vibration bands are the CH_3 stretching and bending bands, lying near 2980 and 1370 cm^{-1} respectively, and a C–O stretching band lying near 1100 cm^{-1}. Detailed wavenumber shifts of two most important bands during heating and reverse cooling of samples can be found in Figure 4 and 5.

Hydrophobic interaction of methyl groups with water exhibits a blueshift, hydrophilic interaction of ether groups with water a redshift of the appropriate infrared bands comparing with the pure compounds. For the copolymer **PE6200** in water, maximum of the CH_3 bending near 1370 cm^{-1} (Figure 4a) and C–O stretching vibration near 1100 cm^{-1} (Figure 4b) shifts for both vibrations on heating to lower and to higher wavenumbers respectively. For the both vibrations, the shift is steepest for the temperature ~35 °C, which corresponds to T_t of this copolymer. At 70 °C, the wavenumber of CH_3 bending vibration decreases almost to the value of respective vibration of the pure copolymer (see Figure 4a). This effect is related with almost complete removing of water from the vicinity of methyl groups. The values of maxima of the C–O stretching band first strongly increase till ~55 °C and then slightly decrease again showing still significant level of C–O groups H-bonded to the vicinal water (see Figure 4b). The slight wavenumber decrease of the C–O stretching band above 55 °C can be explained by some redistribution of water, by its stronger binding to ether group.

For the copolymer **F68** in water, the CH_3 bending vibration behaves on heating analogously as for the copolymer PE6200, with the steepest wavenumber shift appearing

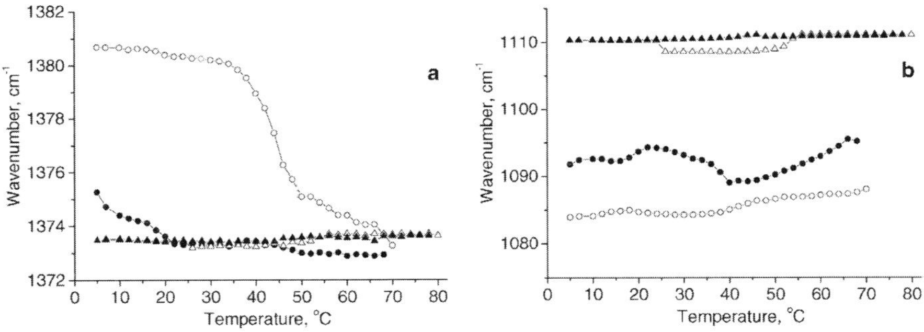

Figure 5.
Wavenumber shifts of methyl deformation (a) and C–O stretching (b) vibration in copolymer F68 with temperature. Water medium, heating (○), Pure, heating (△); Water medium, cooling (●), Pure, cooling (▲).

at the temperature ~45 °C (Figure 5a). The wavenumber of the C–O slightly increases with the increase of temperature; nevertheless its values remain much lower than those for the pure copolymer (see Figure 5b). This can be explained by the fact that the investigated band near 1100 cm^{-1} reflects the C–O vibrations both in PO and EO units. It is apparent that in comparatively long EO blocks of the copolymer F68 the water molecules remain bound even at temperatures between 45 and 70 °C, whereas they were forced out only from the PO blocks at this temperature region.

It is interesting to monitor the wavenumber shifts of the band maxima on **reverse cooling** of the samples. Whereas for PE6200 on cooling only a slight hysteresis appears at increased temperature in the temperature – wavenumber curves for CH_3 bending and C–O stretching vibrations compared with preceding heating (Figure 4a, b), for F68 the previous values of wavenumbers were never reached (Figure 5a, b). This holds especially for the CH_3 bending vibration. Apparently the water molecules, having been expelled from the vicinity of methyl groups in the central PO blocks, could not in this sample penetrate again into the PO parts of the chains. Further, taking into account the results of Figure 3 it follows that the whole sample is not reversibly homogenized on cooling, the local concentration near the diamond crystal remains very large.

Vibrations of Water in Copolymer Neighbourhood

Besides of the vibration bands of copolymers, in Figure 1 and 2 can be also found the bands belonging to the stretching and bending vibrations of water in the vicinity of the polymer material, lying near 3400 and 1650 cm^{-1} respectively. As the origin of the stretching vibrations and the shape of its spectral envelope are rather complicated, only the H–O–H bending vibration in the region 1800–1500 cm^{-1} is used for the further discussion.

A course of the wavenumber in the maxima of the water bands remaining in the difference spectra after subtraction of the curve of bulk water can be found in Figure 6. It can be seen that the wavenumber of the bending bands of water in the vicinity of both copolymers and also of bulk water generally decrease with increasing temperature. This is not surprising, as it reflects a weakening of the hydrogen bonds with temperature. As the maxima of remaining water bands lay for both copolymers at higher wavenumbers than those of bulk

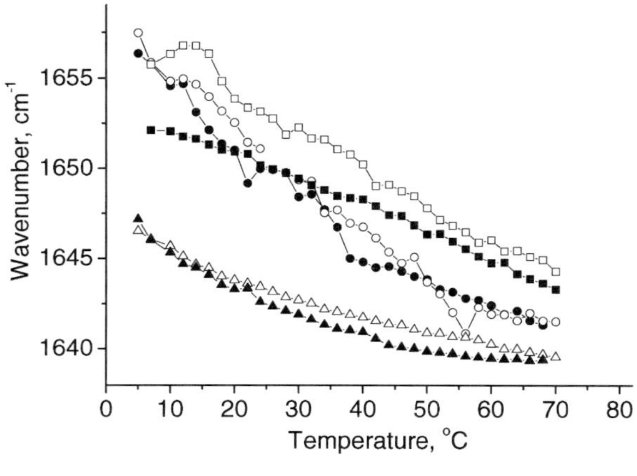

Figure 6.
Wavenumber shifts of H_2O deformation bands with temperature. Water near copolymer PE6200, heating (○), cooling (●); water near copolymer F68, heating (□) cooling (■); bulk water, heating (△), cooling (▲).

water, it can be derived that the average strength of hydrogen bonds in a water envelope around the polymer chains is higher than in bulk water. From this follows that the effect of hydrophilic bonding of vicinal water molecules to ether oxygen apparently predominates over the hydrophobic influence of methyl and methylene groups.

The integral intensities of the H_2O bending bands remaining in difference spectra of both copolymers after subtraction of bulk water were divided by the integral intensities of C–H stretching vibration, so that the concentration effects of sedimentation and other inhomogeneities appearing in the measured layer were partly suppressed. The results are plotted in Figure 7. Supposing the absorption coefficient of bending vibration of water is roughly constant for the measured temperature range, it can be concluded for the sample PE6200 with shorter EO blocks that the relative amount of water in the vicinity of chains diminishes with increasing temperature – water vicinal to copolymer being forced out from the material. During the backward cooling the amount of water near the polymer chains increases again, but below 30 °C the increase is stopped and a decrease begins (see thick line between points in Figure 7). For the sample F68, a copolymer material with a strongly suppressed content of remaining water appeared near the diamond prism at the beginning of backward cooling at 70 °C. This is in accordance with the Figure 3, from which a steep concentration increase of this copolymer near the diamond prism at the beginning of backward cooling can be derived. It is apparent that the vicinal water was expelled from the sedimenting material (compare the arrows in Figure 3 and 7).

A new interesting effect was observed for the copolymer PE6200 in water medium. After heating above transition temperature followed by reverse cooling below 28 °C, a new distinct band with the maximum ~1740 cm^{-1} appeared. Together with its appearance, the bands in the range 3000–2800 cm^{-1} were sharpened and their relative intensities were exchanged (see Figure 8). We assign the new band near 1740 cm^{-1} to a deformation vibration of H_2O molecules forming water bridges connecting EO blocks of two neighbouring copolymer chains. Examples of two PEO segments hydrogen bonded with water molecules are seen in Figure 9. Calculations suggest that the appearance of the band at

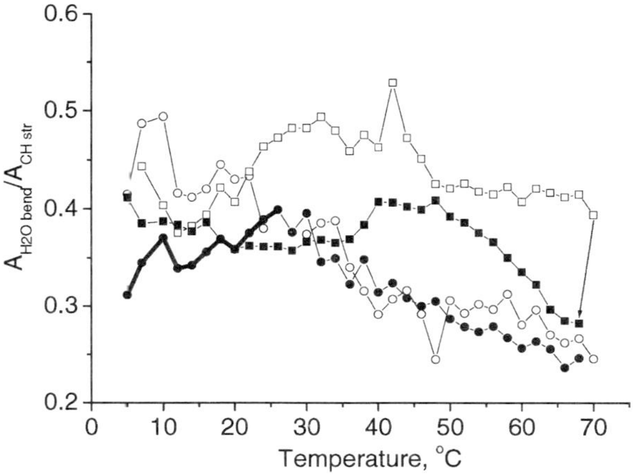

Figure 7.
Changes of relative integrated intensities of deformation band of water near polymer chains ($A_{H2O\ bend}/A_{CH\ str}$). Water near copolymer PE6200, heating (○), cooling (●); thicker line reflects a formation of a new liquid crystalline phase. Water near copolymer F68, heating (□), cooling (■).

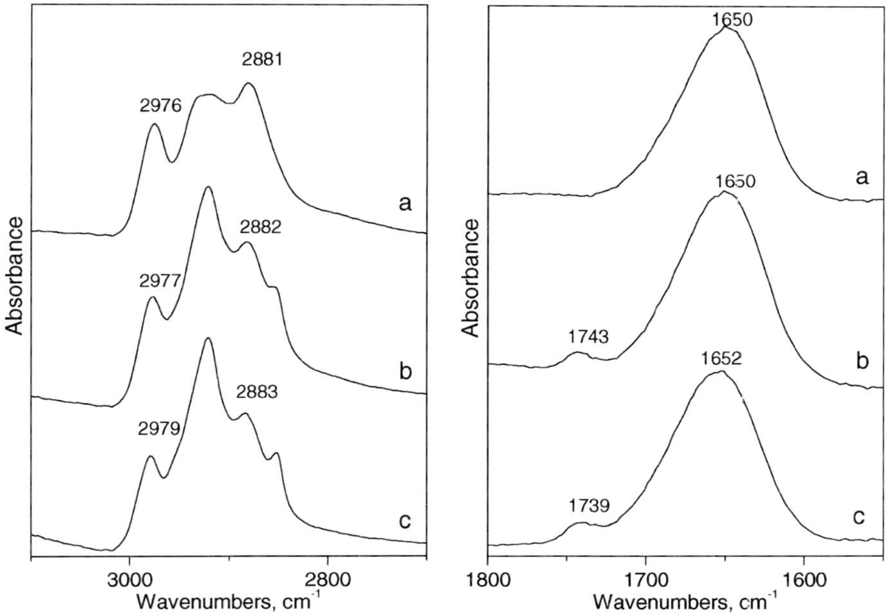

Figure 8.
Changes in infrared spectra during backward cooling of the copolymer PE6200 in water. Solution heated above T_t and cooled to (a) 28, (b) 20, (c) 10 °C.

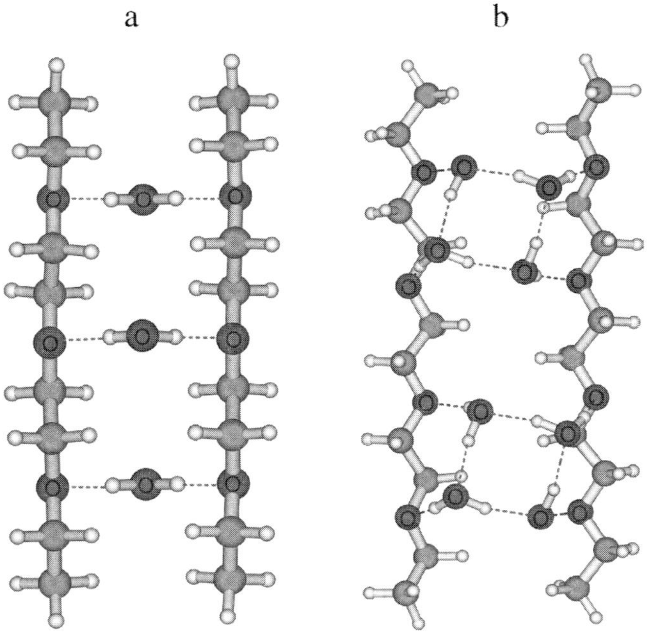

Figure 9.
Optimized geometries of model PEO segments hydrogen bonded through water molecules calculated with the B3LYP/6-31G(d) method.

1740 cm^{-1} can be explained by formation of structures like that shown in Figure 9b. It is supposed that new water bridges between the vicinal EO chains of a micelle with a rigid PO nucleus were formed during heating above the transition temperature, where water redistribution was found (see above the discussion to Figure 4b about redistribution of water). Nevertheless the presence of interchain water bridges in copolymer is visualised by a new band only on backward cooling (below 28 °C) together with a sharpening of the bands in the C–H stretching region. It might be also reflected by a thick line in Figure 7 showing that water ceased to penetrate back among the polymer chains during the end of the reverse cooling. These effects can be explained by the formation of a new liquid crystalline phase. Taking into account that the local weight concentration of the copolymer PE6200 near the ATR prism during the backward cooling reaches ~25% (as can be derived from Figure 3), it is supposed that the phase is identical with that-one shown for this copolymer in.[12] We suppose that the regular water bridges between the vicinal EO blocks are formed only near PO nucleus of the micelles holding the distance between the threads of the EO fringes sufficiently small. This can be the reason why the liquid crystalline phase was detected only in PE6200 having comparatively short EO blocks. As in F68 the EO blocks are roughly 10-times longer, the distance between their outer ends is much larger and the regular water bridges between them can be formed and detected with much lesser probability.

Conclusion

- At low temperatures, unimer chains of copolymers in water are surrounded by water envelopes. Copolymer – water interactions cause the shifts of infrared bands. On increasing the temperature above the appropriate transitions, water molecules are gradually removed from the vicinity of the polymer chains and infrared wavenumbers approach values found for pure copolymers.
- Hydrophobic interaction $H_2O \ldots H_3C$– is weaker than hydrophilic interaction $OH_2 \ldots OC$–. At higher temperatures, water molecules are removed from the central part of the tri-block containing methyl groups and micelles with PO nuclei are formed.
- The transition temperatures increase with the EO molar fraction.
- Above the transition temperature of PE6200, a redistribution of the bound water molecules appears. The polymolecular water clusters between the EO chains are destroyed and water molecules bound by hydrogen bonds form bridges between the ether groups of the vicinal EO chains of the copolymer. During further backward cooling a new liquid crystalline structure is formed.

Acknowledgements: The authors are greatly indebted to the Grant Agency of the Czech Republic (project 203/05/0425) for financial support. Thanks are due to Mrs. M. Brunclíková (IMC) for very helpful technical assistance.

[1] *Nonionic surfactants*, V. M. Nace, Ed., Marcel Dekker, Inc., New York – Basel **1996**.
[2] L. Zheng, C. Guo, J. Wang, X. Liang, P. Bahadur, S. Chen, J. Ma, H. Liu, *Vib. Spectrosc.* **2005**, *39*, 157.
[3] K. L. A. Chan, S. G. Kazarian, *Vib. Spectrosc.* **2004**, *35*, 45.
[4] Y. L. Su, H. Z. Liu, C. Guo, J. Wang, *Mol. Simul.* **2003**, *29*, 803.
[5] Y. L. Su, H. Z. Liu, *J. Phys. Chem. B* **2002**, *106*, 11823.
[6] Y. L. Su, H. Z. Liu, *Macromolecules* **2002**, *35*, 6426.
[7] Y. L. Su, H. Z. Liu, J. Wang, Y., *Langmuir* **2002**, *18*, 865.
[8] Y. L. Su, J. Wang, H. Z. Liu, *Langmuir* **2002**, *18*, 5370.
[9] Y. L. Su, J. Wang, H. Z. Liu, *J Colloid and Interf. Sci.* **2002**, *251*, 417.
[10] C. Guo, J. Wang, H. Z. Liu, J. Y. Chen, *Langmuir* **1999**, *15*, 2703.
[11] C. Guo, H. Z. Liu, J. Wang, J. Y. Chen, *J. Colloid Interface Sci.* **1999**, *209*, 368.
[12] G. Wanka, H. Hoffmann, W. Ulbricht, *Macromolecules* **1994**, *27*, 4145.

Ambiguity in Assignment of Near Infrared Vibrational Bands for Polymerisation Monitoring of Epoxy System

Gilbert Lachenal,[*1] Isabelle Stevenson,[1] Alexandra Durand,[1] G. Seytre,[1] D. Bertrand[2]*

The curing process of epoxy affects the chemical structure of the final network so mechanical and physical properties of the polymeric matrix for a composite may be modified according to the polymerisation conditions. This paper describes the ambiguity in assignment of reference bands to follow the cure of poly-epoxy reactive systems using a laboratory-made system which allows the coupling of dielectric analysis and Fourier Transform Near Infrared Spectroscopy (FTNIR). The dielectric measurements were obtained using interdigitated electrode. *In situ* monitoring of extent of reaction was carried out from room temperature up to 160 °C using fibre-optic FTNIR spectroscopy. For the DGEBA/MCDEA system the epoxy band at 6060 cm^{-1} was chosen in preference to the band at 4530 cm^{-1} as representative of the epoxy function evolution during polymerisation because a small unknown peak probably due to the hardener appears in the 4530 cm^{-1} region. The bands at 4620 and 4680 cm^{-1} assigned to aromatic combination bands and widely used as reference bands are not unique for this formulation hence the band at 5980 cm^{-1} is used as reference. The Principal Components Analysis (PCA) shows clearly also that the bands at 4620 and 4680 cm^{-1} vary during the polymerisation. Surprisingly, the band at 4530 cm^{-1} is equivalent to the one at 6060 cm^{-1} to calculate the conversion rate. It is probably due to the fact that the hardener band near 4530 cm^{-1} follows the same behaviour as the epoxy band at 4530 cm^{-1}.

Keywords: dielectric spectroscopy; in-situ monitoring; near infrared spectroscopy; poly-epoxy reactive system; polymer; PCA

Introduction

Epoxy resins are widely used as adhesive, coating, casting and matrix for glass or carbon composites. They represent an important class of high-performance materials characterized by properties such as good mechanical and thermal behaviours, high resistance to solvents and corrosive agents, outstanding adhesion to various substrate, low shrinkage upon curing and easy processing under a wide range of conditions. These advantages make them attractive in high technology applications such as encapsulation of micro circuitry in the electronic industry, the use as matrices for fibre composites in aerospace applications.

The main reactions during epoxy–amine polymerisation are primary amine epoxy addition, secondary amine addition and etherification (hydroxy-epoxy). These reactions may be catalysed by impurities present in the resin or by hydroxyl.[1] For a stoechiometric molar ratio the etherification is generally low. The homopolymerisation of epoxy groups is generally considered as negligible without Lewis acid or base catalysts.

[1] Université de Lyon, Université Lyon 1, IMP/LMPB, Laboratoire des Matériaux Polymères et des Biomatériaux, Ingénierie des Matériaux Polymères IMP, UMR CNRS 5223, Bât. ISTIL 15, Bld Latarjet 69622 Villeurbanne Cedex – France
E-mail: gilbert.lachenal@univ-lyon1.fr, isabelle.stevenson@univ-lyon1.fr

[2] ENITIAA-INRA, Unité de Sensométrie et Chimiométrie, BP 82225-44322 Nantes Cedex 03 (France)

It is important to monitor polymerisation reactions since the curing process of thermosetting resins i.e. epoxy affects the chemical structure of the final network so mechanical and physical properties of the polymer[1] or the polymeric matrix for a composite may be modified according to the polymerisation temperature.

The main ad-, and disadvantages of alternative methods for reaction control are given as follows:

- Chemical analysis for reaction are time consuming, needs hazardous reagents and leads to sample destruction.
- Differential scanning calorimetry is an usual method for studying cure but is only able to provide the overall extent of reaction.[1]
- HPLC may be used for individual chemical component only before the gel point
- These methods cannot be used for in situ real time monitoring.
- Due to large absorbance coefficient Mid Infrared can be used only for thin sample which often are not representative of industrial samples.[2] In addition, several authors[3,4] warn against using the 915 cm^{-1} band to calculate extent of epoxy reaction because this also originates from another functional group. Alternative methods for monitoring polymerisation reaction are also widely discussed in Billaud et al. [5–6]

Near infrared (NIR) spectroscopy allows to have access to the degree of conversion by following the epoxy band absorption.[3] In situ real-time monitoring of reactive systems by remote fibre optic NIR spectroscopy have been studied and reaction kinetics by coupling NIR spectroscopy with dielectric spectroscopy on model and multifunctional epoxy/amine systems have been compared previously.[7–8] Nevertheless, nobody up to now, has discussed the vibrational band assignment of the reference bands. The extent of reaction calculated using Beer-Lambert's law when considering the peak at 4530 cm^{-1} as representative of the epoxy function evolution (decrease of epoxy functions with polymerisation time) and the band at 4620 cm^{-1} as reference has been successfully presented by several authors.[5–8] This band was chosen instead of the band at 6060 cm^{-1} since it was less overlapped with other neighbouring bands. Unfortunately, a small unknown peak appears in the 4530 cm^{-1} region for the hardener (see Fig. 1), this band was already seen by Min et al.[6]. On the other hand the

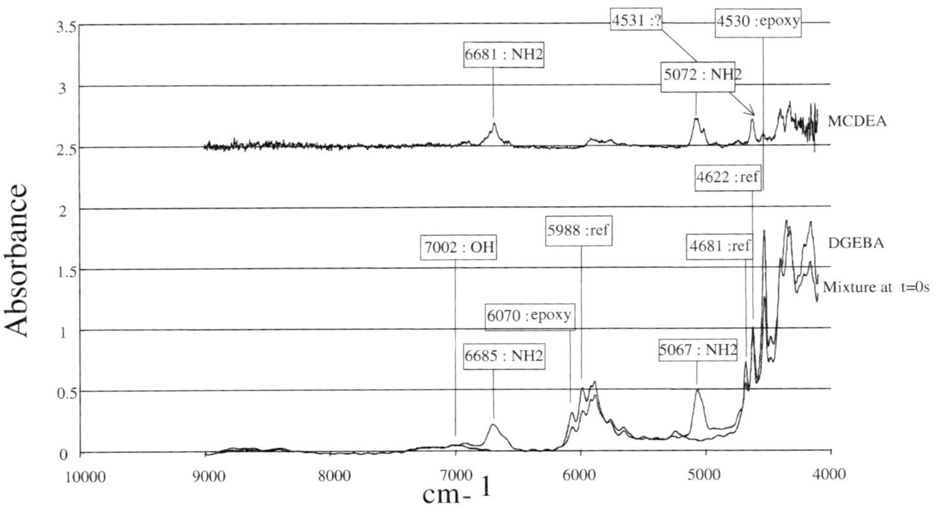

Figure 1.
NIR spectra for the DGEBA/MCDEA system hardener blending before and after polymerisation.

vibrational band at 4620 cm^{-1} has been widely used in the literature for numerous types of epoxy systems but in our case it varies independently of curing which has led us to choose preferably the overlapped vibrational band at 5980 cm^{-1}. The main drawback of the NIR spectroscopy is the difficulty to make classical quantitative analysis for strongly overlapped spectra. Multivariate techniques enable the exploitation of a complex and large set of spectra by improving at the same time selectivity and reliability. These techniques have gained acceptance as routine methods for process analysis. One goal of multivariate techniques is a data reduction without loss of information. The multivariate strategy can lead to new synthetic variables data with higher information content. Chemometric methods are numerous. The Partial Least Squares (PLS) techniques[9,2] are widely used, but need calibrations using primary quantitative analysis techniques. Nevertheless, without reference methods this multivariate analysis fails. On the other hand, Dittmar and Siesler[10] have studied the synthesis of diphenylurethane from phenol and phenyl-isocyanate in chloroform solution, by NIR and multivariate techniques (PLS) and suggested that the familiar Principal Component Analysis[11,12] (PCA) can be appropriate for monitoring the extent of reaction under certain conditions.

PCA can be used to reveal small variations of spectra for samples with differences in chemical (or physical) composition. The main advantage of identification using PCA is that only a limited number of the scores of a spectrum are used. The vectors are usually sorted in order of decreasing eigenvalues, hence, the amount of variance described by the first principal component is greater than the second and so on. The first few principal components (also called factors, latent variables or eigenvectors) depict the largest amount of variance and the noise is contained in the remaining ones. PCA estimates the correlation structure of the variables. Multivariate analysis may provide a base for interpretation of chemical species, identification, classification, or prediction models. Nevertheless, the conversion extent at the final stage must be known for quantification with PCA method.

In this paper, the use of FT-NIR spectroscopy in transmission mode to monitor in situ the curing of epoxy formulation is presented. The band selection is discussed. The extent of reaction versus time obtained from Beer-Lambert analysis and using the PCA will be compared.

Experimental Part

Materials

The thermoset systems used for this study consists of: the polyepoxy resin of DGEBA (Diglycidyl Ether of Bisphenol A) type from CIBA Switzerland-Basel (LY556) with n = 0.15 and a hardener such as MCDEA (4,4′-methylenebis[3-chloro-2,6-diethylaniline] (from LONZA-Levallois-Perret-France) which formulae are given below:

DGEBA

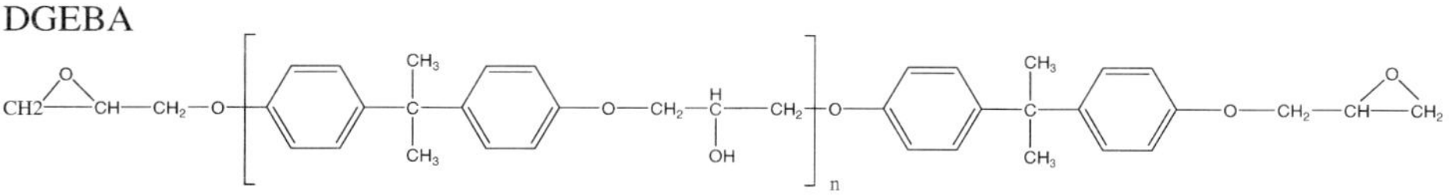

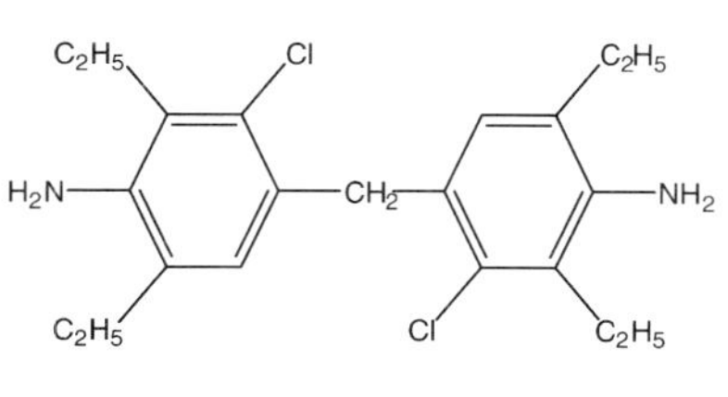

MCDEA

MCDEA hardener is less studied than DDS or DDM hardeners but is less toxic to use which explains mainly why it was chosen for this study.

On PIR spectra, during the polymerisation reaction epoxy/amine can be seen mainly:

- The opening, thus the disappearance of oxirane (epoxy) rings corresponding to the decrease of the bands around 4530, 6060, 8600 cm^{-1}
- Disappearance of primary amine at 5050 et 6670 cm^{-1}
- Appearance of hydroxyl functions at 4700–4800 cm^{-1} and 7 000 cm^{-1}

These components were mixed in a stoechiometric ratio $r = 1$, r being the ratio amine/epoxy functions. The $T_{g\infty}$ of these systems are greater than 150 °C. $T_{g\infty}$ is the glass transition temperature of the totally crosslinked network.

FT-NIR Spectrometer

An ABB NetworkIR spectrometer (ABB Bomem, Canada, Quebec) equipped with GRAMS 5 NIR software (Version 5.2, Galactic, Galactic Industries Corporation, 395 Main Street, Salem, NH 03079, USA) was used for near infrared measurements. All transmission spectra were measured at 8 cm^{-1} resolution using 50 scans between 10000 and 4000 cm^{-1}. The reactive mixture was placed in a sampling quartz cell (from Hellma-France-Paris) which is 1 mm thick, 1 cm wide and 4 cm high. Two optical fibres plugged with SMA connectors and/or an Axiom probe (Irvine-USA) were used. The "in" optical fibre had a diameter of 300 microns and the "out" optical fibre a diameter of 1000 microns, both optical fibres have a length of 2 m. The baseline for area calculations was defined manually from one side to the other side of the vibrational bands chosen.

Dielectric Analysis

Dielectric measurement were obtained using Eumetric Microdielectrometer system III (Micromet-Burlington-USA) equipped with a CHIP interdigitated type electrode which is mounted on a flexible polyimide support and which can be introduced in the NIR quartz cell. This sensor is made of interdigitated electrodes to prevent problems of variation of thickness. This interdigitated electrode consists of several interleaved comb shaped metal strips deposited on a non-conducting substrate and the electric field is generated between the metal strips. The available frequency range extends from 5.10^{-3} Hz to 10^5 Hz. The conductivity measurements range was between $\sigma = 10^{-13}$ to 10^{-3} S/cm. More details on dielectric analysis can be found in.[13–15]

Simultaneous Experiments (Coupling between NIR and Dielectric Spectroscopy)

An experimental system using NIR spectroscopy was carried out simultaneously with dielectric analysis using a quartz cell which could be heated from room temperature to curing temperature (±2 °C) in order to polymerise the thermoset systems. Experimental details can be found in.[2] The isothermal curing temperatures were chosen between 80 to 160 °C for the DGEBA/IPD system and between 105 to 160 °C for the DGEBA/MCDEA system.

Results and Discussion

The quantitative analysis may be performed using either univariate or multivariate approaches. The classic assignment of these bands can be found in Table 1.[16,4] Nevertheless caution had to be taken because many band assignments are not unique (for example there are bands which have not been assigned yet in the 5500–6000 cm^{-1} region), in addition OH and NH bands shift according to the strength of hydrogen bonds.[17]Firstly, the extent of reaction (α) for the DGEBA/MCDEA system at any time t is calculated from the initial areas of the epoxy (4530 cm^{-1}) and reference (4623 cm^{-1}) peaks[5–8] $A_{e,0}$ and $A_{r,0}$ respectively, and their corresponding values at time t $A_{e,t}$ and $A_{r,t}$

Table 1.
Tentative band assignments in the NIR region for epoxy/amine resins.[16,4]

Wavenumber (cm^{-1})	Classic Assignments	Kind of vibration
4060	Aromatic combination bands	
4170	Combination bands of the terminal methylene group	
4350	Combination bands of aliphatic methyl group	
4525	Combination band of the oxyrane (ring C–H)	Conjugated epoxy CH_2 deformation band (1460 cm^{-1}) with the aromatic CH fundamental stretch (3050 cm^{-1})
4550	NH combination band	
4620	Aromatic combination band	Combination band of the aromatic conjugated C=C stretch (1626 cm^{-1}) with the aromatic CH fundamental stretch (3050 cm^{-1})
4680	Aromatic combination band	Combination band of the aromatic conjugated C=C stretch (1626 cm^{-1}) with the aromatic CH fundamental stretch (3050 cm^{-1})
4800–5000	Combination of bands of OH	
5000–5070	NH2 stretching and bending combination bands (aliphatic amine)	
5200	NH2 stretching and bending combination bands (aromatic amine) + water humidity	
5700–5850	Overtones of CH, CH2	First overtones of $-CH_2$ and –CH stretching vibrations
5980	Aromatic C–H stretching first overtone	
6060	First overtone of the terminal C–H in the epoxy ring	First overtone of terminal $-CH_2$, –CH fundamental stretching vibration
6500	NH2 overtone stretching	
6650	NH2 + NH overtone	
7000–7100	OH overtone (hydroxyl + H2O)	
8200	OH combination	
8630	Second overtone of the terminal C–H in the epoxy ring	

according to the Beer-Lambert law.

$$\alpha = 1 - [(A_{e,t})(A_{r,0})]/[(A_{e,0})(A_{r,t})] \qquad (1)$$

As usual to check this result of extent of reaction, the reference band at 4680 cm^{-1} was used[15] since the reference band at 4060 cm^{-1} cannot be used due to the too high absorbance. The same kinetic is not obtained when these two bands (4623, 4680 cm^{-1}) are used separately, thus another functional group must be participating to these bands at 4623 and/or 4680 cm^{-1}. After careful examination of spectra it was observed that the intensity of these classical reference bands change during the polymerisation.

Secondly, as mentioned by Min et al.[6] for the DGEBA/DDS system, a band at 4530 cm^{-1} is present for the amine hardener, this band is also present in MCDEA (see Fig. 1) and can interact with the combination band of the oxirane ring (C–H) which is widely used to follow the kinetics.

Finally, for the DGEBA/MCDEA system the epoxy band at 6060 cm^{-1} was chosen because a small unknown peak appears in the 4530 cm^{-1} region. The band at 5980 cm^{-1} is used as reference. The choice of the baseline can affect the measured areas so that the errors on calculated conversions may be significant and is operator-dependent. The Figure 2 shows the NIR spectra of DGEBA/MCDEA during kinetics of curing at 160 °C.

Band adjustment on the broad 5800–6200 cm^{-1} peak was done using GRAMS software and is shown in Fig. 3 (50/50 Lorentzian/Gaussian contribution, moreover another data treatment such as Voigt

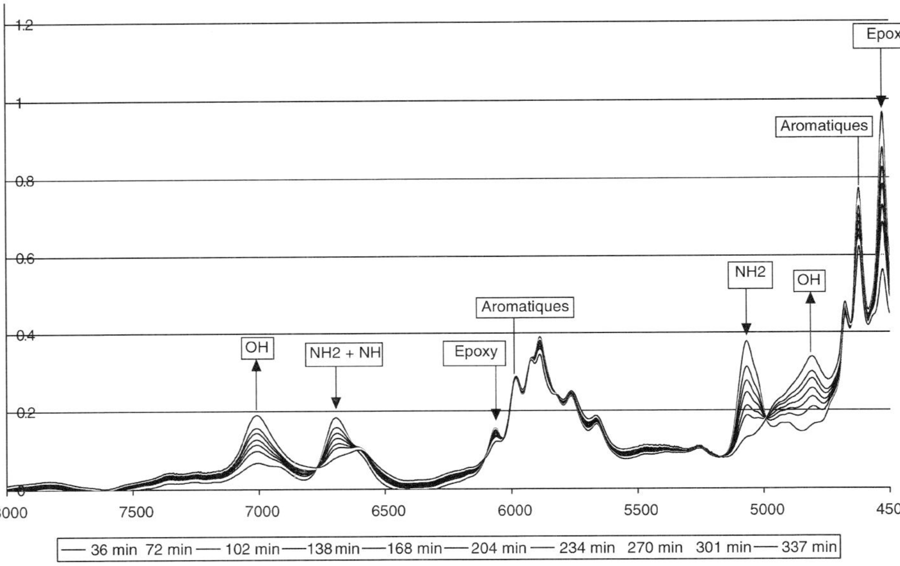

Figure 2.
NIR spectra for the DGEBA/MCDEA system during polymerisation at 140 °C.

functions gave nearly the same peak fitting result). The Fig. 3 also shows that the epoxy peak at 6060 cm^{-1} is not unique, hence a residual area is observed after complete curing reaction. For the calculation of the extent of reaction, this residual spectral area has been removed with the hypothesis that this peak does not change during the polymerisation. It can be noticed that the band at 4530 cm^{-1} has totally disappeared (see Fig. 4) since the curing reaction is completed but the one at 6060 cm^{-1} is still there.

The results given in Figure 4 for DGEBA/MCDEA system show that each functional group can be followed with reaction time.

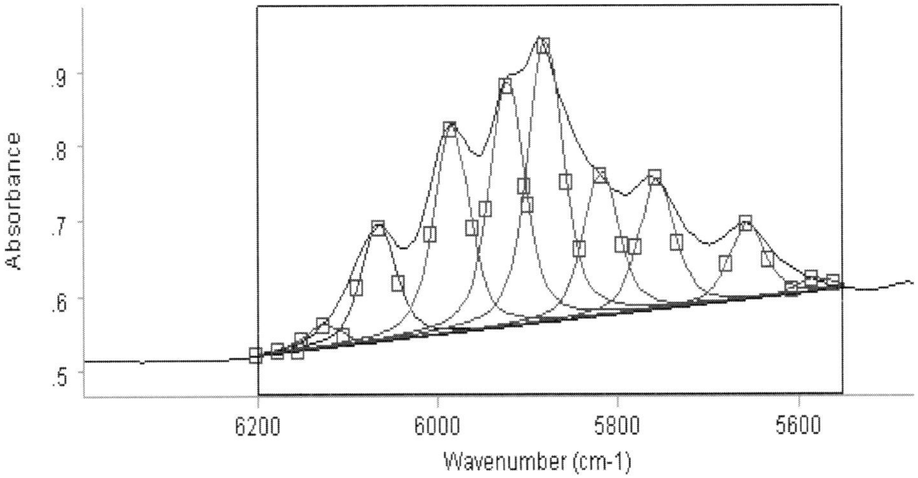

Figure 3.
Band adjustment areas at 6200 and 5550 cm^{-1}.

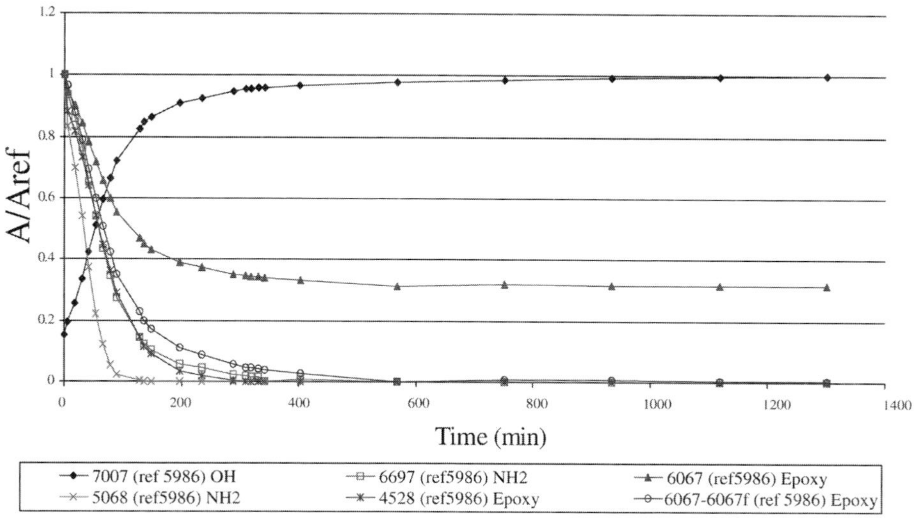

Figure 4.
Normalized absorbance as a function of time for DGEBA/MCDEA during isothermal curing at 160 °C.

OH functions appear and NH_2 as well as epoxy functions disappear.

Another type of quantitative analysis, PCA, can be performed to compare with the previous results. This PCA is first performed on the 4250–9000 cm^{-1} and 4500–6500 cm^{-1} spectral ranges for kinetics above 140 °C where the conversion is full (or seems full). Standard Normal Deviate (SNV) pre-treatment is performed on the unsmoothed spectra. The Standard Normal Deviate transformation normalises each spectrum by dividing the difference between transmittance and the average transmittance by the standard deviation of transmittance.

The first principal component contains about 98–99% of the spectral variance. It was observed that the loading plot of the first factor is linearly correlated to the absorbance changes of the functional groups during the reaction,[9,18–19] hence that it can be used to calculate the conversion by normalising its values between 0 and 1 or the extent final conversion, the conversion extent at the final stage must be known for quantification with PCA method. This kind of result had been already discussed by Maeda and al.[20] They observed that if the spectral variations are very regular, the signal to noise ratio is high and the changes in baseline are small, the loading plot of factor 1 should become very similar to the difference spectrum. This suggests that the plot of this first component as a function of time may be representative of the evolution of the conversion. Such PCA method was successfully applied to evaluate the EVA content,[11] the epoxy extend of cure[21] and more recently for a PU formulation with a reaction induced phase separation.[22] These authors observed a linear correlation between PC1 scores and the extend of reaction measured with a reference method. On the contrary, the second and third components do not seem directly correlated to the conversion. It has to be pointed out that this analysis was carried out on spectra before and after SNV pre-treatment, with no noticeable change.[22]

Figure 5 shows that, at low temperatures (105 °C, 127 °C), the full conversion rate is not achieved. This is in agreement with what is shown in Figure 6 where the vitrification phenomenon can be evidenced by dielectric spectroscopy and not by NIR. When the glass transition temperature of the reactive system reaches the curing temperature, the reaction is limited by the diffusion and the epoxy conversion did not

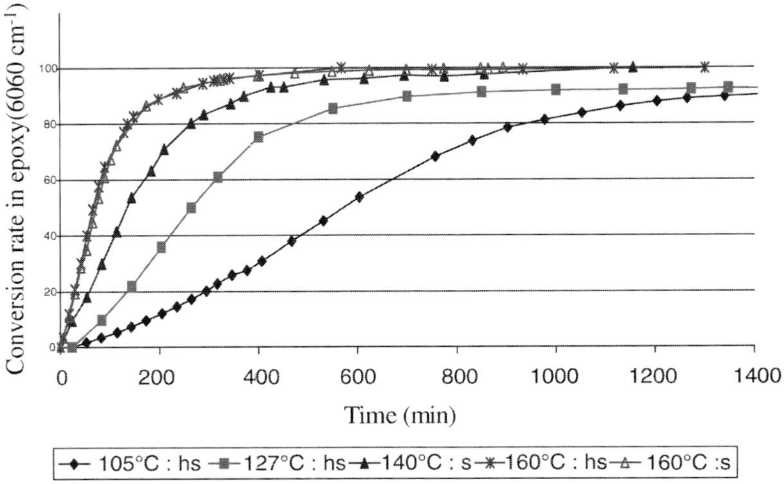

Figure 5.
Conversion rate comparison for different temperatures from 105 °C to 160 °C calculated for the band at 6060 cm^{-1} for DGEBA/MCDEA system. hs = materials stored at −18 °C for 3 months, s = freshly prepared samples.

reach 100% even at long curing times. For 160 °C the two set of points 'hs' and 's' are in good agreement meaning that there is no difference during storage of these samples at low temperature.

The vitrification phenomenon corresponds to the broad peak observed at longer times and the vitrification time is taken to correspond to the peak maximum. The higher the frequency is the earlier appears

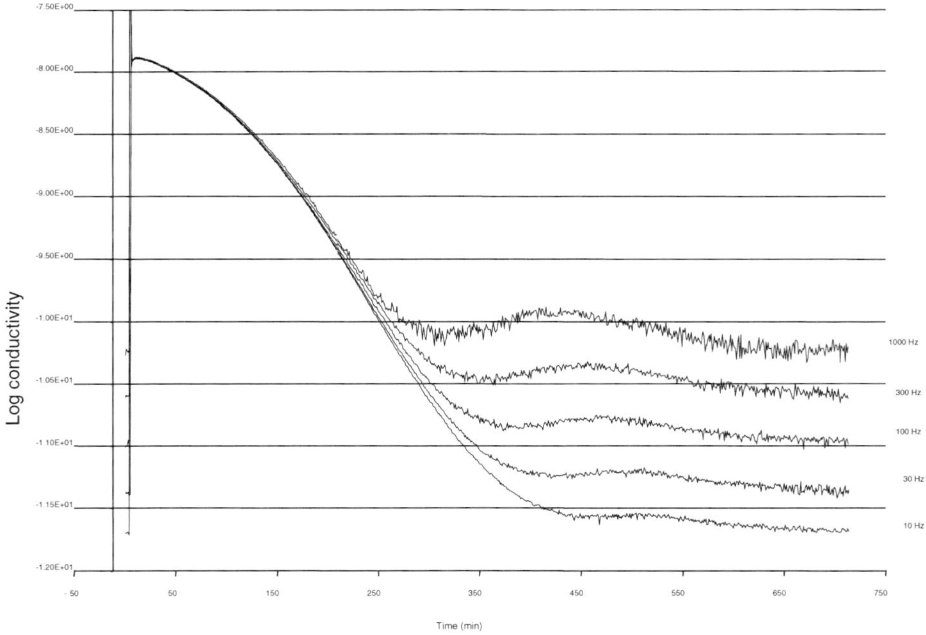

Figure 6.
Dielectric study of the polymerisation kinetics at 127 °C. The same phenomenon occurs at 105 °C.

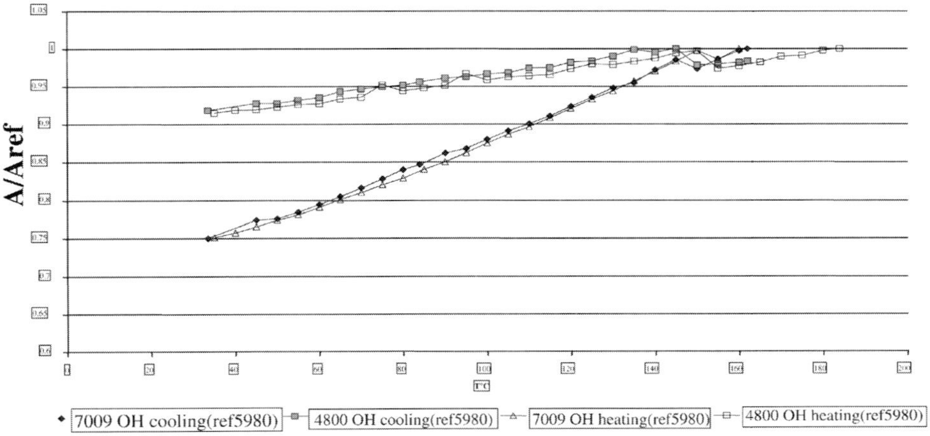

Figure 7.
Band area due to OH vibration at 4800 cm^{-1} and 7009 cm^{-1} which show reversible evolution of hydroxyl bands versus temperature. These measurements were made on a fully cured sample.

the vitrification phenomenon. Vitrification times are respectively: 10 Hz: 530 min, 100Hz: 480 min, 1000 Hz: 425 min.

A strong dependence of OH harmonic band area versus temperature is observed in Fig. 7.

The information contained in the first principal component of the PCA in the OH overtone region may be perturbed by the non-validity of the Beer-Lambert law due to hydrogen bonding. It is well known that hydrogen bonding has a strong influence on

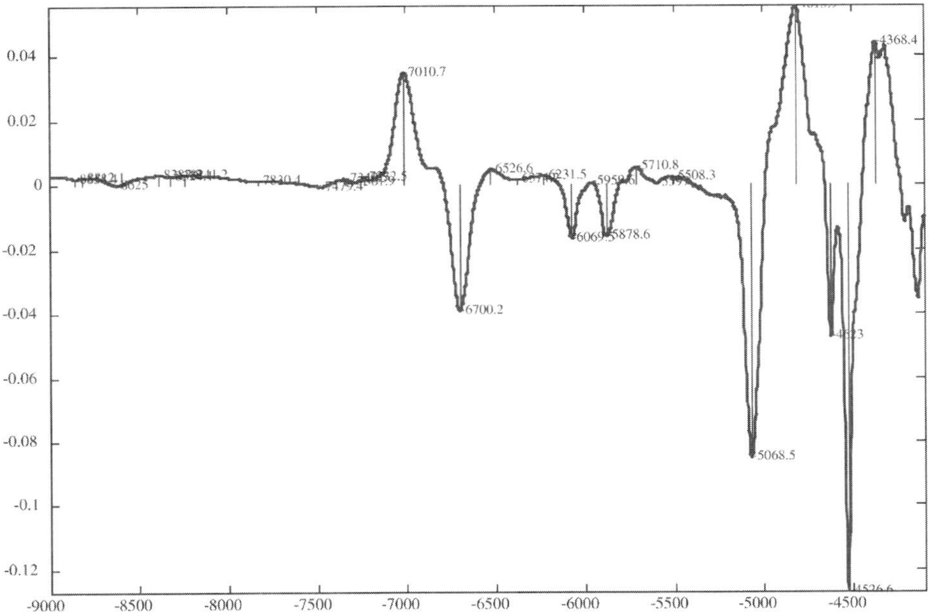

Figure 8.
Principal Component Analysis loading versus wavelength (PC1 98% variability).

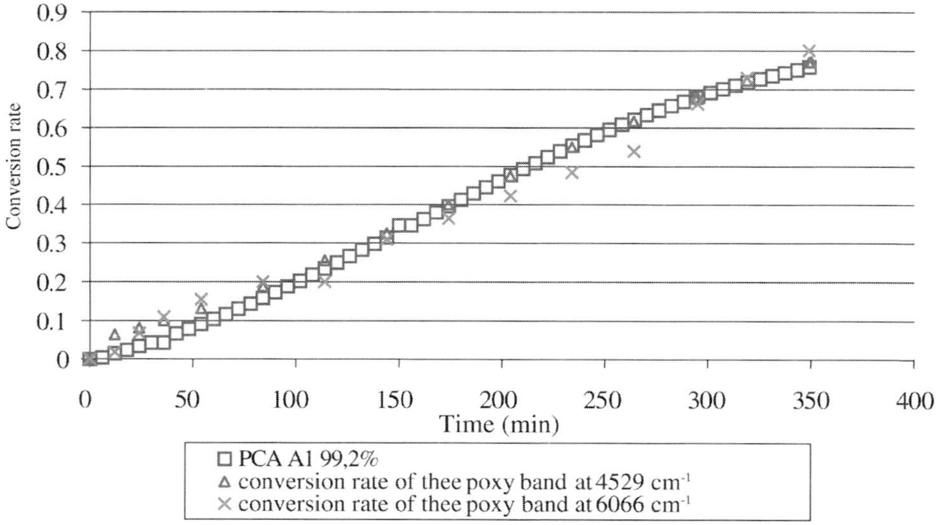

Figure 9.
Comparison between the different data treatments for approximately 6 hours at 140°. It can be noticed that the conversion rates obtained are in relatively good agreement. The epoxy bands at 4530 cm^{-1} and 6060 cm^{-1} give nearly the same extent of reaction using reference band at 5980 cm^{-1}.

NIR spectra.[23] Theoretical and practical studies of the influence of hydrogen bonding on the near infrared spectra, and information concerning the intensity and the position of overtone and combination bands have been presented by Sandorfy.[24] On the other hand the combination band may be shifted according to the strength of hydrogen bonding. However, the stretching bands shift to lower wavenumber while the bending bands shift to higher wavenumber,[24,25] so that this band is not very suitable for quantitative studies of the strength of hydrogen bond formation but on the contrary may be useful for the quantitative analysis.[22] Nevertheless using the 4250–9000 cm^{-1} (including OH and NH harmonic band) and 4500–6500 cm^{-1} spectral ranges for PCA it had been observed very little difference on the first loading of the principal component. The loading of the first PC is similar to difference spectra before and after curing. The choice of bands for the curing monitoring using Beer-Lambert law may be improved by PCA analysis Fig. 8.

Finally, as can be seen in Figure 9, whatever the method (Beer-Lambert law or PCA), whatever the band chosen: nearly equivalent results are obtained.

Conclusion

Everyone has to be very cautious about how to choose the reference band since it can be different for each various system of the same polymer family. Hence, for the DGEBA/MCDEA system the epoxy band at 6060 cm^{-1} was chosen and the band at 5980 cm^{-1} is used as reference. The epoxy band at 6060 cm^{-1} is not unique for this epoxy system. The Principal Components Analysis shows that the band at 4620 cm^{-1} and 4680 cm^{-1} vary. Moreover the band at 4530 cm^{-1} which is at the same place as an hardener band, does not seem to cause any problems, or changes in the same way, however the assignment of this band is not completely elucidated yet and the band at 6060 cm^{-1} can be used also to calculate the conversion rate.

The Principal Component Analysis can give a precious help to choose the right reference band since it allows us to spot clearly the differences in the spectra when reaction kinetics are studied.

The dielectric analysis can evidence the vitrification phenomenon and when coupled simultaneously to near infrared spectroscopy on the same sample and at the same temperature has allowed us to measure simultaneously the spectroscopic extent of reaction and the variation of the electric degree of transformation via the conductivity.

[1] K. Dusek, in: Rubber-Modified Thermoset Resins, C. K. Riew, J. K. Gilham, Eds, AMC, Washington DC, **1989**. or C. J. De Bakker, G. A. George, N. A. St John, P. M. Frederiks, *Spectrochim. Acta* **1993**, **49A/5/6/739**.

[2] D. Bertrand, E. Dufour, La spectroscopie infrarouge et ses applications analytiques, Tec&Doc, Paris, France **2000**.

[3] H. Dannenberg, *SPE Transactions January*, **1963**, *78*.

[4] G. Lachenal, N. Poisson, H. Sautereau, *Macromol. Symp.* **1997**, *119*, 129–136.

[5] C. Billaud, M. Vandeuren, R. Legras, V. Carlier, *Appl. Spectrosc.* **2002**, *56/11*, 1413–1421.

[6] B. G. Min, Z. H. Stachurski, J. H. Hodgkin, G. R. Heath, *Polymer* **1993**, *34*, 3620.

[7] L. Hassi, A. Durand, G. Lachenal, I. Stevenson, G. Seytre, G. Boiteux, *Journal of NIR Spectroscopy* **2006**, *14*, 161–166.

[8] N. Poisson, G. Lachenal, H. Sautereau, *Vibrational Spectros.* **1996**, *12*, 237–247.

[9] H. Martens, T. Naes, Multivariation Calibration, John Wiley & sons. Chichester, UK **1989**.

[10] K. Dittmar, H. W. Siesler, J. Fresenius, Near-infrared light-fiber spectroscopic reaction monitoring of the synthesis of diphenylurethane, *Anal. Chem.* **1998**, *362*, 109.

[11] I. T. Joliffe, Principal Component Analysis, Springer Verlab, N.Y, USA **1996**.

[12] R. E. Aries, J. Sellors, R. A. Spragg, Analytical Applications of spectroscopy II, A. M. C. Davies, C. S. , Eds., RCS, Cambridge, UK **1998**.

[13] K. Friedrich, J. Ulanski, G. Boiteux, G. Seytre, *IEEE Transactions on Dielectrics. and Electrical Insulation.* **2000**, *8/3*, 572–576.

[14] G. P. Johari, D. A. Wasylyshyn, *J. Polym. Sci, Polym. Phys.* **2000**, *38/1*, 122–126.

[15] D. E. Kranbuehl, J. Rogozinski, J. Warner, P. Best, A. Bonnet, J. P. Pascault, H. Sautereau, G. Boiteux, G. Seytre, *Macromol. Symp.* **2003**, *198*(1), 377–388.

[16] N. A. St John, G. A. George, *Polymer* **1992**, *33*, 2679.

[17] G. Lachenal, I. Stevenson, *NIR news* **1996**, *7/6*, 10.

[18] J. F. Aust, K. S. Booksh, C. M. Stellman, R. S. Parnas, M. L. Myrick, Precise Determination of Percent Cure of Epoxide Polymers and Composites via Fibers-Optics Raman Spectroscopy and Multivariate Analysis, *Applied Spectrosc.* **1997**, *51*, 247.

[19] J. Dupuy, S. Benali, A. Maazouz, G. Lachenal, D. Bertrand, *Macromol. Symp.* **2002**, *184*, 249–260.

[20] H. Maeda, Y. Osaki, M. Tanaka, N. Hayashi, T. Kojima, Near infrared spectroscopy and chemometrics studies of temperature-dependent spectral variations of water: relationship between spectral changes and hydrogen bonds, *J. of Near Infrared Spectrosc.* **1995**, *3*, 191–201.

[21] M. G. Hansen, S. Vedula, In-line fiber-optic near-infrared spectroscopy: Monitoring of rheological properties in an extrusion process. Part I., J.A.P.S., **1998**, *68*, 859.

[22] S. Benali, D. Bertrand, J. Dupuy, G. Lachenal, A. Maazouz, *Trans. Inst. Measurement Control* **2007**, In press.

[23] G. Lachenal, I. Stevenson, *NIR news.* **1996**, *7/6*, 10.

[24] C. Sandorfy, Overtones and combination tones: application to the study of molecular associations, in: *Infrared and Raman spectroscopy of biological molecules*, Th. M. Theophanides, D. Reidel, Eds., Dordrecht, Holland **1979**, 305–318

[25] B. Chabert, G. Lachenal, C. Vinh Tung, *Macromol. Symp.* **1995**, *94*, 145–158.

 DOI: 10.1002/masy.200850529

Absorption Study of Norfloxacin – DNA Interaction

L. E. Vijan, M. Conci*

Summary: Norfloxacin, a quinolone antibacterial reagent, has been studied with respect to its binding to calf thymus DNA using absorption spectroscopy. We examined the self-association of the norfloxacin, in order to determine the molar absorption coefficient of the monomer, the molar absorption coefficient of the dimer and the dimerization constant. We also examined the interaction of norfloxacin with DNA by measuring the number of binding sites per DNA segment and the binding constant. Hypochromism, broadening of the envelope and a red-shift in the drug absorption region are apparent for the norfloxacin that bound to DNA. The results were rationalized taking into account both self-association of the drug and the cooperatively effects, in terms of several literature models, Benesi-Hildebrand, Scott, Scatchard, Schwarz and Watanabe-Schwarz.

Keywords: absorption spectroscopy; DNA; norfloxacin; quinolone

Introduction

The term quinolones is used for the quinolonecarboxylic acids or 4-quinolones, which are a group of synthetic antibacterial agents containing a 4-oxo-1,4-dihydroquinoline skeleton. Since the introduction of nalidixic acid (Figure 1) into clinical practice in the early 1960s, a number of structurally related highly potent broad-spectrum antibacterial agents has been isolated.[1]

Modifications of nalidixic acid were made based on structure – activity relationships. It was discovered that a fluorine atom at position 6 and a piperazine ring at position 7 greatly enhance the spectrum of activity. The fluoroquinolones are very active against aerobic Gram-negative microorganisms but less active against aerobic Gram-positive microorganisms.[2] They are useful for the treatment of a variety of infections, including urinary tract infections, respiratory infections, sexually transmitted diseases, acute bronchitis and sinusitis.[1,2]

Faculty of Science, University of Pitesti, Targu din Vale 1, Pitesti, 110040, Romania
E-mail: loredana.vijan@upit.ro

Fluoroquinolones develop its pharmacological action via specific inhibition of sub-unit A of the bacterial gyrase, enzyme that controls DNA shape. The mechanism of this action is unclear but these drugs interact directly with DNA in synergy with gyrase enzyme.[3] Contributions to deeper insight into the mechanism of interaction of this class of antibiotics with DNA would be important for the understanding of their therapeutic efficacy.

Norfloxacin [1-ethyl-6-fluoro-1,4–dihydro-4-oxo-7-(1-piperazinyl) quinoline-3-carboxylic acid] is a synthetic fluoroquinolone antibacterial agent. It has been used for a number of years effectively in humans and animals in the treatments of several bacterial infections, such as *Escherichia coli, Staphylococcus aureus, Citrobacter freundi, Pseudomonas aeruginosa* and *Shigella*.[4] The structure of norfloxacin is shown in Figure 2.

The interaction of norfloxacin with calf thymus DNA has studied by absorption spectroscopy. The study being points to following issues:

- **the self-association of drug**, in order to determine of the molar absorption coefficient of the monomer (ε_M), the molar

WILEY InterScience®

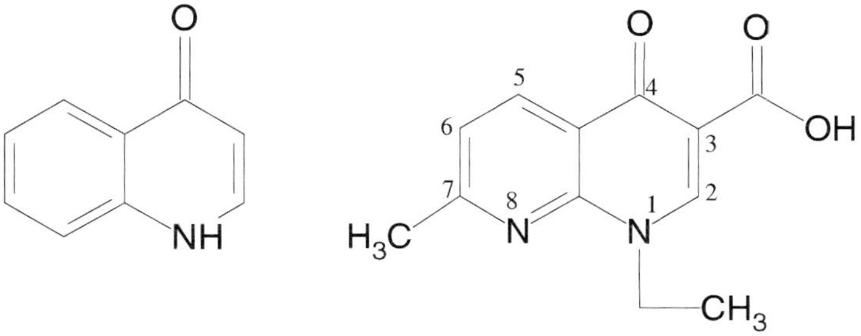

Figure 1.
The formulas of 4-oxo-1,4-dihydroquinoline (left) and nalidixic acid (right).

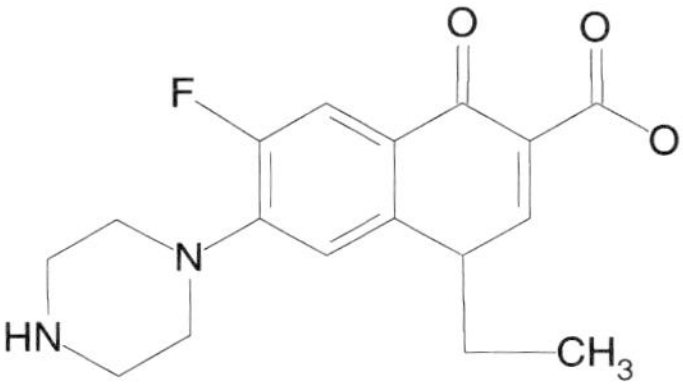

Figure 2.
The formula of norfloxacin.

absorption coefficient of the dimer (ε_D) and the dimerization constant (K_d),
- **the interaction of drug with DNA**, in order to determine of the molar absorption coefficient of stacked drug molecules on the polymer (ε_{st}), the number of binding sites per DNA segment (n) and the binding constant (K).

The results of norfloxacin - DNA system were rationalized taking into account both self-association of the drug and the cooperatively effects, in terms of several literature models, Benesi-Hildebrand, Scott, Scatchard, Schwarz and Watanabe-Schwarz.

Experimental Part

Calf thymus DNA and norfloxacin were obtained from Sigma-Aldrich, USA. The stock solutions of DNA and norfloxacin were prepared by dissolving commercially purchased reagents in doubly distilled water. The concentrations of the stock solutions of reagents were determined by the molar absorption coefficients: ε_{260nm}=6600M^{-1}cm^{-1} for DNA[5] and ε_{275nm}=37500M^{-1}cm^{-1} for norfloxacin.[6] The absorption measurements were performed on a Perkin-Elmer Lambda 25 UV-VIS spectrophotometer using the 1cm optical path length quartz cell, at room temperature.

Results and Discussion

In Figure 3 there is presented the absorption spectra of norfloxacin.

These spectra show a major band centred on 275 nm and two minor bands centred around 320 and 340 nm. The evolution of these spectra with the increase of drug concentration attests the formation of molecular aggregates, by analogy with other compounds[5,7] and allows the following assignment of the bands: the band at 275 nm – monomer and the bands at 320 and 340 nm – higher aggregates.

Although self-association of the drug is more adequately interpreted in terms of indefinite association models,[8] we assumed that in the range of concentrations used in this work, the presence of higher aggregates may be neglected and only the equilibrium between monomer and dimer considered. The literature describes several computation methods for estimating the dimerization constant. Applying two methods, Tipping[9] and Schwarz,[10] we have obtained a value of ~240M^{-1} for the dimerization constant.

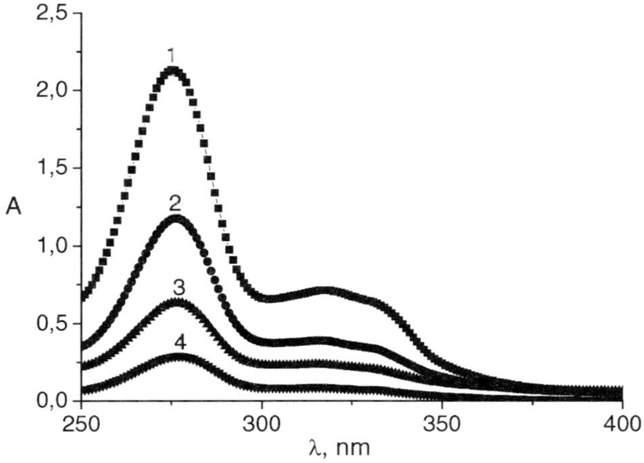

Figure 3.
Absorption spectra of norfloxacin, at different concentrations of drug: (1) $5.59 \cdot 10^{-5}$M, (2) $3.03 \cdot 10^{-5}$M, (3) $1.63 \cdot 10^{-5}$M, (4) $0.73 \cdot 10^{-5}$M.

A family of curves obtained at the titration of norfloxacin solutions of concentrations in the range 10^{-6}–10^{-5}M with calf thymus DNA is presented in Figure 4.

A marked hypochromism (up to ~15%) was observed in the 275 nm and 320 to 340 nm bands for norfloxacin - DNA complex. The interaction of the drug with calf thymus DNA causes a weak red-shift of 2nm for the maximum at 275nm, owing to the perturbation of the complexed chromophore upon binding to DNA bases, while no shift was observed for bands between 320 and 340 nm. The isosbestic point observed in norfloxacin - DNA system, implying the homogeneous conformation of the norfloxacin molecule bound to calf thymus DNA, *i.e.*, the system consists only the DNA free and DNA bound norfloxacin.

The binding constant, for 1:1 norfloxacin - DNA complex can be represented by the following equilibrium:

$$\text{Norfloxacin} + \text{DNA} \overset{K}{\rightleftharpoons} \text{Complex}$$

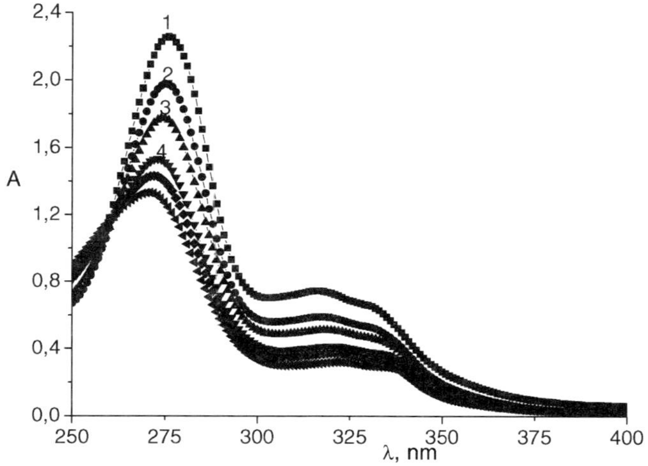

Figure 4.
Absorption spectra of norfloxacin - DNA system, at different polymer/drug $(\frac{P}{D})$ ratios: (1) 0; (2) 0.38; (3) 0.85; (4) 1.54; (5) 2.16; (6) 2.99.

Table 1.
Results of the binding constants for norfloxacin - DNA system.

Method	Equations	K, M^{-1}
Benesi-Hildebrand	$\frac{l}{\Delta A} = \frac{1}{C_D^0 \cdot K \cdot \Delta\varepsilon} \cdot \frac{1}{C_{DNA}} + \frac{1}{C_D^0 \cdot \Delta\varepsilon}$	$1{,}16 \cdot 10^4$
Scott	$\frac{l \cdot C_{DNA}}{\Delta A} = \frac{1}{C_D^0 \cdot \Delta\varepsilon} \cdot C_{DNA} + \frac{1}{C_D^0 \cdot K \cdot \Delta\varepsilon}$	$1{,}06 \cdot 10^4$
Scatchard	$\frac{\Delta A}{l \cdot C_{DNA}} = -\frac{K}{l} \cdot \Delta A + C_D^0 \cdot K \cdot \Delta\varepsilon$	$0{,}89 \cdot 10^4$
	$\frac{r}{C_F} = (n - r) \cdot K$	$0{,}94 \cdot 10^4$
Schwarz	$K = (\gamma_D^o \cdot c_D^o)^{-1} + 2K_d$	$3{,}71 \cdot 10^4$
Watanabe-Schwarz	$\frac{2\theta - 1}{\sqrt{\theta(1-\vartheta)}} = \sqrt{\frac{q}{n}}(K \cdot C_D^o - 1)$	$4{,}52 \cdot 10^4$

The equilibrium constant of the complex formation (K) may be estimated from the changes in the absorbance at 275 nm, using different methods, Benesi-Hildebrand,[11] Scott,[12] Scatchard,[13] Schwarz[14,15] and Watanabe-Schwarz.[16] The equations utilized and the results obtained for norfloxacin - DNA system are summarized in Table 1.

For the interaction of norfloxacin with calf thymus DNA, linear Benesi-Hildebrand, Scott and Scatchard plots were obtained; an example is given in Figure 5.

Considering the simple norfloxacin - DNA equilibrium, the absorbance is assumed to be the sum of the absorbance of the free and bound species, weighted by their respective concentrations:

$$A = f_0 \cdot (C_D^0 - C_B) + f_B \cdot C_B$$

$$A_0 = f_F \cdot C_D^0$$

where A_0 and A are the absorbances of the free drug and that measured at each DNA concentration, respectively; C_D^0 and C_B are the total and bound drug concentrations, respectively.

Under the assumption previously discussed, that the absorption is due only to the free form of the compounds ($f_B = 0$), the concentrations of free and bound drug are given by:

$$C_B = C_D^0 \cdot \frac{A - A_0}{A_0}$$

$$C_F = C_D^0 - C_B$$

The experimental data were also fitted either to the linear Scatchard plot,[7]

$$\frac{r}{C_F} = (n - r) \cdot K$$

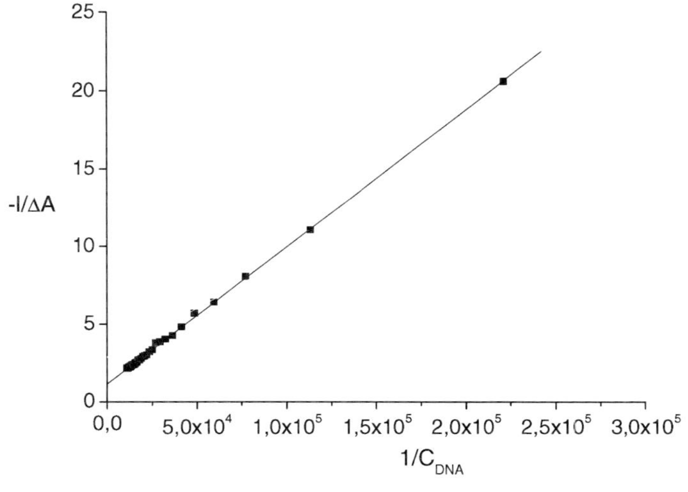

Figure 5.
Benesi-Hildebrand plot for norfloxacin - DNA system.

or to a non-linear regression:

$$r = \frac{n \cdot K \cdot C_F}{1 + K \cdot C_F}$$

corresponding to a single class of non-interacting binding sites that do not exhibit cooperative behaviour. In these relationships, r is the binding ratio ($r = \frac{C_B}{C_{DNA}}$), C_F is the free drug concentration, C_B is the bound drug concentration and n is the number of binding sites.

The Scatchard plots presented in Figure 6, attest the presence of two processes. The solid lines represent the best fit of the linear portion of the plot. This linear part with a negative slope, corresponding to polymer to drug ($\frac{P}{D}$) ratios in the range 0,5–3,5 is characteristic for non-cooperative binding to one class with n equivalent sites. Considering this linear segment, the binding constant $K = 0{,}94 \cdot 10^4 M^{-1}$ and the number of binding sites $n = \sim 1$ were obtained.

The form of the binding curves at the absorption maximum of the monomer ($\varepsilon_{app} = f\left(\frac{P}{D}\right)$, at 275 nm) presents the features of a cooperative binding caused by stacking interaction of neighbouring bound drug molecules and may be interpreted in the frame of a theory developed by G. Schwarz.[10,14,15] This theory defines the fraction of free drug by:

$$\gamma_D^* = \frac{C_F}{C_D^0} = \frac{\varepsilon_{app} - \varepsilon_{st}}{\varepsilon_M - \varepsilon_{st}}$$

Plotting the apparent absorption coefficient ε_{app}, at 275 nm, versus the reciprocal value of the total weighing-in concentration of the drug $\left(\frac{1}{C_D^0}\right)$, at different constant polymer to drug $\left(\frac{P}{D}\right)$ ratios lead to straight lines which converge to ε_{st} (Figure 7).

Extrapolation to $\frac{1}{C_D^0} \rightarrow 0$ yields the molar absorption coefficient of bound and stacked drug molecules ε_{st} as being the intercept on the ordinate axis, value practically the same for all $\frac{P}{D}$ ratios within our experimental errors $28000(\pm 250) M^{-1} cm^{-1}$.

We may determine the fraction of free drug γ_D^* as a function of polymer to drug ratio $\left(\frac{P}{D}\right)$, at constant concentration of norfloxacin. This is carried out by evaluating the decrease of ε_{app} due to increasing the concentration of the polymer. An example is given in Figure 8.

Extrapolation of the limiting straight line to the axis of abscissas yields the reciprocal value of n, which stands for the number of binding sites per DNA segment (this yields $n = 0{,}67$ in our case). The second auxiliary straight line in the diagram has half the slope of the first one and corresponds

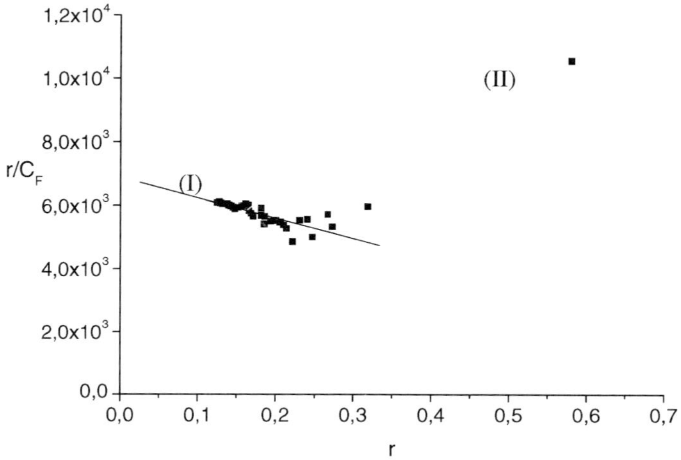

Figure 6.
Scatchard plot for norfloxacin - DNA system.

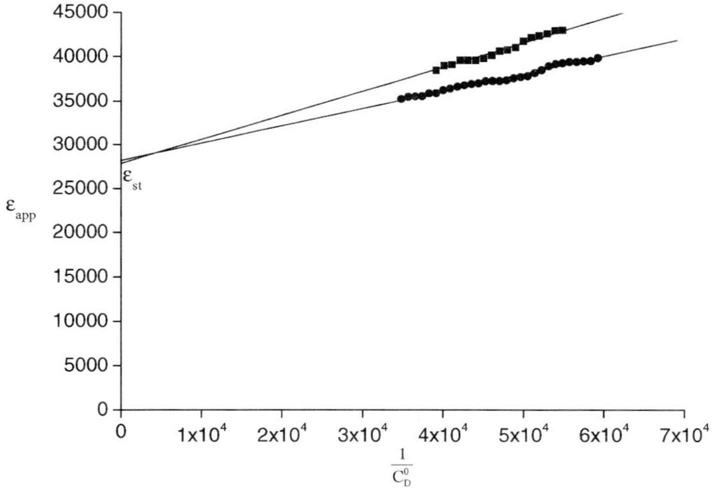

Figure 7.
The apparent absorption coefficient as a function of reciprocal value of the total weighing-in concentration of the norfloxacin.

to $\theta = \frac{1}{2}$, i.e. to the limiting case where no cooperativity is observed (the binding sites being alternatively occupied and not occupied).

The value of γ_D^* at the intersection point with the experimental curve, γ_D^o, may be used to calculate the binding constant K according to the equation[10]:

$$K = (\gamma_D^o \cdot c_D^o)^{-1} + 2K_d$$

valid if binding to the polymer is stronger than the dimerization tendency of the drug, $K \gg K_d$. In our case, we find $K \gg K_d$ so this is a good approximation.

Another useful experimental approach to evaluate binding data at large polymer to drug ratios was developed by Schwarz and Watanabe.[16] In any experiment, binding is usually measured by means of optical absorbance that changes when the free drug is bound to the polymer. One may

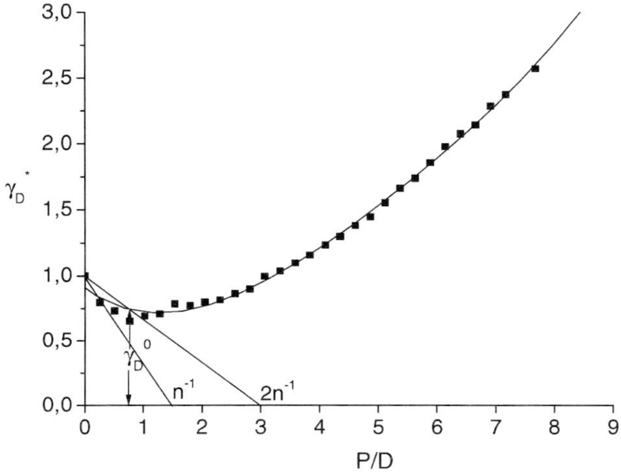

Figure 8.
The plot of γ_D^* versus $\frac{P}{D}$ ratio for norfloxacin - DNA system.

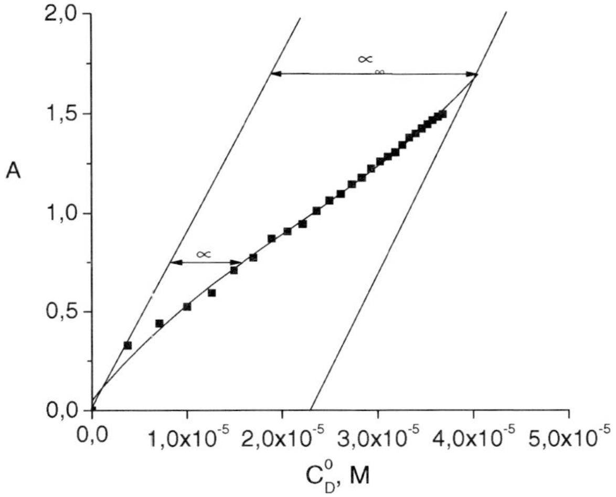

Figure 9.
Plot of absorbance versus total norfloxacin concentration.

ordinarily assume that absorbance depends linearly on the concentrations of free and bound drug, respectively. Thus, we write:

$$A = (\varepsilon_A \cdot C_A + \varepsilon_a \cdot C_a)l$$

with the molar absorption coefficient ε_A, ε_a referring to the two states of the drug.

The quantity ε_A can conveniently be obtained from the concentration dependence of A in a polymer free solution of the drug. An analogous determination of ε_a requires the availability of solution where all drug molecules are bound. In practice it may be difficult to realize this condition. A method of measurement and data processing starts from a drug free solution of polymer to which the drug is added bit by bit. In other words, the absorbance will be measured at constant polymer concentration and gradually increasing total drug concentration. A corresponding plot of absorbance versus total norfloxacin concentration is presented in Figure 9. If the polymer concentration is equal to zero then the absorbance follows the zero-line. The deviations from this zero-line indicate binding of drug. When norfloxacin concentration has finally been increased so much that all available polymer is saturated the curve runs parallel to zero-line. There we have a constant $\propto$, denoted $\propto_\infty$. Thus the degree of binding θ is equal to:

$$\theta = \frac{\alpha}{\alpha_\infty}$$

where $\propto$ and $\propto_\infty$ are defined in the Figure 9.

We applied the method of Schwarz and Watanabe[16] to determine the cooperative binding constant (K) and the cooperative interaction parameter (q), from absorption measurements by using the equation:

$$\frac{2\theta - 1}{\sqrt{\theta(1-\vartheta)}} = \sqrt{\frac{q}{n}}(K \cdot C_D^0 - 1)$$

From the linear plot of $\frac{2\theta-1}{\sqrt{\theta(1-\theta)}}$ agains- drug concentration (Figure 10), we obtained a cooperative binding constant equal to $4{,}52 \cdot 10^4 M^{-1}$ and a value of 6–8 for the cooperative interaction parameter.

where l is the path length, ΔA is the change in the absorbance, $\Delta\varepsilon$ is the molar absorptivity difference ($\Delta\varepsilon = \varepsilon_B - \varepsilon_F$), ε_F and ε_B are the free and bound drug absorption coefficients, C_D^0 is the total concentration of drug, C_F and C_B are the free and bound drug concentrations, C_{DNA} is the concentration of DNA, K is the binding constant, r is the binding ratio, n is the number of binding sites.

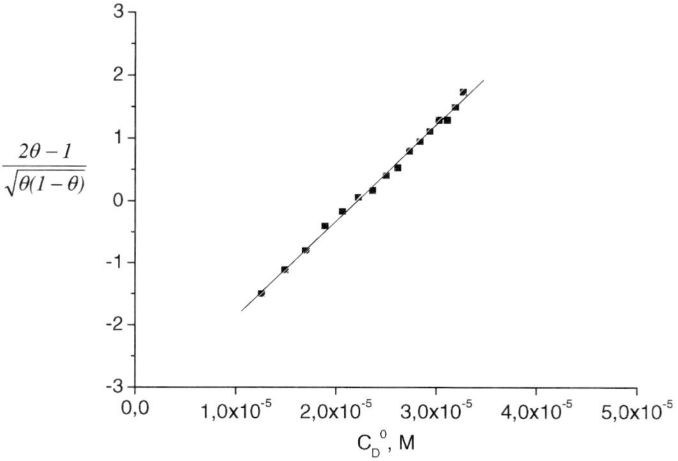

Figure 10.
Plot of $\frac{2\theta-1}{\sqrt{\theta(1-\theta)}}$ versus total norfloxacin concentration.

Conclusion

The binding of norfloxacin to calf thymus DNA produces hypochromism, broadening of the envelope and a red-shift in the drug absorption region, the characteristics that indicated that norfloxacin interacts with DNA. Based on the isosbestic point observed in norfloxacin - DNA system and the assumption that at a relatively low concentration of norfloxacin and DNA the formation of higher order complexes is unlikely, it can be concluded that norfloxacin forms a 1:1 homogeneous ground state complex with DNA.

The analysis of the norfloxacin interaction with calf thymus DNA, using several methods points out two binding types for the norfloxacin - DNA system: a non-electrostatic (internal) type consisting of the intercalation of the drug between the base-pairs from DNA and an external type, cooperative where the electrostatic interactions with the phosphate groups of DNA are predominant. The first binding process analysed by Benesi-Hildebrand, Scott and Scatchard models suppose a 1:1 binding ratio and do not account explicitly for either the dimerization of the drug or cooperativity effects on the binding. The second binding process analysed by Schwarz and Watanabe-Schwarz methods, supposes a linear lattice of equivalent binding sites with nearest neighbour cooperativity.

[1] D. T. W. Chu, P. B. Fernandes, in: B. Testa, Ed., *Advances in Drug Research*, Vol. 21, Academic Press, London **1991**, p. 39.
[2] O. Paulsen, *Drugs of today* **1987**, 23, 269.
[3] I. Turel, *Coordination Chemistry Reviews* **2002**, 232, 27.
[4] G. Song, Y. He, Z. Cai, *Canadian Journal of Analytical Sciences and Spectroscopy* **2004**, 49, 203.
[5] E. Volanschi, L. E. Vijan, *Romanian J. Biophys.* **2000**, 10, 1.
[6] H. J. Hwangbo, Y. A. Lee, J. H. Park, Y. R. Lee, J. M. Kim, S. Y. Yi, S. K. Kim, *Bull. Korean Chem. Soc.* **2003**, 24, 579.
[7] L. E. Vijan, E. Volanschi, M. Hillebrand, *Annals of West University of Timisoara* **2001**, 10, 711.
[8] J. B. Chaires, N. Dattagupta, D. M. Crothers, *Biochemistry* **1982**, 21, 3927.
[9] E. Tipping, B. Ketterer, P. Poskelo, *Biochem. J.* **1978**, 169, 509.
[10] G. Schwarz, S. Klose, W. Balthasar, *Eur. J. Biochem.* **1970**, 12, 454.
[11] H. Benesi, J. H. Hildebrand, *J. Am. Chem. Soc.* **1949**, 71, 2703.
[12] R. L. Scott, *Rec. Trav. Chim.* **1956**, 75, 787.
[13] G. Scatchard, *Ann. N. Y. Acad. Sci.* **1949**, 51, 660.
[14] G. Schwarz, *Eur. J. Biochem.* **1970**, 12, 461.
[15] G. Schwarz, *Eur. J. Biochem.* **1970**, 12, 442.
[16] F. Watanabe, G. Schwarz, *J. Mol. Biol.* **1983**, 163, 467.

 DOI: 10.1002/masy.200850530

Spectroscopic Study of Hedamycin – DNA Interaction

L.E. Vijan, A. Raducu*

Summary: Binding of hedamycin, a member of the antitumour antibiotics pluramycin class, to calf thymus DNA has been studied using UV-Vis absorption spectroscopy. The results have been rationalized in terms of several literature models: Wolfe, Benesi-Hildebrand, Scott and Scatchard.

Keywords: calf thymus DNA; hedamycin; pluramycin antibiotic; UV-Vis absorption spectroscopy

Introduction

Nucleic acids have an important function in the life processes and their study has become an important research field of life sciences. A variety of drugs can bind to DNA and interfere in processes like transcription and replication.

Different classes of anticancer drugs that interact with DNA in different ways have been developed. There are non-covalent interactions by intercalation (e.g. the anthracycline antibiotics[1]: doxorubicin, epirubicin, daunomycin) or minor groove binders (e.g. the aureolic acids[2–4]: mithramycin, chromomycin A_3, olivomycin). Other drugs bind covalently to DNA, including mitomycins, anthramycin and related antibiotics, and some cause backbone cleavages, e.g. bleomycin,[5,6] streptonigrin. Many of these drugs show sequence selectivity, which make them potential targets for DNA from different sources. For example, the minor groove binders preferentially bind to AT-rich sequences whereas the intercalators have been proposed to prefer GC-rich sequences.[1–8]

The interaction of anticancer drugs with DNA is generally highly specific but not necessarily selective. DNA has many specific sites for these interactions, including the polyanionic phosphate, sugar backbone and various hydrogen acceptors and donors of bases from minor and major grooves. For example, the exocyclic N2 atom of guanine in the minor groove and the N7 atoms of guanine and adenine in the major groove are binding sites for alkylating agents. The N3 atoms of both adenine and guanine in the minor groove are also drug-binding sites.[9]

Drugs can be classified into the categories: intercalators, minor or major groove binders and alkylating agents, depending on their mode of the interaction with DNA. In some cases drugs bind via more than one mode (e.g. intercalating alkylators such as the pluramycin antibiotics). The pluramycin antibiotics (altromycin, hedamycin) are a group of the DNA-reactive agents that represent a range of 4H-anthra[1,2]-β-pyran-4,7,12-trione structures with attached carbohydrate and epoxide moieties on the corners of their planar anthrapyrantrione chromophores. Following intercalation of the chromophore into DNA, the epoxide side chain is located in the major groove, allowing selective alkylation of N7 atom of guanine.[10,11]

Hedamycin was originally isolated from fermentation broths of *Streptomyces griseoruber*. Interest in this compound arose from its potent antibacterial action and also its ability to inhibit the growth of HeLa cells in culture and some transplanted rodent tumours.[12] Early studies directed towards elucidating the mechanism of action of hedamycin, revealed that it binds to double-stranded DNA, produces substantial increases

Faculty of Science, University of Pitesti, Targu din Vale 1, Pitesti, 110040, Romania;
E-mail: loredana.vijan@upit.ro

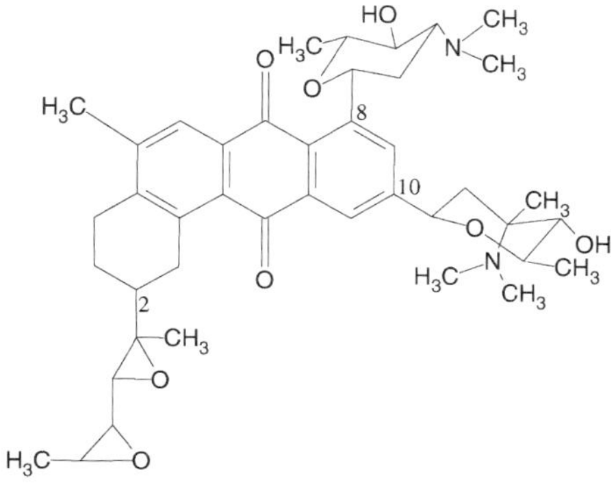

Figure 1.
The chemical structure of hedamycin.

in DNA melting temperature and inhibits DNA and RNA polymerases.[13,14]

Hedamycin (Figure 1) consists of an anthrapyrantrione chromophores to which are attached amino sugar rings at carbons 8 and 10 and a six carbon, bis(epoxide)-containing side chain at carbon 2.

In this paper, we present the results of the interaction of hedamycin with calf thymus DNA, obtained by the UV-Vis absorption spectroscopy. The results have been rationalised in terms of several literature methods: Wolfe,[15] Benesi-Hildebrand,[16] Scott[17] and Scatchard.[18] We have determined the binding constant (K) and the number of binding sites per DNA segment (n).

Experimental Part

Calf thymus DNA was obtained from Sigma-Aldrich, USA. Hedamycin supplied by the National Cancer Institute (NCI), National Institutes of Health (NIH), USA, was generously donated by the Developmental Therapeutics Program, Division of Cancer Treatment and Diagnosis. The stock solutions were prepared by dissolving the reagents in doubly distilled water. The concentrations of the stock solutions of reagents were determined by the molar absorption coefficients $\varepsilon_{260nm}=6600 M^{-1}cm^{-1}$ for DNA[1–6] and $\varepsilon_{428nm}=10000 M^{-1}cm^{-1}$ for hedamycin.[19]

The absorption measurements were performed on a Perkin-Elmer Lambda 25 UV-Vis spectrophotometer using a 1cm optical path length quartz cell, at room temperature.

Results and Discussion

Figure 2a presents three absorption spectra of hedamycin, which shows a major band centred at 428 nm and a shoulder at 340 nm.

The influence of DNA on hedamycin is presented in Figure 2b by a family of curves obtained at the titration of hedamycin solutions of concentrations in the range 10^{-6}–10^{-5}M with calf thymus DNA. It may be observed that at small polymer to drug ratios ($\frac{P}{D}$) the changes of the drug absorption spectrum are similar to those observed at increasing concentration of drug.

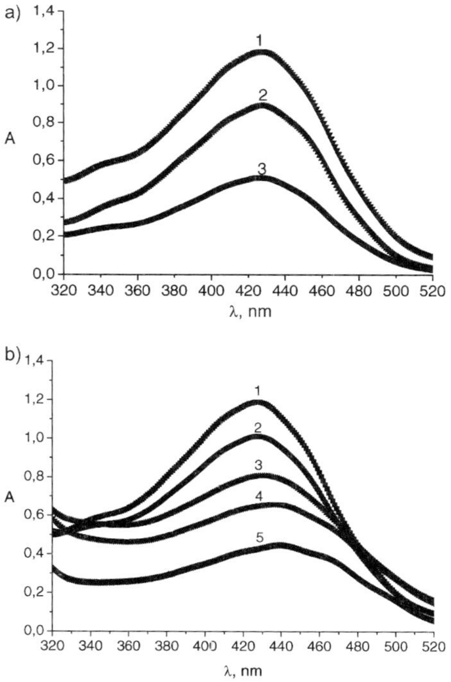

Figure 2.
(a) The absorption spectra of hedamycin at concentrations: (1) 11,89 · 10^{-5} M, (2) 8,95 · 10^{-5} M, (3) 5,12 · 10^{-5} M. (b) Absorption spectra of hedamycin - DNA system. The polymer to drug ratios are: (1) 0; (2) 0,39; (3) 0,94; (4) 1,44; (5) 2,22.

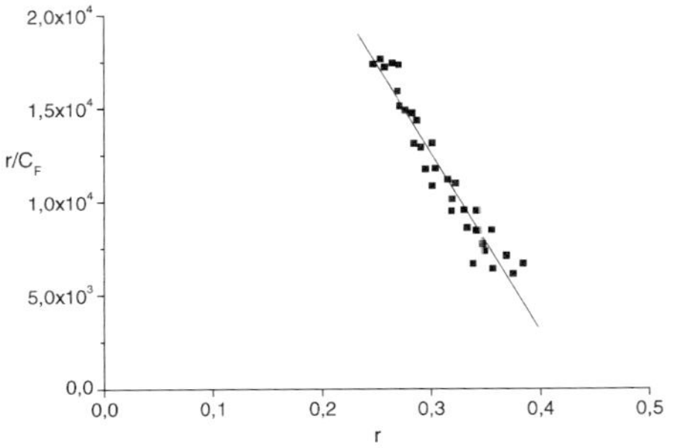

Figure 3.
The plot of $\frac{r}{C_F}$ versus the binding ratio r for hedamycin - DNA interaction.

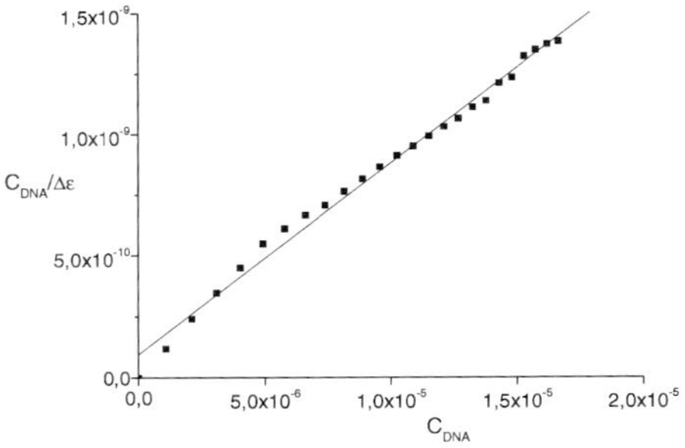

Figure 4.
Wolfe plot of hedamycin - DNA interaction.

On the basis of the equilibrium between hedamycin, DNA and hedamycin - DNA complex:

$$\text{Hedamycin} + \text{DNA} \rightleftharpoons \text{complex}$$

the absorbance is assumed to be the sum of the absorbance of the free and bound species, weighted by their respective concentrations:

$$A = f_0 \cdot (C_D^0 - C_B) + f_B \cdot C_B$$

$$A_0 = f_F \cdot C_D^0$$

where A_0 is the absorbance of free drug, A - the absorbance of drug measured at each DNA concentration, C_D^0 - the total drug concentration, C_B - the bound drug concentration. On the assumption of the absorption is due only to the free form of drug ($f_B = 0$), the concentrations of free

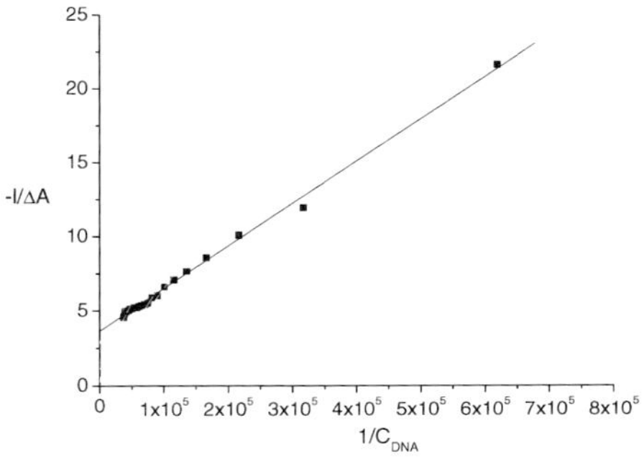

Figure 5.
Benesi-Hildebrand plot of hedamycin - DNA interaction.

and bound drug are given by:

$$C_B = C_D^0 \cdot \frac{A - A_0}{A_0}$$

$$C_F = C_D^0 - C_B$$

Thus, the experimental data (Figure 3) may be fitted on the basis of Scatchard

Table 1.
Results of the hedamycin - DNA interaction.

Method	Equations	K, M^{-1}
Wolfe	$\frac{C_{DNA}}{\Delta\varepsilon_{app}} = \frac{C_{DNA}}{\Delta\varepsilon} + \frac{1}{K \cdot \Delta\varepsilon}$	$3{,}36 \cdot 10^5\ M^{-1}$
Benesi-Hildebrand	$\frac{l}{\Delta A} = \frac{1}{C_D^0 \cdot K \cdot \Delta\varepsilon} \cdot \frac{1}{C_{DNA}} + \frac{1}{C_D^0 \cdot \Delta\varepsilon}$	$1{,}28 \cdot 10^5\ M^{-1}$
Scott	$\frac{l \cdot C_{DNA}}{\Delta A} = \frac{1}{C_D^0 \cdot \Delta\varepsilon} \cdot C_{DNA} + \frac{1}{C_D^0 \cdot K \cdot \Delta\varepsilon}$	$1{,}36 \cdot 10^5\ M^{-1}$
Scatchard	$\frac{\Delta A}{l \cdot C_{DNA}} = -\frac{K}{l} \cdot \Delta A + C_D^0 \cdot K \cdot \Delta\varepsilon$	$1{,}29 \cdot 10^5\ M^{-1}$
	$\frac{r}{C_F} = (n - r) \cdot K$	$1{,}07 \cdot 10^5\ M^{-1}$

where ε_{app}, ε_F and ε_B are the apparent, free and bound drug absorption coefficients, l is path length, ΔA is the change in absorbance at a given wavelength, C_D^0 is the total concentration of drug, C_{DNA} is the concentration of calf thymus DNA and $\Delta\varepsilon$ is the molar absorptive difference.

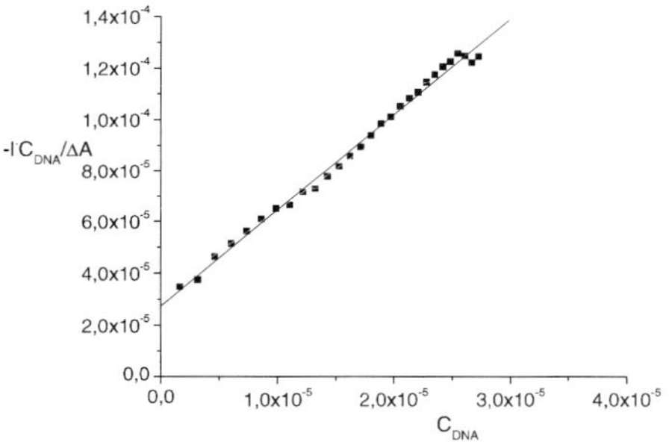

Figure 6.
Scott plot of hedamycin - DNA interaction.

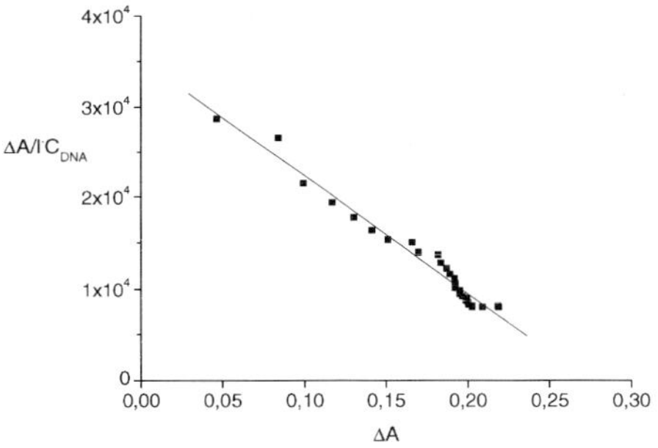

Figure 7.
Scatchard plot of hedamycin - DNA interaction.

equation:[18]

$$\frac{r}{C_F} = (n - r) \cdot K$$

corresponding to a single class of the binding sites. In this relationship, C_F is the free drug concentration, n - the number of binding sites and r is the binding ratio:

$$r = \frac{C_B}{C_{DNA}}.$$

Therefore, the binding constant $K = 1{,}07 \cdot 10^5\ M^{-1}$ and the number of binding sites n = ~0,4 were obtained.

The binding constant may be also evaluated by the methods proposed by Wolfe,[15] Benesi-Hildebrand,[16] Scott[17] and Scatchard.[18] The equations utilized and the results obtained are summarized in Table 1. In the Figures 4–7 are presented the used plots.

Acknowledgements: The authors gratefully acknowledge the financial support by the Education and Research Minister, Romania (Excellence research project - young researchers, no. 56/2006-2008). The authors are also grateful to Jill Johnson from National Cancer Institute (NCI) for the generous gift of hedamycin used in this work.

[1] E. Volanschi, L. E. Vîjan, *Revue Roumaine de Chimie* **2001**, *46*, 163.
[2] L. E. Vîjan, *Rev. Chim. - Bucharest* **2005**, *56*, 527.
[3] L. E. Vîjan, *Rev. Chim. - Bucharest* **2005**, *56*, 655.
[4] L. E. Vîjan, *Rev. Chim. - Bucharest* **2005**, *56*, 735.
[5] L. E. Vîjan, A. Tase, *Rev. Chim. - Bucharest* **2007**, *58*, 199.
[6] L. E. Vîjan, A. Tase, *Rev. Chim. - Bucharest* **2007**, *58*, 628.
[7] L. E. Vîjan, E. Volanschi, M. Hillebrand, *Progr Colloid Polym. Sci.* **2003**, *122*, 67.
[8] M. J. Waring, *Ann. Rev. Biochem.* **1981**, *50*, 159.
[9] X. L. Yang, A. H. J. Wang, *Pharmacology & Therapeutics* **1999**, *83*, 181.
[10] F. Charmantray, A. Duflos, J. Lhomme, M. Demeunynck, *J. Chem. Soc., Perkin Trans.* **2001**, *1*, 2962.
[11] S. J. Lee, L. H. Hurley, *J. Am. Chem. Soc.* **1999**, *121*, 8971.
[12] W. T. Bradner, B. Heinemann, A. Gourevitch, *Antimicrobial Agents Chemother.* **1966**, *6*, 613.
[13] H. L. White, J. R. White, *Biochemistry* **1969**, *8*, 1020.
[14] P. B. Joel, I. H. Goldberg, *Biochim. Biophys. Acta* **1970**, *224*, 361.
[15] A. Wolfe, G. H. Shimer, T. Meehan, *Biochemistry* **1987**, *26*, 6392.
[16] H. Benesi, J. H. Hildebrand, *J. Am. Chem. Soc.* **1949**, *71*, 2703.
[17] R. L. Scott, *Rec. Trav. Chim.* **1956**, *75*, 787.
[18] G. Scatchard, *Ann. N. Y. Acad. Sci.* **1949**, *51*, 660.
[19] G. N. Bennett, *Nucleic acids Research* **1982**, *10*, 4581.

 DOI: 10.1002/masy.200850531

Investigations on Multilayer Films: Electron Microscopy and Infrared Spectroscopy – Possibilities and Limitations

B. Chernev, M. R. Belegratis, E. Ingolič*

Summary: The study of multilayer films has become an important issue, since every year improved materials are produced. They have to be optimally designed in order to enable the maintenance of aroma, taste and nutritional value of the goods. This is primarily achieved by the use of special laminated films, additives and oxygen barriers. In addition, ecological and economical factors force the industry forwards to create more environmental-friendly and cheaper films. In this study, two multilayer packaging films differing in the layer and barrier composition are examined with light microscopy (LM), transmission electron microscopy (TEM) and Fourier transformed infrared (FTIR) microspectroscopy: LM gives the number and the thickness of the layers, but there is no information about the chemical composition and the coatings. FTIR and Raman Spectroscopy enable the chemical characterization of the identified layers, but the lateral resolution is restricted in the μm range. TEM visualizes the important characteristics (coatings, filler particles) and enables the detection of inorganic fillers (EELS, EDX), but the organic components cannot be analyzed. Therefore, the use of complementary techniques is crucial in order to achieve a complete analysis of multilayer foils.
Special attention was paid on the preparation procedure, since ultrathin sections are an important prerequisite for TEM observation and there is little known about the "perfect" preparation of multilayer films.

Keywords: FTIR imaging; microtomy; multilayer

Introduction

Multilayer films are composed of several layers, which vary in thickness, width and coatings. Each layer performs a different function: layers based on polyolefines (i.e. polypropylene, polyethylene, polyamide) convince due to their excellent flexibility and stability, whereas layers of EVOH, PVDC, aluminium or special polyamides provide impermeability and sealing ability.[1]

The final application of multilayered materials is mainly in the area of packaging. Every year, improved multilayer films are demanded to enlarge the range of material application. The desired properties of the final product depend on the composition of the different layers.[2] Most multilayer films use aluminium as a barrier layer. Although aluminium has proved to be an important ally in the fight against waste, energy consumption and environmental damage compared to other materials, there is still a tremendous amount of energy needed for its production. An alternative to aluminium is found in the area of special polyamides or coated polyethylene terephthalates (PET).

In the study of multilayer materials, FTIR spectroscopy is used to identify the different layers, whereas the important characteristics of the structure are examined by TEM, which has generally been proved to be a

Research Institute for Electron Microscopy, Graz University of Technology and Austrian Centre for Electron Microscopy and Nanoanalysis Graz, Steyrergasse 17, 8010 Graz, Austria
E-mail: boril.chernev@felmi-zfe.at

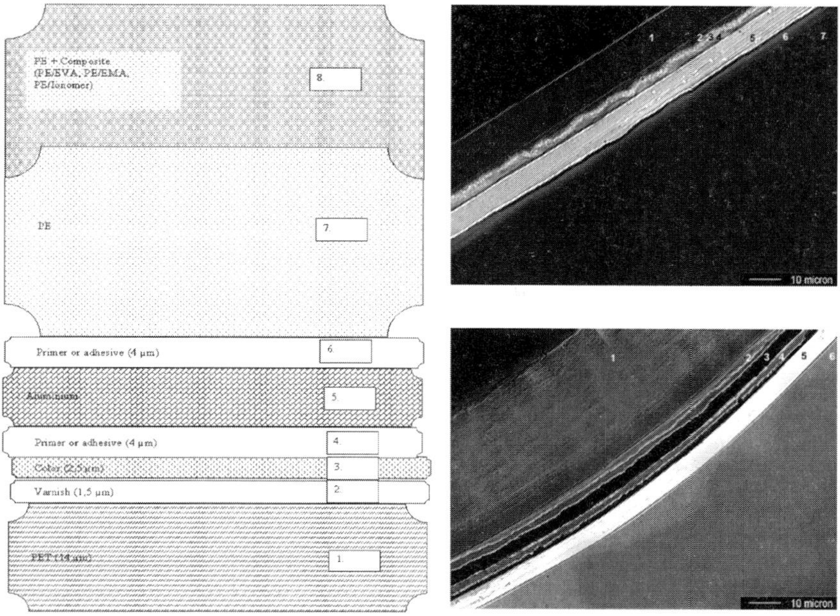

Figure 1.
Schematic representation of the multilayered architecture of the investigated sample (MLF1) and white light micrograph.

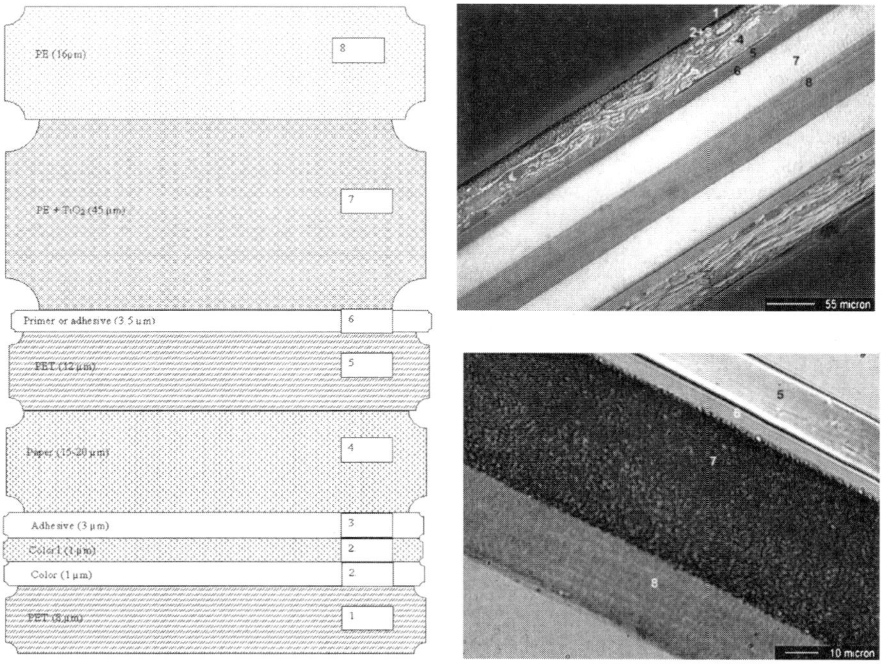

Figure 2.
Schematic representation of the multilayered architecture of the investigated sample (MLF2) and white light micrograph.

promising tool revealing the finest details of internal material structure.[3]

Transmission studies with FTIR spectroscopy require flat semithin cross-sections (2–20 μm thick), whereas specimen preparation for TEM, which is an essential requirement, is a difficult venture. Normally, a series of ultra-thin sections of an equal thickness are produced with an ultramicrotome. However, in the case of multilayer films microtomy encounters its limits. The main problems result from the fact that it is almost impossible to "ideally" cut several adjoined layers of different composition simultaneously, since each layer requires its own cutting requirements. Therefore, even if the ideal embedding material is used and the knife is accurately adjusted for one layer, the other layers are sliced too thickly, might break, curl or peel away. As a consequence, it is almost impossible to view all layers at once.

With FTIR-imaging the problems become a bit more complicated. In order to achieve clear chemical images, the neighbouring layers should consist of components with different characteristic vibrations. Moreover, since the spatial resolution with imaging is

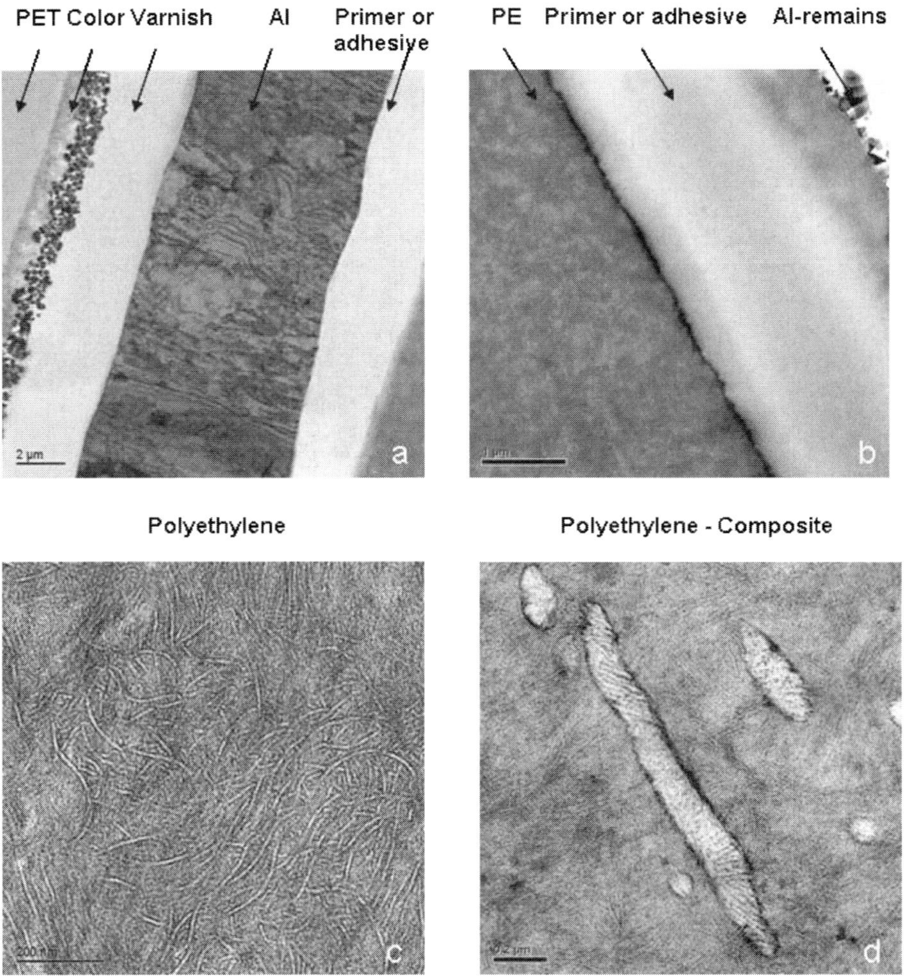

Figure 3.
Transmission electron micrographs from the sample MLF1: a – cryo-ultrathin section (Al-layer present); b – ultrathin section (Al-layer corroded due to RuO_4-staining); c and d – ultrathin sections after RuO_4-staining.

bound to the characteristic wavelength, used to create the chemical image, very thin layers (5 μm and below) are quite difficult to represent and/or identify, even by using the ATR-imaging technique.

Experimental Part

Two different multilayer packaging films (MLF) were tested, varying in thickness and (barrier) layer composition: MLF 1 with aluminium as barrier layer, MLF 2 without aluminium.

Stripes were embedded in two component epoxy embedding resin Epofix®, which was cured for 8 hours at 50 °C. The hardened samples were trimmed and the multilayer materials were cut with an Ultracut E Leica ultramicrotome, using a 45 ° Diatome diamond knife at an angle of 6 °. In cases of difficulties (i.e. detaching or curling foils), the samples were cut with a Leica FCS cryo-ultramicrotome, using a 35 ° diamond knife.

Surface sections and semithin sections were observed by light microscopy (Zeiss Axioplan).

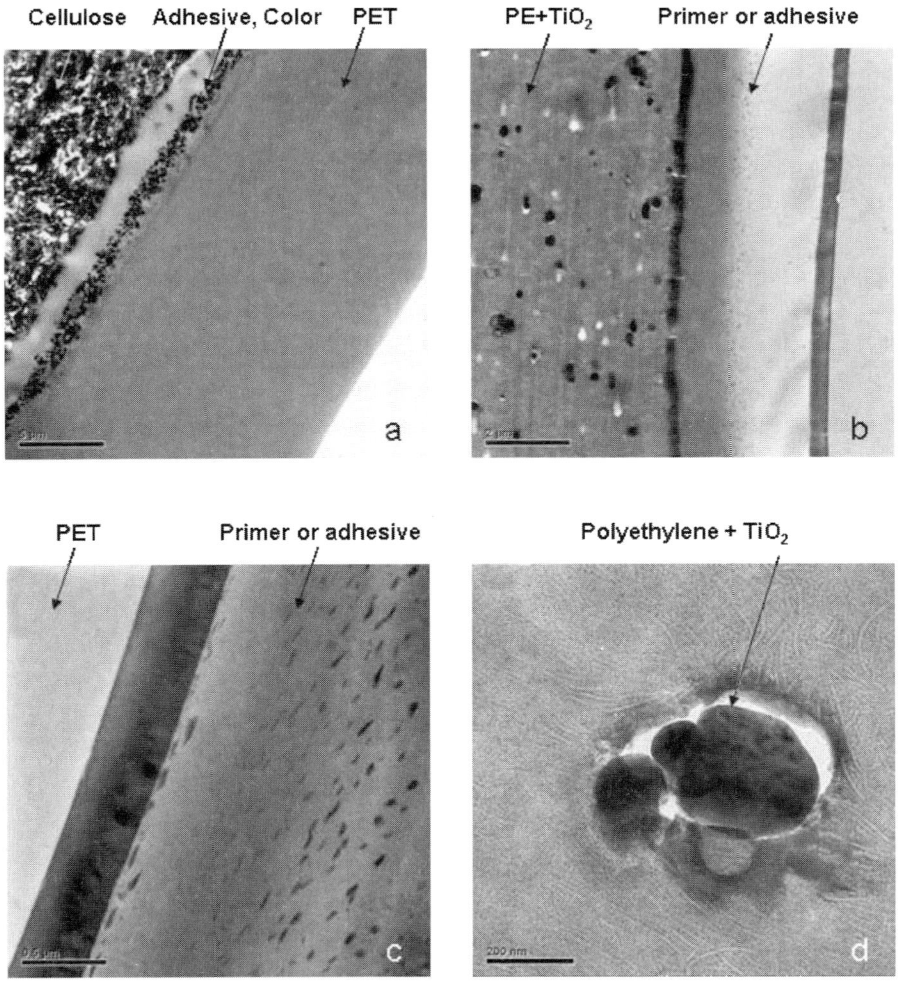

Figure 4.
Transmission electron micrographs from the sample MLF2: a–d – ultrathin sections after RuO_4-staining.

Ultrathin sections were observed with TEM using a FEI Tecnai G^2 electron micro-scope with Gatan Bioscan camera (Model 792).

Semithin sections were analyzed by means of FTIR transmission- or ATR- imaging. The experiments were carried out using an Equinox 55 scientific FTIR spectrometer, coupled to a Bruker Hyperion 3000 FTIR microscope. A cassegrainian x15 mirror objective was used for the transmission experiments and a x20 ATR- objective with Ge-crystal was used for the ATR-imaging experiments. The imaging detector was a 64×64 FPA MCT detector, cooled with liquid nitrogen. Thus, the investigated sample area was 270×270 μm^2 in transmission and $50 \times 50\,\mu m^2$ in ATR mode. Images were collected with typically 16fold accumulation and 4 cm^{-1} spectral resolution in the 4000–900 cm^{-1} spectral range; measuring time per image including background was about 30 minutes. Data evaluation was carried out by using the OPUS® 6.0 software package (Bruker).

Raman spectra were obtained using a Renishaw Ramascope 2000 system, coupled to a Leica DMLM scientific microscope. The used objective was Olympus NPlan 50x NA = 0.75, dispersive grating with 1800 lines/mm, corresponding to a spectral resolution of 4 cm^{-1}. A He-Ne laser (633 nm) with 17 mW output power was used. Typical accumulation time of the spectra was 3×10 seconds.

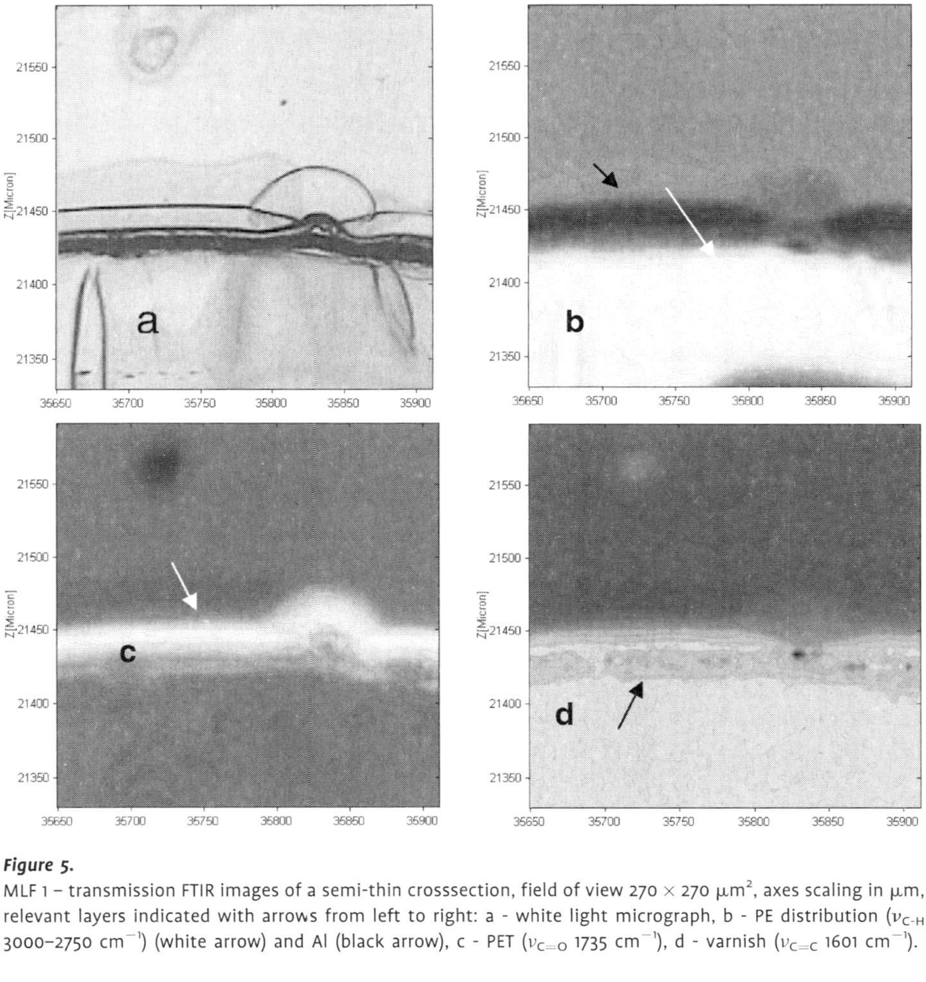

Figure 5.
MLF 1 – transmission FTIR images of a semi-thin crosssection, field of view 270×270 μm^2, axes scaling in μm, relevant layers indicated with arrows from left to right: a - white light micrograph, b - PE distribution (ν_{C-H} 3000–2750 cm^{-1}) (white arrow) and Al (black arrow), c - PET ($\nu_{C=O}$ 1735 cm^{-1}), d - varnish ($\nu_{C=C}$ 1601 cm^{-1}).

Results and Discussion

Light microscopy gives the number and the thickness of the layers, but there is no information about the chemical composition and the coatings. Both multilayer foils consist of 8 layers (Figure 1).

TEM visualizes the important characteristics (coatings, filler particles) and enables the detection of inorganic fillers (EELS, EDX), but the organic components cannot be identified. In addition to the cutting problems and the detached from the adjacent epoxy-embedding, the samples

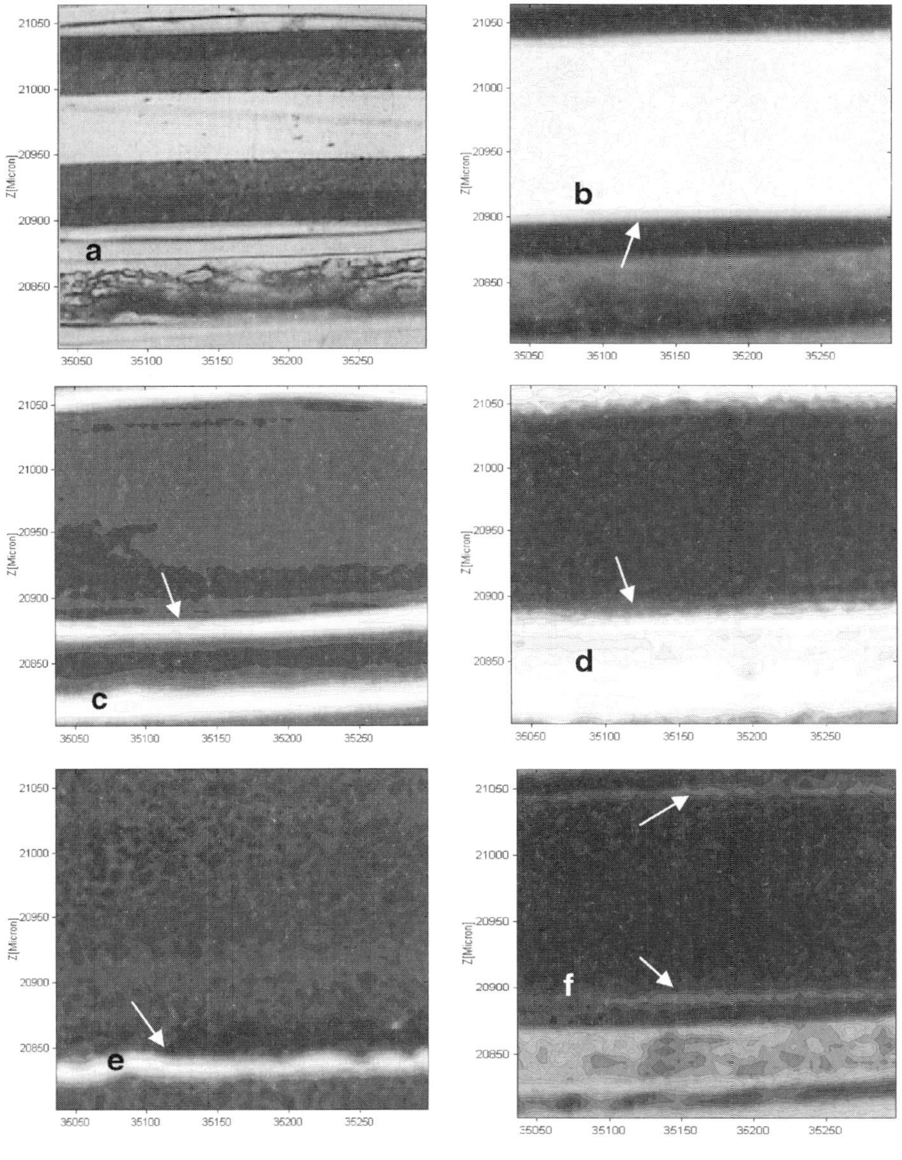

Figure 6.
MLF 1– transmission FTIR images of a semi-thin crosssection, field of view 270 × 270 μm^2, axes scaling in μm^2, relevant layers indicated with arrows from left to right: a – light micrograph, b - PE and PE + TiO_2 (ν_{C-H} 3000–2750 cm^{-1}), c - PET ($\nu_{C=O}$ 1735 cm^{-1}), d - cellulose (ν_{O-H} 3400 cm^{-1}), e - SiO_x (ν_{O-H} 3550 cm^{-1}), f - varnish ($\nu_{C=C}$ 1601 cm^{-1}).

have to be stained with RuO_4 in order to visualize their ultrastructure in TEM. In the case of MLF 1 the aluminium is destroyed. Samples with aluminium are therefore cut unstained with the cryo-ultramicrotome (Figure 3 and 4).

FTIR and Raman microspectroscopy enable the chemical characterization of the identified layers, but the lateral resolution is restricted (5–10 μm for FTIR and 1–2 μm for Raman). Most of the layers in the samples could be resolved in transmission imaging mode, as it is shown in Figure 5 and 6. The overall image quality is sufficiently high, as well as the quality of the extracted spectra, which were used for the materials' identification. It is noted here, that the FTIR imaging is an appropriate tool for layers identification and/or representation, requiring however good quality micrographs, which serve as a help and reference. In this case TEM possesses an unrivalled spatial resolution and serves as a reference for the determination of layers thicknesses

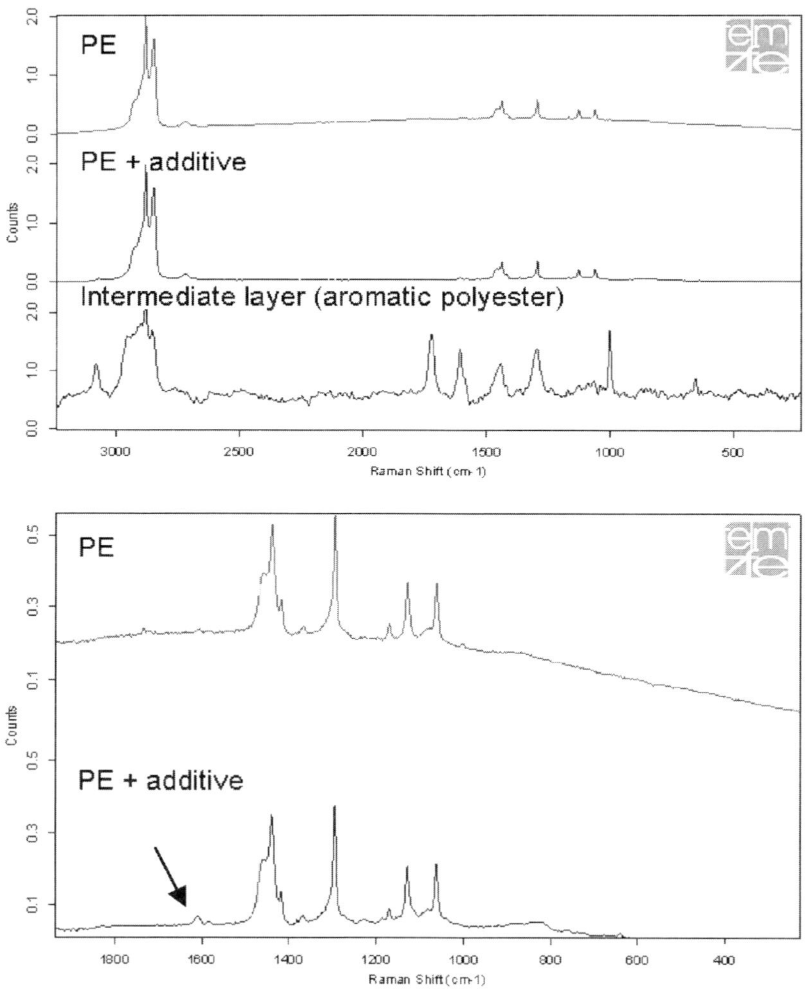

Figure 7.
MLF1: Raman spectra, showing the difference between the two PE-layers note the weak band at 1600 cm^{-1} in the zoomed spectral region on the lower figure, indicated with an arrow; and the identification of the intermediate layers 4 and 6.

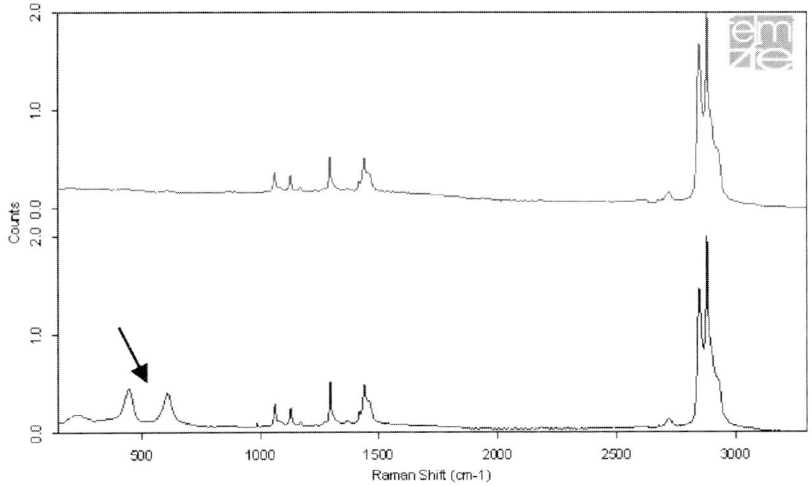

Figure 8.
MLF2: Raman spectra from the PE-layers, proving the presence of rutile (TiO_2); in the lower spectrum there are Raman bands at 612, 449 and 240 cm^{-1} (indicated with arrow).

or the location of certain layer. The latter is quite useful when components with similar spectral features are present in the sample next to each other, like in the case of MLF1 – PET and aromatic polyester as neighbouring layers.

Raman microspectroscopy was useful for the identification of the composite PE material in MLF1 – small band at 1608 cm^{-1}, obviously an additive and/or maleic anhydride grafting – Figure 7. It was already shown by TEM, that in one of the PE-layers of MLF2 a TiO_2 phase was present; this turned out to be the rutile modification – Figure 8.

Since the thinner layers in MLF1 and 2 couldn't be clearly distinguished in the transmission FTIR images, ATR imaging experiments were carried out. The results (Figure 9) show clearly two distinctive layers

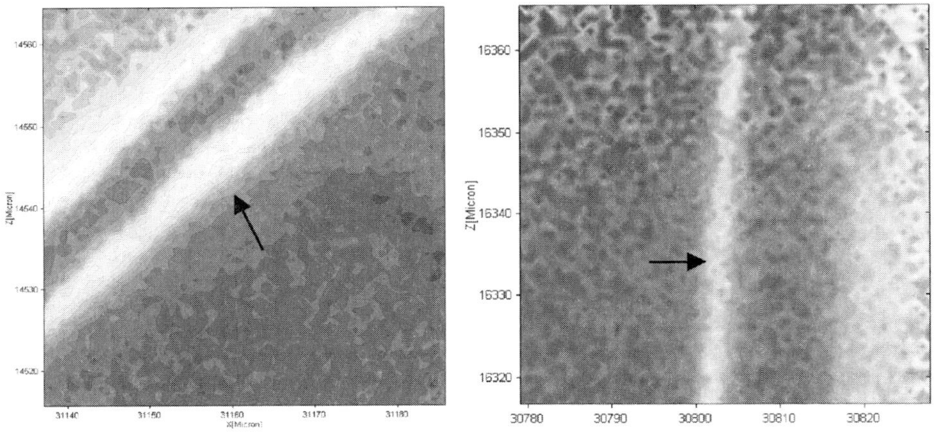

Figure 9.
FTIR-ATR images from the sample MLF1 and MLF2, field of view 50 × 50 μm^2, axes scaling in μm. The relevant layers (layer numbered with 6, cf. Figure 1 and 2) are marked with arrows.

for MLF1 with the Al in-between. The thinner layer consists of aromatic polyester and/or polyurethane (cf. Figure 3 – "Varnish") and is about 5 μm thick; the thicker layer in the figure summarizes two layers (cf. Figure 3 – "Varnish" and "Color") and its thickness is about 9 μm, according to the ATR data. In the ATR-image of MLF2 there is a thin layer of around 3.5 μm thickness; next to the PET layer (cf. Figure 4 "Primer or adhesive"). This is in good agreement with the TEM data.

In conclusion, we were able to show that the investigation and characterization of multilayered materials is quite complicated with respect to the visualization and identification. White light microscopy is an uncomplicated approach, giving enough preliminary information about the layered architecture of the sample. However, TEM is superior and the information thereof is much more detailed and, therefore, helps within the further investigation steps. The spatial resolution of the vibrational spectroscopic methods in their classical form is the limiting factor in that case. The theoretical achievable spatial resolution with ATR-imaging is about 1–2 μm. This would be the case with an "ideal" sample e.g. perfect contact between the crystal and the sample and high spectra heterogeneity in terms of distinctive bands in the low wavelengths region (3000 cm^{-1} and above). Thus, layers with thickness as low as 3 μm, could successfully be resolved with ATR-FTIR imaging. The thinnest layers in Figure 1–3 are too thin and/or possess no distinctive spectral features with respect to their environment, making their identification with imaging spectroscopy an impossible task. The attempt to use Raman microscopy for that purpose did not bring any results because of strong fluorescence. It is however possible that at other excitation wavelengths or even by using FT- Rman spectroscopy the Raman spectra could become strong enough and/or the fluorescence could drop down, however we did not try that because of lack of instrumentation.

[1] J. L. White, D. D. Choi, *Polyolefines. Processing, Structure Development and Properties*, Carl Hanser Verlag, Munich **2004**.
[2] J. Nentwig, *Kunststoff-Folien. Herstellung – Eigenschaften – Anwendung*, Carl Hanser Verlag, München Wien **2006**.
[3] D. S. Tucker, E. J. Jemkins, J. J. Hren, *Journal of Electron Microscopy Technique* **1985**, 2, 29–33.
[4] P. Wilhelm, B. S. Chernev, P. Pölt, G. Kothleitner, K.-J. Eichhorn, G. Pompe, N. Johner, A. Piry, *Spectroscopy Europe* **2004**, *16*(2), 14–19.

Application of FTIR Microscopy in Combinatorial Experimentation on Polymer Blends

Roy l'Abee,[1,2] *Weizhen Li,*[1] *Han Goossens,*[*1,2] *Martin van Duin*[1,3]

Summary: A novel combinatorial, high-throughput experimentation (HTE) setup has been developed, which allows for rapid mapping of the phase behavior of blends of homopolymers and block copolymers. The principle is based on the preparation of composition (ϕ)-temperature (T) gradient films. Linear ϕ gradients were obtained over a large composition range, as shown by FTIR microscopy. The applicability of this combinatorial approach was demonstrated by studying the phase behavior of a poly(styrene-*co*-acrylonitrile) (SAN)/poly(methyl methacrylate-*co*-ethyl acrylate) (PMMA-EA) blend with varying EA content and a poly(styrene-*b*-butadiene-*b*-methyl methacrylate) (SBM) triblock copolymer.

Keywords: block copolymers; FTIR microscopy; high-throughput experimentation; phase behavior; polymer blends

Introduction

Pure polymers are rarely used in practical applications. Instead, polymeric materials generally consist of mixtures of several components, such as polymers, fillers, plasticizers and stabilizers. Blending two or more polymers is a convenient route to develop new polymeric materials, since synergy between the blend components can lead to materials with enhanced properties.[1,2] The material properties of polymer blends are the result of a complex interplay between the individual component properties and the blend morphology. Generally, polymer blends are classified into three categories depending on the thermodynamic miscibility, i.e. immiscible, partially miscible and fully miscible. Although most polymer blends are immiscible, a significant number of polymer systems is known to be partially miscible.[1–3] A key issue in partially miscible polymer blends is the control of the morphology, which is directly related to the thermally-induced demixing behavior of the blend.

It is known that small changes in the molecular structure of the blend components can have a significant influence on the miscibility of polymer blends.[3] Lee et al. showed that by varying just the end group of poly(dimethylsiloxane) (PDMS) from a methyl to an amide group, the upper critical solution temperature of poly(isoprene)/PDMS blends decreases by 165 °C.[4] Van Durme et al. showed that the miscibility of poly(vinyl methyl ether) (PVME) with water is highly influenced by the PVME end group. Compared to methoxy-terminated PVME, hydroxy-terminated PVME shifts the miscibility gap to higher temperatures, whereas bromine-terminated PVME is immiscible with water at room temperature.[5] These are only two of many examples that show the importance of the polymer chain architecture to control the miscibility of polymer blends.

The miscibility of polymer blends is typically studied by light scattering techniques to

[1] Department of Chemical Engineering and Chemistry, Eindhoven University of Technology, PO Box 513, 5600 MB Eindhoven, The Netherlands
E-mail: j.g.p.goossens@tue.nl
[2] Dutch Polymer Institute, PO Box 902, 5600 AX Eindhoven, The Netherlands
[3] DSM Research, PO Box 18, 6160 MD Geleen, The Netherlands

determine the cloud points.[2] However, cloud point measurements are generally time-consuming and fine-tuning of the miscibility gap of polymer blends by small variations in the chemical structure of the blend components is highly impracticable. Because of the expected time reduction, combinatorial methods for the evaluation of polymer blend phase behavior have recently received increasing attention.[6–12] A convenient approach to study the miscibility of polymer blends is based on the preparation of composition-temperature (ϕ-T) gradient films, where one single film provides extensive information on the phase diagram and the phase-separated morphology of the mixture.[6,8,11] The main limitations of the currently reported techniques lie in the achievable ϕ range, which is typically from ~15 to ~65 wt%, the reproducibility of the ϕ gradients, due to manual sampling and deposition methods, and the restriction to use polymers that dissolve at room temperature, which makes the preparation of ϕ gradients of most semi-crystalline polymers impossible.

This paper describes a method to rapidly screen the phase behavior of polymer blends, utilizing a high-throughput experimentation (HTE) setup for the preparation of ϕ-T gradient film libraries. The setup described in this work enables the preparation of ϕ-T gradients over the whole composition range and a temperature gradient between 25 and 300 °C. The experimental reproducibility is excellent, since the sampling and deposition processes are fully automated. The setup and the procedure for the preparation of ϕ-T gradient films are evaluated by using blends of poly(styrene-*co*-acrylonitrile) (SAN) and poly(methyl methacrylate) (PMMA). Subsequently, the method is used to study the phase behavior of SAN/P(MMA-EA) blends with different levels of ethyl acrylate (EA) comonomer and of a poly(styrene*b*-butadiene-*b*-methyl methacrylate) (SBM) triblock copolymer. Especially in blends with one of the homopolymers three regimes of phase separation can be distinguished: microphase separation, macrophase separation, and a combination of both. In the microphase separation regime, phase separation of the blocks of the triblock copolymer occurs upon mixing of one block with the homopolymer. In the macrophase separation regime, the homopolymer is separated from the triblock copolymer, which itself may be in the disordered state. The macro/microphase separation regime can be described by the coexistence of the microphase separated phase and the macrophase separated homopolymer.

Experimental Part

Materials

SAN with 28 wt% acrylonitrile (AN) was obtained from the DOW Chemical Company ($M_n = 41$ kg/mol and a polydispersity of 2.2). P(MMA-EA_x) copolymers were provided by Arkema (France), where x indicates the wt% of ethyl acrylate (EA) comonomer. The samples contained 0.5, 5, 15 and 25 wt% of EA, with a M_n ranging from 40 to 55 kg/mol and a PDI ranging from 2.1 to 3.3. Two SBM triblock copolymers were supplied by Arkema, France, i.e. $S_{20}B_{25}M_{55}$ and $S_{60}B_{10}M_{30}$, where the subscripts represent the weight fraction of the respective blocks. Acetone and methyl ethyl ketone (MEK) were obtained from Sigma Aldrich and used as received.

Characterization Techniques

The composition of the library films was studied by Fourier transform infrared (FTIR) microscopy. Spectra were recorded with a BioRad UMA 500 microscope coupled to a BioRad FTS6000 spectrometer, equipped with a slide-on attenuated total reflection (ATR) crystal (germanium). For each spectrum 100 scans were taken at a resolution of 4 cm^{-1} over a spectral range of 600 to 5000 cm^{-1}. In order to quantitatively determine the composition of the library films, the ratio in peak intensity of the phenyl group of SAN at 698 cm^{-1} and the carbonyl band of P(MMA-$EA_{0.5}$) at 1740 cm^{-1} was compared to a calibration curve.

Light scattering was used to determine the cloud point curves of the SAN/P(MMA-$EA_{0.5}$) system. The set-up consisted of a 1 mW HeNe laser ($\lambda = 632.8$ nm) of which the light was directed through the sample positioned on a Linkam THMS600 hot-stage for temperature control. A light-sensitive diode was placed under a fixed angle of 20 ° with respect to the incident light. The output signal of the diode, which was coupled to a *xy*-recorder, registered the scattered light.

AFM investigations of the morphology of the block copolymer blends were performed by using a Smena P47H (NT-MDT Ltd, Moscow, Russia). The AFM was operated in semicontact mode in air using silicon cantilevers (NSG 11 NT-MDT), coated with a gold layer for a higher laser beam reflectivity. The resonance frequencies applied were 250–270 kHz. The neat copolymer was dissolved in MEK and cast on a glass substrate by evaporating the solvent at room temperature for 2 days. The samples were further dried at 130 °C under vacuum and N_2 for 24 h to ensure complete removal of the solvent.

Results and Discussion

High-throughput Experimentation (HTE) Setup

The preparation of the ϕ gradient films by using the HTE setup (Figure 1) involves several steps. First, each supply vessel is filled with a solution of the respective polymer dissolved in a common solvent. The two syringes can be filled with the solutions, since each syringe is connected to a supply vessel by means of a 6-way valve. After switching the position of the valve, the solutions are transferred from the filled syringes and are deposited from the deposition tip onto a glass slide of 70×70 mm, which is positioned on a movable stage. The speed of both syringes is individually controlled by an in-house developed computer program based on Labview to prepare a gradient in polymer composition during deposition. Simultaneously, the stage moves the glass slide under the deposition tip over a distance of 70 mm in the x-direction. The thin 70 mm long stripe prepared via the deposition procedure is spread out as a film over a distance of 70 mm by moving the

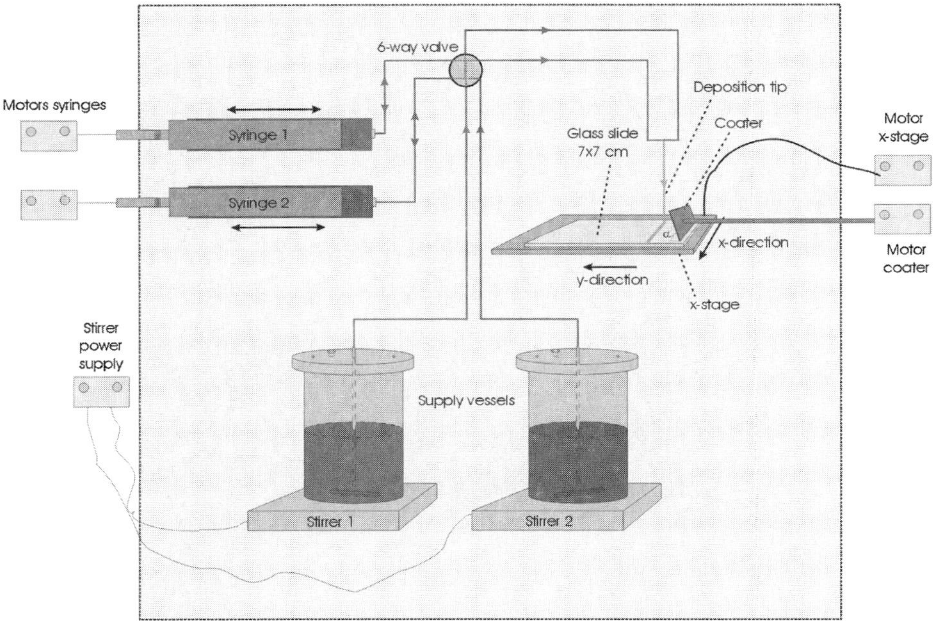

Figure 1.
Schematic representation of the HTE setup.

coater at a constant speed orthogonal to the ϕ gradient direction. The speed of the coater is typically in the order of 10 mm/s, to avoid extensive evaporation of the solvent before completion of the film formation. The distance and angle between the coater and the substrate are adjustable and are typically 200–600 μm and 75–85°, respectively. After evaporation of the solvent, a thin film with a gradient in polymer composition is obtained. Residual solvent is removed by annealing at elevated temperature under a N_2 flow for 1-2 hours. To increase the robustness of the setup, several additional features have been implemented. A delay in the composition gradient can be applied, which reduces the length of the gradient to less than 70 mm. This feature is convenient to exclude disturbances in the gradient close to the edges of the glass slide. The setup is designed to operate at temperatures up to 180 °C, which enables the use of semi-crystalline polymers that do not readily dissolve at room temperature, such as poly(ethylene) and poly(propylene). Furthermore, to enhance mixing of the two components, the dispension tip can be replaced by a static micromixer.

Calibration of HTE Setup

Many parameters affect the quality of the prepared ϕ gradient films, including software settings (e.g. speed of the syringes, x-stage and coater), hardware settings (e.g. coater distance and angle), as well as the properties of the polymer solutions (e.g. type of solvent, polymer concentration and solution viscosity). Optimization of those parameters was performed by preparing ϕ gradient films based on mixtures of SAN and P(MMA-$EA_{0.5}$). FTIR microscopy was used to determine the exact composition of the ϕ gradient. A calibration curve was constructed using SAN/P(MMA-$EA_{0.5}$) mixtures with pre-determined compositions.

Figure 2 shows the FTIR spectra of the SAN/P(MMA-$EA_{0.5}$) blends with increasing SAN content. The spectra show a clear decrease in the absorption band at 1725 cm^{-1}, which is assigned to the carbonyl (C=O) stretching vibration of the methacrylate group in P(MMA-$EA_{0.5}$), and an increase in the absorption band at 698 cm^{-1}, which is assigned to the out-of-plane ring bending vibration of the phenyl group of SAN. The inset shows the peak positions of the carbonyl

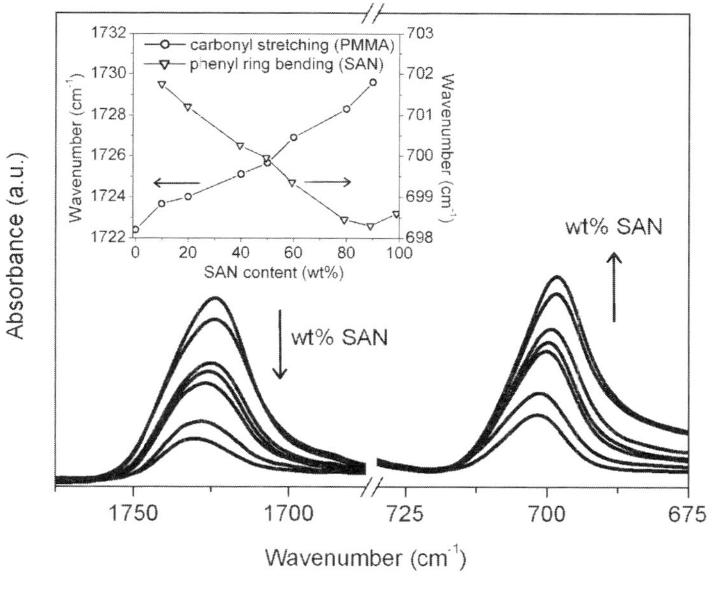

Figure 2.
FTIR spectra of SAN/P(MMA-$EA_{0.5}$) blends with increasing SAN content from 10, 20, 40, 50, 60, 80 to 90 wt%, indicated by the arrow.

stretching vibration of PMMA and the phenyl ring bending vibration of SAN, both as a function of the blend composition. Upon increasing the P(MMA-EA$_{0.5}$) content in the blend, the carbonyl peak shifts to lower wavenumbers, while the phenyl peak shifts to higher wavenumbers, indicating the existence of electron transfer interactions between the phenyl groups in SAN and the carbonyl groups in PMMA, which may be partially responsible for the miscibility of the blend.[13,14] This interaction is formed partially at the expense of the disruption of electron transfer complexes formed between the styrene and acrylonitrile comonomers within SAN, since the C≡N stretching vibration at 2238 cm^{-1} shifts to slightly higher wavenumbers.[15]

Based on the data presented in Figure 2, a calibration curve was constructed. The blend composition was plotted as a function of the intensity ratio of the representative SAN and P(MMA-EA$_{0.5}$) absorption bands (inset Figure 3). The experimental data can be reasonably well fitted by an allometric-type function.[16] Based on this calibration curve, the ϕ gradients of SAN/P(MMA-EA$_{0.5}$) as prepared with the HTE setup can now be determined. The HTE setup was used with the optimized settings to prepare ϕ gradient coatings, based on 8 wt% solutions of SAN and P(MMA-EA$_{0.5}$) in acetone. A maximum syringe speed of 0.85 mm/s, a table speed of 5.5 mm/s and a coater speed of 10 mm/s were used, in combination with a delay of 5 mm. By using these settings, the whole film preparation takes less than 20 seconds. Figure 3 shows that a fairly linear composition gradient is obtained in the x-direction. FTIR microscopy showed that the composition gradient is the same in the y-direction, which indicates that no diffusion or turbulent mixing occurs during the coating procedure. As determined by interferometry, the thickness of the coating typically varied from 1 to 1.8 μm in the direction orthogonal to the ϕ gradient.

Application of the HTE Setup to Blends of SAN/P(MMA-EA$_x$)

While traditional scattering experiments are very time-consuming, combinatorial experimentation is a much more effective tool to rapidly map the phase behavior of

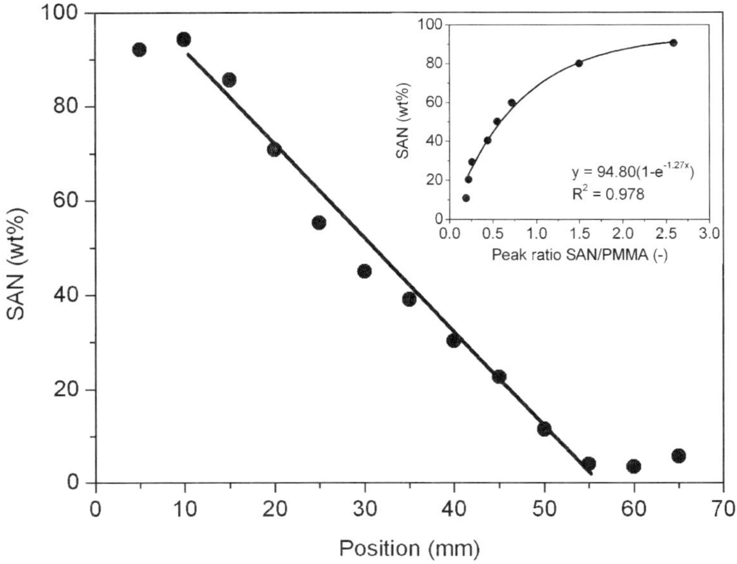

Figure 3.
ϕ–gradient of SAN/P(MMA-EA$_{0.5}$) prepared with the HTE setup. The inset shows the calibration curve based on discrete SAN/P(MMA-EA$_{0.5}$) blends.

polymer blends. Since the phase behavior of SAN/PMMA homopolymer blends was extensively studied in the past, it is a convenient system to validate the HTE setup.[17,18] The SAN/PMMA blend is miscible at room temperature and phase separates at elevated temperature, i.e. the system exhibits a lower critical solution temperature (LCST) behavior. Phase separation typically occurs far above the glass transition temperature (T_g) of the two components. This is convenient, since the phase-separated structure is frozen in after decreasing the temperature below T_g, which facilitates the off-line study of the phase-separated morphology as a function of temperature and composition. To demonstrate the applicability of the HTE setup, the phase behavior of SAN/P(MMA-EA_x) blends with a variation in EA content (x) is presented. Small amounts of EA are typically incorporated into PMMA to prevent unzipping of the polymer at elevated temperatures, but the influence of the EA content on the miscibility with SAN has not been documented.

70 × 70 mm coatings with a ϕ gradient in one direction were prepared. Subsequently, the films are annealed on a temperature (T) gradient heating stage, consisting of a metal block that is heated on one side and cooled on the other side. This yields a linear T gradient with a maximum temperature range from 25 to 300 °C. Upon placing the glass slide with the ϕ gradient film orthogonally onto the T gradient stage, a two-dimensional ϕ-T library is created. Typically, a T gradient over a range of 80 °C is applied under a N_2 flow to prevent oxidative degradation of the polymers. Subsequently, the phase-separated morphology is studied by optical microscopy. Figure 4 shows the phase diagram of SAN/P(MMA-EA_x) as obtained from the ϕ-T gradient films. Traditional cloud point measurements (CPMs) were performed on SAN/P(MMA-$EA_{0.5}$) blends to verify the phase behavior as determined from the ϕ-T gradient films. Figure 4 shows that the LCST behavior of the ϕ-T gradient library convincingly matches the CPMs and fully agree with data presented in literature for SAN/PMMA blends.[17,18] Incorporation of EA into PMMA leads to a significant shift of the binodal to lower temperatures. This result is consistent with the work of Paul et al., who visually determined the decreased

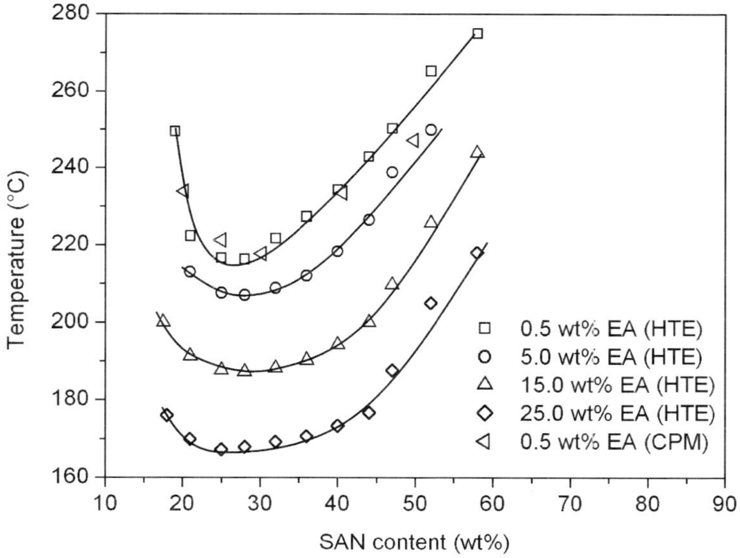

Figure 4.
Phase diagrams of SAN/P(MMA-EA_x) blends as determined from ϕ-T gradient films.

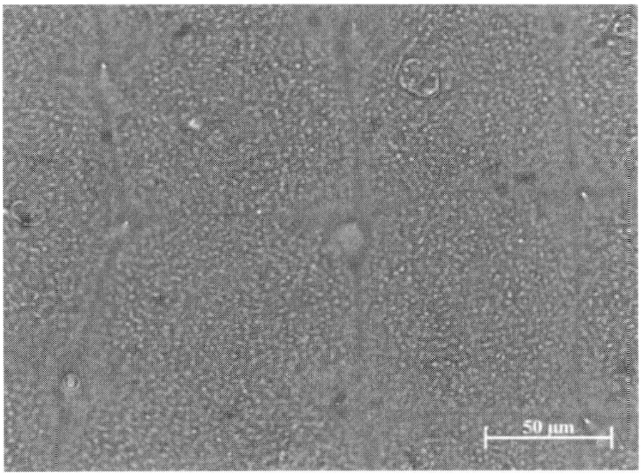

Figure 5.
OM image of a PMMA/$S_{20}B_{25}M_{55}$ blend prepared with the HTE setup.

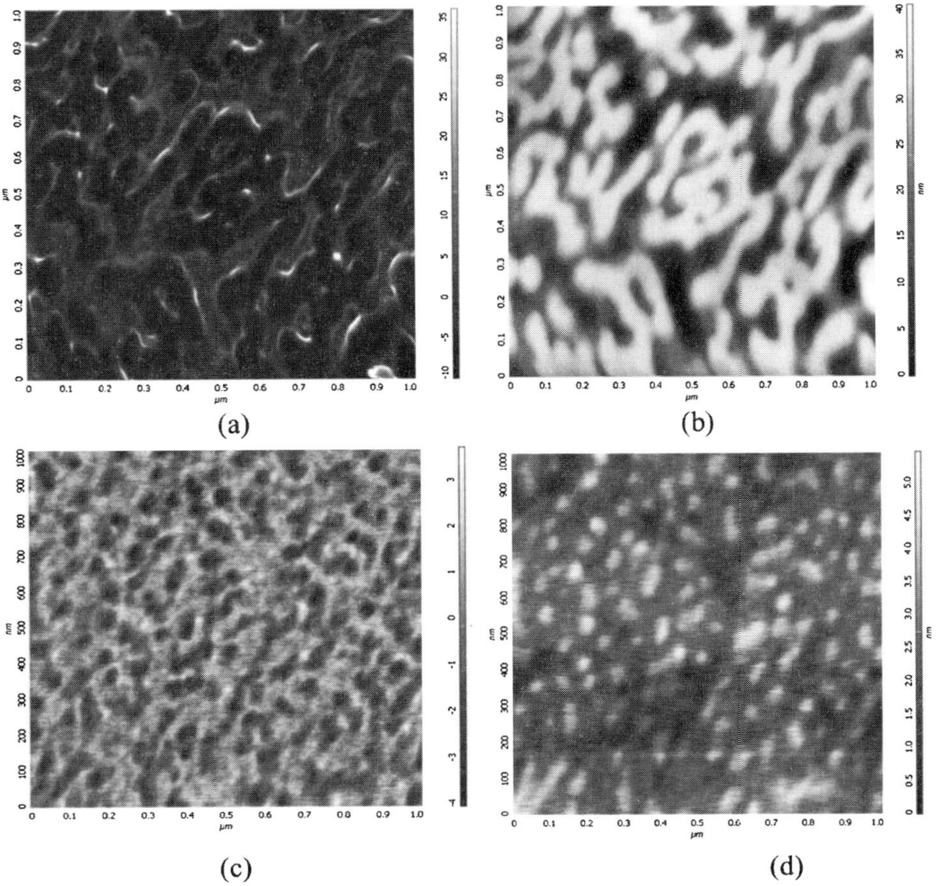

Figure 6.
AFM images of $S_{20}B_{25}M_{55}$: a) phase, b) height and $S_{60}B_{10}M_{30}$: c) phase, d) height.

miscibility of SAN/PMMA upon incorporation of EA from discrete polymer blends.[19]

Application of the HTE Setup to SBM Triblock Copolymer Systems

The HTE setup was also used to study the morphology of triblock copolymer blends. In Figure 5, the OM image of a PMMA/ $S_{20}B_{25}M_{55}$ blend is shown. It is evident that the film is phase-separated. The spatial resolution of OM limits the observation to the level of macrophase separation. To get insight in the other phase separation regimes (micro- and micro-/macrophase separation), techniques with a higher spatial resolution are required, such as TEM or AFM, of which the latter has the advantage of being a surface technique. However, both techniques lack the possibility of chemical identification. The linearity of the ϕ-gradient over the 70 mm-film was checked by using FTIR microscopy using the ratio of the carbonyl (C=O) stretching vibration of the methacrylate group and the out-of-plane ring bending vibration of the phenyl group of PS for quantification.

The morphology, visualized by AFM, of the as-received $S_{20}B_{25}M_{55}$ copolymer (Figures 6a and b) shows an imperfect lamellar morphology, because of the presence of SB-diblock copolymer impurities.[20] For the $S_{60}B_{10}M_{30}$ system, PS forms the matrix and the PMMA cylinders are covered by little PB spheres (see Figure 6c and d). Future work will be focused on the influence of ϕ and T on the micro- and macrophase separation of homopolymer/ triblock copolymer (PMMA/SBM) blends by applying the HTE setup.

Conclusions

A HTE setup has been developed to prepare ϕ gradient films of polymer blends. In combination with a T gradient heating stage, ϕ-T gradient films were prepared, yielding fast and detailed information on miscibility, demixing processes and final morphologies of polymer blends. The HTE setup has several advantages over the high-throughput techniques reported in literature so far: larger composition and temperature ranges, full automation of the preparation of the ϕ-T gradients and the possibility to extend the gradient preparation to elevated temperatures. FTIR microscopy has been used to show that linear ϕ gradients are obtained. FTIR microscopy is also the preferred technique to study the composition of the different phases after the temperature-induced phase separation, although time consuming. The recently developed macroATR accessory might be interesting to enable a much faster screening.[21] For the block copolymer systems, the spatial resolution of FTIR microscopy is far from sufficient, but the recent developments in combining AFM with either Raman spectroscopy[22] or with scanning near-field optical microscopy (SNOM)[23] are promising. Although this paper only discusses the phase behavior of polymer blends, the HTE setup is a promising tool to study other systems such as reactive polymer blends, block copolymer (dis)assembly processes and (nano)particle-controlled phase behavior of polymer blends.

Acknowledgements: This work is part of the Research Programme of the Dutch Polymer Institute (DPI) under project number 537 and the Dutch Technology Foundation (STW) (project 07057).

[1] D. R. Paul, C. B. Bucknall, "*Polymer Blends*", Wiley, New York **2000**.
[2] L. A. Utracki, "*Polymer Blends Handbook*", Kluwer Academic Publishers, Dordrecht **2002**.
[3] K. F. Feed, J. Dudowicz, in "*Phase Behavior of Polymer Blends*", K. F. Feed, Ed., Springer, Berlin **2005**, p. 63–126.
[4] M. H. Lee, C. A. Fleischer, A. N. Morales, J. T. Koberstein, R. Koningsveld, *Polymer* **2001**, *42*, 9163.
[5] K. van Durme, B. van Mele, K. V. Bernaerts, B. Verdonck, F. E. Du Prez, *J. Polym. Sci. Part B.* **2006**, *44*, 461.
[6] J. C. Meredith, A. Karim, E. J. Amis, *Macromolecules* **2000**, *33*, 5760.
[7] J. T. Cabral, A. Karim, *Meas. Sci. Technol.* **2005**, *16*, 191.
[8] J. C. Meredith, A. Karim, E. J. Amis, *MRS Bull.* **2002**, *27*, 330.

[9] A. P. Smith, J. F. Douglas, J. C. Meredith, E. J. Amis, A. Karim, *J. Polym. Sci. Part B.* **2001**, *39*, 2141.
[10] A. P. Smith, J. F. Douglas, J. C. Meredith, E. J. Amis, A. Karim, *Phys. Rev. Lett.* **2001**, *87*, 15503.
[11] T. A. Dickinson, D. R. Walt, J. White, J. S. Kauer, *Anal. Chem.* **1997**, *69*, 3413.
[12] J. W. Gilman, S. Bourbigot, J. R. Shields, M. Nyden, T. Kashiwagi, R. D. Davis, D. L. Vanderhart, W. Demory, C. A. Wilkie, A. B. Morgan, J. Harris, R. E. Lyon, *J. Mat. Sci.* **2003**, *38*, 4451.
[13] K. Naito, G. E. Johnson, D. L. Allara, T. K. Kwei, *Macromolecules* **1978**, *11*, 1260.
[14] G. Das, A. N. Banerjee, *J. Appl. Polym. Sci.* **1997**, *63*, 991.
[15] H. Feng, L. Shen, Z. Feng, *Eur. Polym. J.* **1995**, *31*, 243.
[16] F. Picchioni, J. G. P. Goossens, M. van Duin, P. Magusin, *J. Appl. Polym. Sci.* **2003**, *89*, 3279.
[17] M. Suess, J. Kressler, H. W. Kammer, *Polymer* **1987**, *28*, 957.
[18] M. Nishimoto, H. Keskkula, D. R. Paul, *Polymer* **1989**, *30*, 1279.
[19] J. H. Chu, D. R. Paul, *Polymer* **1999**, *40*, 2687.
[20] S. Ritzenthaler, F. Court, E. Girard-Reydet, L. Leibler, J. P. Pascault, *Macromolecules* **2003**, *36*, 118.
[21] S. G. Kazarian, K. L. A. Chan, *Prog. Coll. Pol. Sci.* **2006**, *132*, 1.
[22] S. Nie, S. R. Emory, *Science* **1997**, *275*, 1102.
[23] D. W. Pohl, *Thin Solid Films* **1995**, *264*, 250.

 DOI: 10.1002/masy.200850533

An FTIR Study on the Solid-State Copolymerization of bis(2-hydroxyethyl)terephthalate and Poly(butylene terephthalate) and the Resulting Copolymers

Maya Ziari,[1,2,3] *Otto van Asselen,*[2] *Martijn Jansen,*[2,3] *Han Goossens,**[2]
Peter Schoenmakers[1]

Summary: The first aim of this work was to study the solid-state copolymerization (SSP) of bis(2-hydroxyethyl)terephthalate (BHET) with poly(butylene terephthalate) (PBT) by FTIR spectroscopy. The development of the chemical microstructure during the SSP-reaction was examined as a function of the BHET content, showing the different regimes. The thermal behaviour of the resulting copolymers with different BHET contents was also investigated during cooling using infra-red dynamic spectra. For low BHET-concentrations, only crystallization of PBT-sequences was observed, while for high BHET-concentrations, only crystallization of PET-sequences was detectable with a cross-over behaviour for intermediate concentrations.

Keywords: FTIR spectroscopy; polyesters; solid-state modification

Introduction

Poly(butylene terephthalate) (PBT) and poly(ethylene terephthalate) (PET) are two semi-crystalline polymers that are used in many engineering applications. The main advantage of PBT is its high crystallization rate, making it suitable for injection molding applications with short molding cycle times,[1] while PET has a considerably lower crystallization rate and is therefore mainly used for fiber applications and packaging.[2] The glass transition temperature of PBT is rather low compared to PET ($T_g \approx 45\,^\circ C$ for PBT and $T_g \approx 80\,^\circ C$ for PET). To obtain a T_g higher than 45 °C, PBT and PET can be reactively blended in the melt. The concomitant ester-interchange reactions occurring in the melt first result in the formation of block copolymers, but as the reaction proceeds, random PBT-PET copolymers are obtained.[3–5] These random PBT-PET copolymers have a T_g in between that of PBT and PET homopolymer. However, the shorter and more irregular homopolymer sequences of these random copolymers consequently lead to a lower melting temperature, crystallization rate and crystallinity with respect to pure PBT. It is therefore desirable to synthesize PBT-PET copolymers having a T_g higher than PBT homopolymer, but with a crystallization behavior comparable to that of PBT. Therefore, a copolymerization method should be used, which enables to retain large crystallizable homopolymer PBT blocks with a non-random chemical microstructure.

In our previous work,[6] PBT-PET copolymers were made by copolymerization in the melt (MP) and the solid state (SSP) of bis(2-hydroxyethyl)terephthalate (BHET) with PBT. The chemical microstructure of the synthesized PBT-PET copolymers was examined by ^{13}C-NMR spectroscopy sequence distribution analysis. As expected, the chemical microstructure of the PBT-PET

[1] Polymer-Analysis group, Van 't Hoff Institute for Molecular Sciences, University of Amsterdam, Nieuwe Achtergracht 166, 1018 WV Amsterdam, The Netherlands
[2] Department of Chemical Engineering and Chemistry, Eindhoven University of Technology, PO Box 513, 5600 MB Eindhoven, The Netherlands
E-mail: j.g.p.goossens@tue.nl
[3] Dutch Polymer Institute, PO Box 902, 5600 AX Eindhoven, The Netherlands

copolymers obtained by MP was fully random. However, when BHET was incorporated into PBT via SSP, a non-random blocky chemical microstructure was obtained. When the fraction of BHET used for incorporation in PBT was increased, the blocky character of the chemical microstructure became more pronounced, suggesting that BHET may react by self-condensation to form homopolymer PET blocks.[7,8] The ^{13}C-NMR sequence distribution analysis also showed that transesterification reactions occurred between BHET monomer and PBT. It was not clear if the formed homopolymer PET blocks are present as a separate phase or if these blocks reside in the amorphous phase of PBT.

In this paper, the incorporation of BHET into PBT via SSP is studied in more detail. The kinetics of the incorporation via SSP is studied under isothermal conditions by using FTIR spectroscopy. During the SSP process, a competition between annealing and dissolution of the PBT-crystals, incorporation of BHET in the amorphous PBT and BHET homopolymerization exists, of which the relative rates depend on the ratio of PBT/BHET. In a previous paper, the miscibility of the BHET with PBT and the thermal properties of the resultant BD_xEG_y-copolymers obtained via SSP were investigated using DSC.[9] BD_x denotes the mol% of PBT (expressed in 1,4-butanediol units), whereas EG_y denotes the mol% of BHET units. It was observed that BHET is partially miscible in the amorphous PBT, depending on the PBT/BHET ratio. The crystallization behaviour of the resultant BD_xEG_y copolymers, where BD_x represents the initial mol% PBT and EG_y the mol% of hydroxyl end groups per mole BHET, was studied by DSC and it was observed that when more BHET is incorporated the onset of crystallization shifts to lower temperatures, while the crystallization exotherm becomes broader.[9] However, the $(BD_{30}EG_{70})_{SSP}$ copolymer displayed a much sharper crystallisation exotherm with a slightly higher onset. A similar observation was made by Misra and al.[10] However, DSC lacks the possibility to investigate the crystallization behaviour of the individual components if the crystallization exotherms display a large overlap. In this respect, temperature-dependent FTIR spectroscopy is much more discriminating by following absorption bands, which can either be assigned to the crystalline or amorphous phase of PBT or PET.

The SSP kinetics was investigated under isothermal conditions using two compositions. In that way, the development of the chemical microstructure during the SSP reaction could be examined together with the miscibility of BHET and PBT. Temperature-dependent experiments were performed on three different BD_xEG_y copolymers and compared to PBT using the dynamic infra-red spectra obtained through the 2D IR correlation algorithm developed by Noda et al.[11,12]

Experimental Part

The BDEG copolymers were obtained by incorporation of BHET into PBT by using SSP.[6,9]

At first, the kinetics of the SSP reactions was investigated with attenuated total reflection (ATR)-FTIR spectroscopy. A BioRad FTS3000 spectrometer equipped with a MCT detector was used to record the spectra. The samples were pressed onto the diamond window of a SPECAC Golden Gate ATR accessory. Next, the sample was heated to 120 °C, where the initial spectrum was recorded. Upon starting the experiment, the sample was heated to 180 °C using a heating rate of 30 °C/min. Spectra were recorded during the reaction with a time interval of 60 seconds using a spectral resolution of 4 cm^{-1}, co-adding 62 scans. At 180 °C, the feed consists of a semi-crystalline fraction of PBT and a liquid fraction of BHET. During the initial stage of the SSP-reaction, the fraction of the liquid BHET in contact with the crystal increases, while the fraction of PBT decreases. It becomes then very difficult to monitor the reaction quantitatively.

Therefore, it was decided to perform all experiments in the transmission mode. Then, a more or less constant cross-section of the sample is monitored. These experiments were performed using a BioRad UMA500 IR microscope coupled to a BioRad FTS6000 spectrometer. A Linkam TMS600 hot stage was used for temperature control. To obtain a thin sample, the powder samples were cold pressed by a diamond anvil cell. Next, the sample was placed in between two zinc selenide (ZnSe)-windows and transferred to the Linkam hot stage. Spectra were recorded with a resolution of 2 cm^{-1} co-adding 250 scans.

The crystallization behaviour of three BDEG copolymers were also followed using the same set-up as described above with a cooling rate of 10 °C/min. The samples, however, were not covered with a second ZnSe window. For these experiments, spectra were recorded using a resolution of 2 cm^{-1} co-adding 100 scans. The 2D-Pocha software, developed at the Kwansei Gakuin University (Japan), was used for data treatment. The spectrum at 180 °C was taken as the mean spectrum for the dynamic spectra calculations.

Results and Discussion

Assignment of the Absorption Bands of the Pure Components

Table 1 shows the assignment of most relevant absorption bands of PBT, PET and BHET. For crystalline PBT, a band at 1454 cm^{-1} (CH_2 bending) is observed that for amorphous PBT shifts to 1468 cm^{-1} together with absorption bands at 1320 cm^{-1} (ring ester in plane) and 1389 cm^{-1} (CH_2 wagging). Crystalline PET shows characteristic absorption bands at 1471 cm^{-1} (CH_2 bending), 1342 cm^{-1} (CH_2 wagging), and at 990 cm^{-1}, of which the latter band is only observed if chain folding occurs. Typical bands of amorphous PET can be observed at 1371 cm^{-1} (CH_2 wagging), 1458 cm^{-1} (CH_2 bending). For BHET, a lot of absorption bands overlap with PBT or PET. To follow the copolymerization, the absorption band at 3550 cm^{-1} (OH stretching) can be used.

Kinetics of the SSP Reaction Studied by FTIR Spectroscopy

Since the spectra of PBT, BHET and PET show overlapping absorption bands, we decided to focus on the four following absorption bands: the crystalline PET absorption band at 1342 cm^{-1}, the absorption band at 1371 cm^{-1} of amorphous PET or BHET, the crystalline PBT band at 1454 cm^{-1} and the OH-stretch vibration band of BHET at 3550 cm^{-1}.

The SSP-kinetics at 180 °C was followed for the $(BD_{70}EG_{30})_{feed}$ and $(BD_{30}EG_{70})_{feed}$ systems. Figures 1a and 1b show the individual spectra before and after the SSP-reactions, while the kinetic results are shown in Figures 2a and 2b, respectively. The disappearance of the absorption band at 3550 cm^{-1} shows that in both cases the BHET is converted. For the $(BD_{70}EG_{30})_{feed}$ system (Figure 2a), the intensity decreases gradually and the behaviour is typical for a series of two irreversible consecutive first-order reactions. This was also observed for the incorporation of 2,2-bis[4-(2-hydroxyethoxy)phenyl]-propane (Dianol 220).[14] The $(BD_{30}EG_{70})_{feed}$ system (Figure 2b) shows a

Table 1.
Assignment of the absorption bands of the pure components.[13]

	PBT (cm^{-1})		PET (cm^{-1})		BHET (cm^{-1})
	amorphous	crystalline	amorphous	crystalline	
OH stretching					3550
CH_2 bending	1468	1458/1454	1458/1440	1471	1453
Ring-in-plane def	1408	1409	1409	1409	1407
Ring CH in-plane def				1380	
CH_2 wagging	1389	1389	1371	1342	1372
Ring CCH			1315		
Chain folding		1320		990	

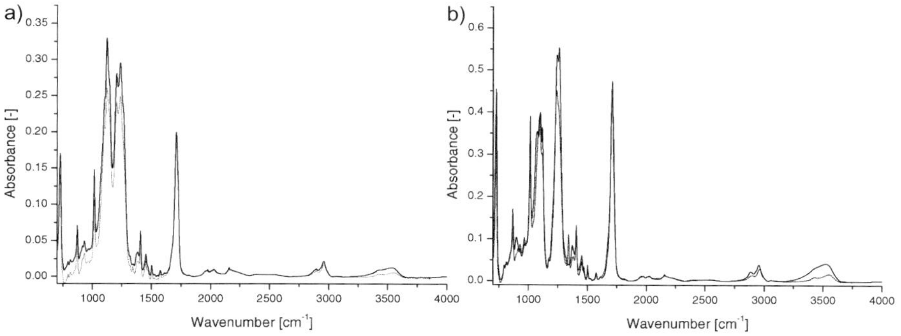

Figure 1.
Spectra before (-) and after (..) the SSP-reaction: a) $(BD_{70}EG_{30})_{feed}$ and b) $(BD_{30}EG_{70})_{feed}$.

different behaviour. After a similar gradual decrease of the intensity up to approx. 20 min, the reaction accelerates and finally reaches a plateau level after approx. 50 min. In both figures, the absorption band of the PET CH_2 wagging mode in the crystalline state is also depicted, although the band position changes from at 1340 to 1342 cm^{-1} on going from $(BD_{70}EG_{30})_{feed}$ to $(BD_{30}EG_{70})_{feed}$. Cole et al. studied the microstructure of PET homopolymer using reflection and transmission FTIR spectroscopy and showed that the ethylene glycol segments for PET in the crystalline phase are always in the mutual trans orientation (T), while in the amorphous phase a mutual gauche orientation predominantly exists (G).[13,15] They also observed that different conformation states can also exist for the C–O bond of the glycol group, which can likewise exist in a trans (t) or gauche (g) conformation as well as the terephthalic acid segment. If the latter is in a non-trans conformation and the ethylene glycol segment is in a trans (T) conformation, a "rigid" amorphous band will be observed at 1340–1338 cm^{-1}. X-ray diffraction measurements showed that all three groups adopt the trans conformation (T, t, T_b) in the crystalline phase. Then, an absorption band at 1342 cm^{-1} of the glycol C-C unit will appear. The change in peak position is a strong indication that for the $(BD_{70}EG_{30})_{feed}$ system the BHET is incorporated in the PBT chain, while for the $(BD_{30}EG_{70})_{feed}$ system separate PET crystals can be formed. The difference in behaviour is also found for the 1371 cm^{-1} band. For the $(BD_{30}EG_{70})_{feed}$ system, the same intensity profile is found as for the OH-stretching vibration band of BHET

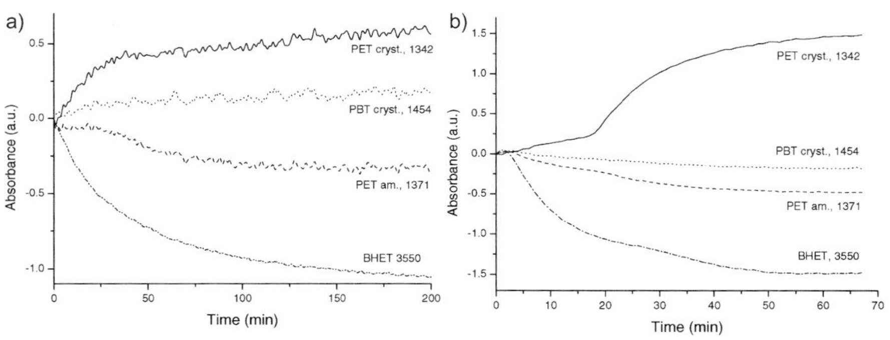

Figure 2.
Relative intensities of the indicated absorption bands of crystalline PET, amorphous PET, PBT and BHET during SSP of: a) $(BD_{70}EG_{30})_{feed}$ and b) $(BD_{30}EG_{70})_{feed}$.

(3550 cm^{-1}), but $(BD_{70}EG_{30})_{feed}$ system remains at the same absorption level for the first 30 min and then starts to decrease up to approx. 40 min followed by a plateau. This might indicate that during the first 30 min, the decrease of the intensity due to the conversion of BHET is compensated by the formation of amorphous PET, which also has an absorption band at that wavenumber.

For the $(BD_{30}EG_{70})_{feed}$ system, the crystalline PBT band at 1454 cm^{-1} decreases, probably because of the decrease of the BHET that also has an absorption band at 1454 cm^{-1}, although this decrease may also be attributed to partial dissolution of the PBT-crystals that may occur because of the high mole fraction of BHET acting as a solvent for PBT. On the contrary, for the $(BD_{70}EG_{30})_{feed}$ system, the band at 1454 cm^{-1} increases. This behaviour can only be explained by the occurrence of annealing and perfectioning of the PBT-crystals due to the high mobility during the SSP-reaction.

Temperature Dependent Experiments

Changes in the FTIR-spectra were monitored as a function of temperature. Figure 3 shows the dynamic spectra of BDEG copolymers with different PBT/BHET ratios cooled from the melt in a temperature range from 240 to 170 °C. For the PBT homopolymer, positive bands at 1454, 1386 and 1320 cm^{-1} develop during cooling. These bands can be assigned to crystalline PBT (CH_2 bending and wagging modes). Also the positive band at 1408 cm^{-1} can be assigned to amorphous PBT (See Table 1). In the spectra of the $(BD_{70}EG_{30})_{SSP}$ system, beside the positive bands of crystalline PBT, also a positive band is observed at 1338 cm^{-1},

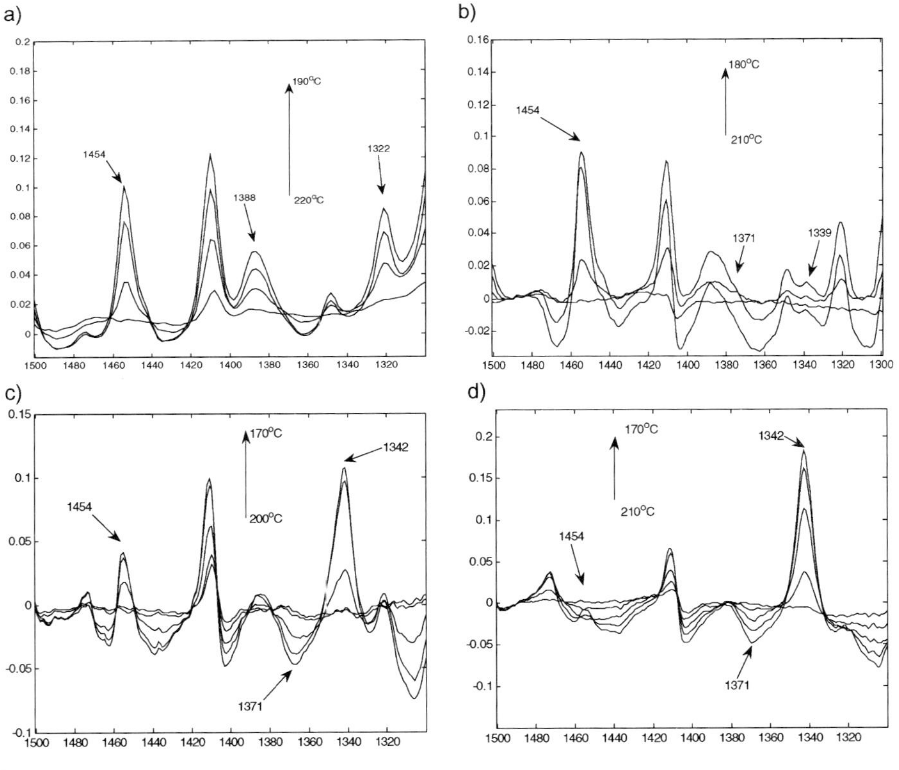

Figure 3.
Dynamic spectra as a function of temperature: 210–170 °C with 10 °C steps: a) PBT, b) $BD_{70}EG_{30}$, c) $BD_{50}EG_{50}$, and d) $BD_{30}EG_{70}$.

which can be assigned to CH_2 wagging of an ethylene segment (trans conformation) with the terephthalate segment in a non-trans configuration. This is an indication that ethylene segments are incorporated into the PBT chain.

In the spectra corresponding to sample $(BD_{50}EG_{50})_{SSP}$ the positive bands of PBT are weaker and positive bands at 1471 and 1342 cm^{-1} corresponding to crystalline PET appear. Negative bands can be observed at 1315 and 1371 cm^{-1}, which can be assigned to amorphous PET, as well as a band at 1465 cm^{-1}, which can be assigned to amorphous PBT. These spectral features clearly indicate that PET crystals are formed. Furthermore, the crystalline PBT and PET bands appear almost at the same time. The limited miscibility of BHET with PBT promotes the formation of separate PET-crystals. However, no visible absorption band at 990 cm^{-1} related to PET chain folding is found, suggesting that the crystalline PET chain segments are not long enough.

In the spectra corresponding to the $(BD_{30}EG_{70})_{SSP}$ system weak spectral features of crystalline PBT are present together with strong positive and negative features of PET. If we have a closer look at the bands at 1320 and 1454 cm^{-1} and the band at 1342 cm^{-1} it can be noticed that PET crystallizes much earlier, indicating that PBT and PET crystallize independently and are mostly like in a separate phase. This is confirmed by the presence of an absorption band at 990 cm^{-1} that can be assigned to chain folding of PET.

These observations are in line with the previously reported DSC-results.[9] When more BHET is incorporated, the onset of crystallization shifts to lower temperatures, while the crystallization exotherm becomes broader, mainly because the PBT crystallization is disturbed by the presence of short PET sequences in PBT. However, the exotherm of the $(BD_{30}EG_{70})_{SSP}$ copolymer, which is much sharper with a slightly higher onset, is due to crystallization of PET, while at lower temperatures only a small PBT crystalline fraction is formed. Most likely, because of the limited miscibility of BHET in the amorphous phase of PBT and the concomitant formation of PET homopolymer by self-condensation, the large interfacial area leads to a strong nucleation effect.

Conclusions

FTIR spectroscopy in transmission can be used to follow the kinetics of BHET-incorporation in PBT during SSP and to study the microstructure of the resulting BDEG copolymer. These results show that depending on the ratio of PBT/BHET, a competition exists between annealing of PBT, incorporation of BHET in amorphous PBT and BHET homopolymerization. For the $(BD_{70}EG_{30})_{feed}$ system, BHET is incorporated in PBT forming a non-random copolymer, while for high BHET concentrations a separate PET phase is found.

The dynamic spectra measured during cooling from the melt clearly showed that for low and intermediate BHET-concentrations, crystallization of PBT is responsible for the crystallization exotherm, as observed by DSC, while for high BHET-concentrations the DSC crystallization exotherm is caused by crystallization of PET.

Acknowledgements: This work is part of the research program of the Dutch Polymer Institute (DPI) under project numbers 143 and 337.

[1] H.-J. Radusch, in Handbook of Thermoplastic Polyesters, S. Fakirov, Ed., Wiley-VCH, Weinheim **2002**, Chapter 8, 389.
[2] V. B. Gupta, Z. Bashir, in Handbook of Thermoplastic Polyesters, S. Fakirov, Ed., Wiley-VCH, Weinheim **2002**, Chapter 7, 317.
[3] S. C. E. Backson, A. M. Kenwright, R. W. Richards, *Polymer* **1995**, *36*, 1991.
[4] B. Jacques, J. Devaux, R. Legras, E. Nield, *J. Polym. Sci., Polym. Chem. Ed.*, **1996**, *34*, 1189.
[5] J. H. Kim, W. S. Lyoo, W. S. Ha, *J. Appl. Polym. Sci.* **2001**, *82*, 159.
[6] M. A. G. Jansen, J. G. P. Goossens, G. de Wit, C. Bailly, C. E. Koning, *Anal. Chem. Acta* **2006**, *557*, 19.

[7] K. Tomita, *Polymer* **1973**, *14*, 50.
[8] C.-C. Lin, S. Baliga, *J. Appl. Polym. Sci.* **1986**, *31*, 2483.
[9] M. A. G. Jansen, L. H. Wu, J. G. P. Goossens, G. de Wit, C. Bailly, C. E. Koning, *J. Pol. Sci., Pol. Chem. Ed.*, **2007**, 45, 882.
[10] A. Misra, S. N. Garg, *J. Pol. Sci., Pol. Phys. Ed.*, **1986**, 24, 983.
[11] I. Noda, *Applied Spectroscopy* **1990**, 4, 44.
[12] P. Harrington, A. Urbas, P. J. Tandler, *Chemometrics and Intelligent Laboratory Systems* **2000**, *50*, 149.
[13] K. C. Cole, A. Ajji, E. Pellerin, *Macromolecules* **2002**, 35, 770.
[14] M. A. G. Jansen, J. G. P. Goossens, G. de Wit, C. Bailly, C. E. Koning, *Macromolecules* **2005**, *38*, 2659.
[15] K. C. Cole, A. Ajji, E. Pellerin, *Macromol. Symp.* **2002**, *184*, 1.

